International Perspectives on Aging

Volume 9

Series editors

Jason L. Powell, Preston, Lancashire, UK
Sheying Chen, New York, USA

For further volumes:
http://www.springer.com/series/8818

Halaevalu F. Ofahengaue Vakalahi
Gaynell M. Simpson · Nancy Giunta
Editors

The Collective Spirit of Aging Across Cultures

Springer

Editors
Halaevalu F. Ofahengaue Vakalahi
School of Social Work
Morgan State University
Baltimore, MD
USA

Gaynell M. Simpson
School of Social Work
University of Alabama
Tuscaloosa, AL
USA

Nancy Giunta
Silberman School of Social Work at Hunter College
The City University of New York
New York, NY
USA

ISSN 2197-5841 ISSN 2197-585X (electronic)
ISBN 978-94-017-8593-8 ISBN 978-94-017-8594-5 (eBook)
DOI 10.1007/978-94-017-8594-5
Springer Dordrecht Heidelberg New York London

Library of Congress Control Number: 2014930186

Printed on acid-free paper

Springer is part of Springer Science+Business Media (www.springer.com)

To all of my earthly angels who carried me to this place and to my spiritual ancestors may your "truth" be witness in my future generation of nephews and nieces—Kirsten G. Simpson-Winston and Alex Seddon

—Gaynell M. Simpson

To my grandparents and parents, whose shoulders I stand on as they supported my aspiration for education beyond their own; to my partner Bill, whose support in this fleeting journey of life is beyond measure; and finally, to my siblings, nieces, and nephews who embody the optimism I have for the future

—Nancy Giunta

To Mom and Dad—I thank God every day for you, for your constant inspiration, and life examples of the collective meaning of "living well" which you have instilled in me and my siblings as an existence that only occurs when we live our lives inclusive of others. To all of my elders from across so many cultures and communities in which I've lived, I thank God for you and your investment of your wisdom in me, may your legacies live on for generations to come

—Halaevalu F. Ofahengaue Vakalahi

Foreword

Given my commitment to the study of aging and diversity, it is truly an honor to be invited to write the foreword for this extraordinary and innovative book that focuses on aging and old age from varied perspectives and gives voice to people who have been overlooked in the past. The underlying premise of this text is that integrating a chorus of voices will enrich readers' knowledge, understanding, and practices by highlighting the commonalities and differences across diverse segments of the older population in the contemporary United States.

As the distinguished contributors to the volume have demonstrated, different elements of diversity—such as gender, race or ethnicity, religious affiliation, social class, community location, and sexual orientation—are not independent of one another. In day-to-day life, as well as within institutions, each of these elements interacts with the others and adds to the complexity of individual and collective experiences in our society. Focusing on just one element of diversity in isolation can create an inaccurate and incomplete picture of lived realities. The intersections between social categories create a complex and dynamic set of experiences, strengths, and challenges for individuals and social groups which may be completely missed if we focus our analysis and practices on a single social category (such as race or age only). Considering the experiences of older adults through a more complex and intersectional perspective, as is illustrated in the chapters that follow, has the potential to deepen readers' understanding of diverse elders and strengthen interventions for these populations.

Instead of reinforcing the "multiple jeopardy" perspective, this book emphasizes strengths and resources, as well as challenges facing older populations. To be sure, it is essential to recognize the problems faced by those coping with the effects of disadvantaged status. But, it is equally important for the readers to develop an understanding and appreciation of how diverse elders create meaning in their lives, how they draw upon their past experiences, cultural traditions, and values, and how they demonstrate resilience and strengths in addressing barriers such as those based on gender, race, class, sexual orientation, and other social identities. Thus, reading the large variety of chapters included in this timely book will deepen readers' understanding of aging and old age in our pluralistic society and this, in turn, would influence the research they undertake, the programs they design, and the services they provide for older adults from diverse ethnic and cultural groups.

Another strength of this volume is that the contributors have taken a multidisciplinary approach to the study of aging and diversity. This approach is indeed consistent with gerontology's focus on biological, psychological, and social aspects of aging. Today's gerontologists include researchers, instructors, and practitioners in such diverse fields as biology, medicine, nursing, dentistry, social work, physical therapy, sociology, psychology, economics, physical science, pharmacy, architecture, and anthropology. Given their commitment to teaching, research, and service, scholars from these disciplines would benefit from engaging with the ideas that the authors have presented in their chapters. Having reviewed these chapters, it is my assessment that this book would be of interest to a wide range of audiences such as social service providers, health professionals, policy makers, faculty members, and students interested in aging and diversity. They will find that the book includes relevant topics, incorporates current research and scholarship, and presents the material in an easy-to-understand language and style.

Reading the *Collective Spirit of Aging Across Cultures* and listening to the voices of the contributors who themselves represent different elements of diversity and bring varied viewpoints is likely to yield new insights, create a unique fusion of horizons, and provide new lenses through which to view multiple social worlds that exist within our communities and/or places of work. The editors and the contributors hope that providing multiple perspectives can help illuminate significant aspects of social reality and experience that may have been overlooked in the past.

Congratulations to the editors and the chapter authors for writing a book that students, faculty, service providers, and other professionals will want to keep close at hand for regular consultation and use. If this book helps generate new research and affects how diverse elders are viewed and served, it will have achieved its purpose.

Chandra M. Mehrotra

Chandra M. Mehrotra is a professor of psychology and dean of special projects at the College of St. Scholastica in Duluth, Minnesota. He is a fellow of the American Psychological Association (APA) and the Gerontological Society of America (GSA). He received the mentor award from APA Division 20 in 2003 and Outstanding Mentorship Award from the GSA Task Force on Minority Aging in 2011. With support provided by the National Institute on Aging, and the Office of Behavioral and Social Sciences Research, National Institutes of Health and the John A. Hartford Foundation, he directs faculty training programs in aging research.

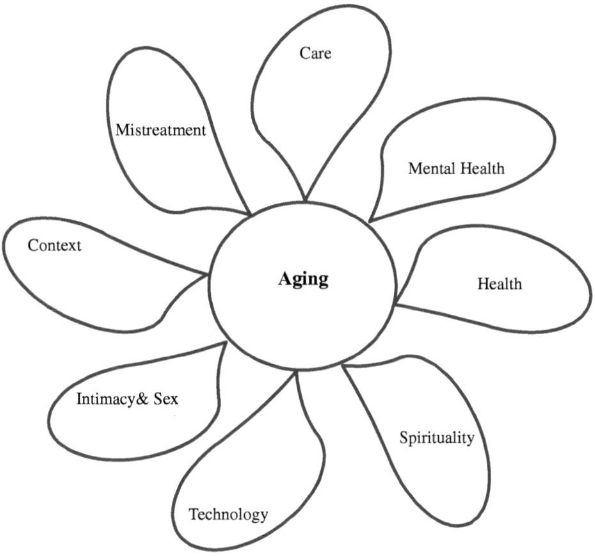
Care
Mistreatment
Mental Health
Context
Aging
Health
Intimacy& Sex
Spirituality
Technology

Contents

4th Generational Theories of Intersection: Multicultural Aging

Gaynell M. Simpson, Nancy Giunta and Halaevalu F. Ofahengaue Vakalahi

Abstract In this chapter, authors asserted that the multifarious reality of older adults' social positions offer unique opportunities and challenges to our traditional gerontological theories. Authors presented intersectionality as an analytical framework to move readers beyond their disciplinary boundaries towards a *4th generation of theories* for co-creation of innovative theories, methods, and interventions. In the co-edited book, *The Collective Spirit of Aging Across Cultures*, we critically reflect on how chapter authors integrated gerontological theories guided by the philosophical assumptions of intersectionality for addressing multicultural issues of aging, including caregiving, intimacy, elder mistreatment, financial crisis, housing transitions, and technology that span and interact with all social identities at multiple levels. We have invited readers to embrace a lens of intersectionality, and provided a discourse towards a *4th generation of theories* to inform trans-disciplinary professionals on how to co-create with collectivist communities aimed at addressing social justice, health and well-being of multicultural older adults.

In the United States and globally, empirical data indicate older adults are the fastest growing population, and within this population, racial and ethnic groups, are projected to increase at even a faster rate (Vincent and Velkoff 2010). The emerging racial and ethnic population growth of older adults may even be larger as census reports may not account for inter-racial marriage, persons with mixed ancestry, and immigrants (Perez and Hirschman 2009). In this book, authors recognize that older adults are not stagnant classifications but represent "multi-dimensionality" of lived experiences (Crenshaw 1989). Older adults' multiple

G. M. Simpson (✉)
University of Alabama, Tuscaloosa, USA
e-mail: simpson.gaynell@gmail.com

N. Giunta
Hunter College, City University of New York, New York, USA

H. F. O. Vakalahi
Morgan State University, Jenkins Building, Baltimore 21251, USA

H. F. O. Vakalahi et al. (eds.), *The Collective Spirit of Aging Across Cultures*,
International Perspectives on Aging 9, DOI: 10.1007/978-94-017-8594-5_1,

social positions, diverse cultures and values, historical and political cohort experiences, challenges our traditional approaches (i.e., methods) and explanations (i.e., theories) to address "…difference while also strategically mobilizing the language of commonality… in the service of constructing a coherent theoretical and political agenda" (Nash 2008, 4).

In this book, *The Collective Spirit of Aging Across Cultures*, authors challenge the boundaries of "third generation" theories (see Bengtson et al. 1997) by "bridging" (see Marshall 1995) aging phenomena at *both* the micro-level (individual, group, and family) *and* macro-level (social structures and social constructs) of analysis. The following authors address the call by Gans et al. (2009) that multiple disciplines must collaborate to co-create different theories, methods and interventions to address the bio-psycho-social-spiritual life-long transitions of older adults. This is done by moving the discourse towards a *4th generation of theories* through integration of multiple gerontological theories guided by *intersectionality* as an analytical framework for deconstructing *multiple* social categories, at *multiple* levels, simultaneously (Crenshaw 1989, 1991; McCall 2005). This collection of works, reflect a multidisciplinary perspective guided by the following philosophical assumptions of intersectionality: (1) understand that older adults come from a diverse and varied background; (2) refrain from viewing any one category of social identity as primary over any other; (3) all elders from the same social group do not have equal or the same shared experiences as we are all shaped differently by cohort historical events; (4) elder's analytical framework must include 'age' at the axis of two or more areas of intersection (e.g., age, ethnicity, and sexuality), at the same time; recognizing they are inseparable, interconnected and linked; (5) researchers recognize their own position of power and status, particularly with individuals we seek to undertake research, as well as, critically reflect on our own personal and collective reasons (implicit or explicit) for engaging a community and/or individual in seeking scientific 'truth'; and (6) we actively engage in collaborative relationships with elders and various community stakeholders for co-creating social change; thus we must make sure that our engagement with all elders do not end when funding ends and that we remain committed to addressing needs of elders' communities (see Collins 2000; Crenshaw 1989, 1991; McCall 2005).

Three chapters have been selected to represent the spirit of this book as they move us beyond the binary boundaries of disciplines and static social categories toward a *4th generation of theories* through integration of multiple gerontological theories guided by an i*ntersectional* analytical framework: *A Profile of Rural African American Lesbian Elders: Meeting Their Needs* (Harley, Stansbury, Nelson and Espinosa); *LATIN@*[1] *ELDERS: Securing Healthy Aging In Spite of Health and Mental Health Disparities* (Medina and Negroi); and *City Life—What*

---

[1] As cited in Medina and Negroi referenced chapter, the @ sign is the deference to the quite recent determination to develop and use nonsexist language and to identify with Latin America (Wallerstein 2005).

a Wonderful Way of Life: Aging in the Urban Environment (Yancy, Baldwin, Saran, and Vakalahi).

In the chapter, *A Profile of Rural African American Lesbian Elders: Meeting Their Needs*, Harley and colleagues integrates different theoretical concepts from multiple disciplines to examine inseparable and linked axes of difference, such as age, race, gender, sexual orientation and geographic location. They do not yield to an easier interpretation of assuming that all rural, African American, lesbian elders who hold these multiple spaces have a shared same experience nor do they give primacy to gender or race. Instead, they recognize how these mutually interdependent and interactive positions often relegate rural, African American lesbian elders to positions of "intersectional invisibility" (Purdie-Vaugns and Eibach 2008) by critiquing the social, historical, and political contextual experiences. These authors provide supportive evidence to demonstrate how the mechanism of 'intersectional invisibility' have removed the concerns of rural, elder, African American lesbians from being adequately addressed in the fields of gerontology, healthcare, education, public policy, and many other organizations and disciplines. Harley and colleagues critically analyze historical and current contextual influences that required older African American lesbians to constantly negotiate their multiple identities and struggle for acceptance *within* and *across* race, age, sexual orientation, and geographical locations. In their conclusion, authors exemplify serving as social change agents as they call upon all of us to engage in educating each other, and ourselves by providing resources to guide us in this journey of aging.

Medina and Negroi continue with the intersectionality of social determinants of health for advancing *Latin@ elders' right to Securing Healthy Aging In Spite of Health and Mental Health Disparities.* These authors critique and analyze how the socio-historical–geographical and cohort experiences impact Latin@ elders' health and well-being. This analysis was achieved through an intersectionality framework that is integrated from social justice, health disparity, and acculturation to demonstrate the complex interplay of how both micro- and macro- levels of influence can create challenges or opportunities for healthy aging of Latin@ elders. Medina and Negroi suggest that the social health determinants for promoting healthy aging of Latin@ elders will not be successful unless we extend our lenses to their unique and varied experiences, including consideration of an elder's country of origin, socio-political context, colonization history, ethnic identities, and generational status. They highlight how both their country of origin and United States' context interacts with Latin@ elders' cultural and life transitions. Medina and Negroi conclude with examples of how Latin@ families and communities work with inter-disciplinary collaborations to positively shape the physical and emotional well-being of Latin@ elders.

Yancey, Baldwin, Saran, and Vakalahi's chapter, *City Life—What a Wonderful Way of Life: Aging in the Urban Environment*, invite readers into the world of elders who are indigenous to urban communities through their use of the elders' voices and depictions that represent how elders indigenous to an urban community make *meaning* of their environment. Authors transform our worldview about what it means to live longer and healthier lives in an urban environment through integration and analysis of theories and conceptual frameworks from Intersectionality, Systems/Ecological, Strengths Perspective and resilience. They help readers understand that despite how 'outsiders' may view an urban environment, which is often conflated with negative characteristics, exclusionary and marginalize language, urban elders are *defining* their own "spaces" as they work collaboratively to embrace the positive spaces of an urban environment while at the same time employ collective action on challenges from within and external to urban communities. Yancey and colleagues re-position our critical lenses to uncover the ways that indigenous urban elders negotiate micro- and macro-level structural constraints and *opportunities* by analyzing *from, within* and *across* these multidimensional levels. These authors conclude with indigenous elders' world view of the urban environment as a collective, inclusive, community enriched with strengths and extraordinary people. Their recommended strategies for working with the elders in multicultural urban environments draw on intersectionality theory by maximizing resilience and collective wisdom and minimizing the use of dominant cultural approaches to service delivery.

Each chapter in this book, although not all discussed here, clearly have cross-cultural implications to consider, particularly around issues that touch all elders, including caregiving, intimacy, elder mistreatment, and technology which are issues that do not choose one social identity or culture over another, but span and interact with all social identities at multiple levels. The application of intersectionality is illustrated throughout this book as authors integrate different theoretical perspectives from various disciplinary backgrounds. Authors throughout the book recognize that the *real intention* of cross-fertilization of various theories and disciplines is to create points of interventions and future directions. They move us from theorizing or acknowledging aging issues to recommending points of intervention that are multidimensional, cross-disciplinary and collaborative. They answer questions that are often neglected after engaging in multidisciplinary research, questions such as: Where do we intervene? And how do we co-create with community to create culturally, appropriate interventions?

Readers of this collection are invited to embrace a lens of intersectionality. Explore the multiple dimensions at play in the lives of our elders. Use their stories to hold multiple perspectives and build knowledge using their collective wisdom by piecing together these intersecting narratives. It is through this collective process that we will begin to build the fourth generation of gerontology theories that truly embodies multicultural aging.

References

Bengtson, V. L., Burgess, E. O., & Parrott, T. M. (1997). Theory, explanation, and a third generation of theoretical development in social gerontology. *The Journals of Gerontology, 52B*(2), S72–S88.

Collins, P. H. (2000). *Black Feminist Thought* (2nd ed.). New York: Routledge.

Crenshaw, K. (1989). Demarginalizing the intersection of race and sex: A black feminist critique of antidiscrimination doctrine, feminist theory, and antiracist politics. *University of Chicago Legal Forum*, 139.

Crenshaw, K. (1991). Mapping the margins: Intersectionality, identity politics, and violence against women of color. *Stanford Law Review, 43*(6), 1241–1299.

Gans, D., Putney, N. M., Bengtson, V. L., & Silverstein, M. (2009). The future of theories of aging. In V. L. Bengtson, D. Gans, N. M. Putney, & M. Silverstein (Eds.), *Handbook of theories of aging* (2nd ed., pp. 3–23). New York: Springer.

Marshall, V. W. (1995). Social models of aging. *Canadian Journal on Aging, 14*, 12–34.

McCall, L. (2005). The complexity of intersectionality. *Signs, 30*(3), 1771–1800.

Nash, J. C. (2008). Re-thinking intersectionality. *Feminist Review, 89*, 1–15.

Perez, A. D., & Hirschman, C. (2009). The changing racial and ethnic composition of the U.S. population: Emerging American identities. *Population and Development Review, 35*(1), 1–51. doi:10.1111/j.1728-4457.2009.00260.x

Purdie-Vaugns, V., & Eibach, R. (2008). Intersectional invisibility: The distinctive advantages and disadvantages of multiple subordinate-group identities. *Sex Roles, 59*(5–6), 377.

Vincent, G. K., & Velkoff, V. A. (2010). *The next four decades, The older population in the United States: 2010 to 2050. Current Population Reports, P25-1138*. Washington, DC: U.S. Census Bureau.

Wallerstein, I. (2005). Latin@s: What's in a name? In R. Grosfoguel et al., *Latin@s in the World-System: Decolonization struggles in the twenty-first century U.S. empire*. (pp. 31–39). Boulder: Paradigm Press.

A Historical Perspective in Aging and Gerontology

Patsy R. Smith

Abstract The chapter provides the reader with an historical view of aging and gerontology in the United States by focusing on pioneers in the field of geriatrics and groundbreaking contributions to the field. Research leaders who explore how older adults find success in managing the aging process are presented with exemplars of the work that has catapulted gerontology to the forefront of social, political, and healthcare agendas. Highlights capture the work of pioneers across multiple disciplines and presents select accomplishments in research, education, and policy. Milestones include development of organizations focused on aging and gerontology resulting in a proliferation of forums for research support and dissemination of findings. Brief descriptions of the contributions of such organizations provide an overview of growth in the field of aging and gerontology with transcendence of race and ethnicity, economic or political status, or physical ability. A timeline presents notable events in recognition of the legacy established by the collective contributions of others, and inspiration toward the achievements of future generations that focus on aging and gerontology.

Introduction

Providing a foundation upon which to build the diverse content and multidisciplinary relevance of this book, this chapter provides historical perspective on aging and gerontology, highlighting examples of pioneers and groundbreaking work in the field. This chapter intentionally broadens the focus beyond geriatrics (the health and care of older adults, the ability of older adults to function in daily life, and the quality of daily life for older adults) to applications of research about

P. R. Smith (✉)
The University of Oklahoma Health Sciences Center, Oklahoma, USA
e-mail: patsy-smith@ouhsc.edu

H. F. O. Vakalahi et al. (eds.), *The Collective Spirit of Aging Across Cultures*, International Perspectives on Aging 9, DOI: 10.1007/978-94-017-8594-5_2,

the aging process and how older adults manage. This chapter offers no lists of interventions or expected outcomes for specific pathological conditions; instead, the focus is on gerontology as purposed by the National Institute of Aging, Clinical Gerontology Branch (http://www.nia.nih.gov/research/dgcg/clinical-gerontology-branch). Important aspects of gerontology include the influences of how change occurs within a disease or physiologic condition and its relationship to the characteristics of that condition in later years. Gerontology integrates the biology of aging, the psychology of coping, and social science of living in an environment that is unique for each individual (Ruiz 1990). Such variability occurs across and within populations, and may be related to genetic conditions, early-life influences or other factors and may remain asymptomatic until later years. Despite the unknowns of aging, the population of centenarians rapidly increases while contributions to the science of aging and gerontology continue to develop.

The study of aging cells and genomics has produced an evolutionary portrait of germ plasm, cell division, mutation accumulation, late life mortality, and other key research findings (Ljubuncic 2009). The pioneering work of scientists who develop evolutionary theories of aging provides the underpinning for understanding longevity (Ljubuncic 2009). Notwithstanding those developments, the aging of populations across the world provides a ripe field for study.

Examinations of gerontological publications reflect a need for greater use of theory to support research (Alley et al. 2010; Fry 1992; Hendricks et al. 2010). Theoretical foundations for research in aging and gerontology appeared approximately 12 % more frequently as the years progressed through publications that were examined in a study of relevant publications in social gerontology from 1990 to 2004 (Alley et al. 2010). These researchers' review of studies in eight major journals revealed an increase of at least 20 % explicit sociology of aging theory use in four of the journals, and a decrease or no change in sociology of aging theory in two of the journals. The most frequently used theories were life course perspective theory and life-span developmental theories. Models of aging were also used in the studies examined, including successful aging models and Andersen's behavioral model of health services use. These researchers discussed the importance of using theory to guide research, explained the findings of research studies, and helped researchers develop interventions (Alley et al. 2010). Study findings supported the work of other researchers (Hendricks et al. 2010) who recognized and recommended greater use of theory to support research that expands understanding and improving of the lives of older adults.

Similar advice was recommended in the early 1990s for counselors preparing to work with older adults (Fry 1992; Qualls 1992). The advice included sharpening knowledge of the developmental stages of adulthood, including knowledge of aging, assessment, and intervention. The focus of Fry and Qualls attests to the heterogeneity of older adults and encourages health workers to engage in learning about what happens among older adults throughout the aging process: aging occurs differently among different populations and within populations. Another important aspect of Fry's work was acknowledgement of the contribution that activity and interactivity make toward life satisfaction among older adults. Fry concludes by espousing socio-environmental theory as encompassing three factors that influence adjustment to older adulthood: physical living space, psychological abilities and coping mechanisms, and economic resources which support life choices.

The more recent work of Hendricks et al. (2010) supports the continuing need for researchers to include theory as a foundation for gerontological research. The authors examined literature published over a 5-year span in three journals pertaining to the inclusion of theory-based research. Findings indicate the need for application of theoretical foundations in social gerontology research. Hence, the authors called for researchers and journal reviewers to conduct and disseminate theory-based research that clearly demonstrates application to practice. Such has been the focus of the pioneers and leaders in ethnogeriatric research.

Select Highlights of Pioneers and Policy

Leaders in social gerontology, psychology, medicine, education, and nursing set the stage for groundbreaking work to enhance knowledge in gerontology. The progress in the field is the result of the groundbreaking and phenomenal work of many pioneers both living and those who have passed on. A few contemporary pioneers in the field are highlighted in this chapter only as an initial gesture of celebrating and honoring those whose shoulders this generation stands. These are but a few of the pioneers who have paved the way for a focus on the intersections of aging, culture, and interdisciplinary practice. These highlights are not presented in any particular format or importance rather they are presented as a starting point. The author of this chapter encourages each reader to consider who could be added to this list of leaders and pioneers.

Author and educator, Gwen Yeo is Senior Research Scholar Emerita of Stanford University School of Medicine (http://med.stanford.edu/profiles/Gwen_Yeo/). Her pioneering work in ethnogeriatrics encourages the delivery of culturally relevant care. She emphasizes that the experience of aging must be recognized in context with the unique influences of each person's culture and environment (Yeo 2003, 2009). Noting that the population of older adults is increasing rapidly, Dr. Yeo and colleagues at the Stanford Geriatric Education Center urge interdisciplinary and interprofessional collaboration to embrace the increasing ethnic diversity among older adults by recognizing implications for older adult services in a way that is culturally meaningful. Geriatric Education Centers (GECs), awarded through the U.S. Department of Health and Human Services (HHS), Health Resources and Services Administration, support education and training in geriatrics. The GECs promote engagement with community members to develop care processes and information delivery to enhance understanding, improving the likelihood that health decisions are well informed.

Barbara Berkman is an iconic leader in geriatric social work having been Associate Director of the Geriatric Education Center in the Division on Aging at Harvard Medical School, and having led the Geriatric Social Work Faculty

Scholars Program funded by the John A. Hartford Foundation (http://www.gswi.org/programs/Barbara_Berkman.html). Through programs such as the GEC and Faculty Scholars, Dr. Berkman has mentored and launched numerous social work scholars and educators in aging. Another prominent contribution to the field, Dr. Berkman mobilized and connected a host of experts to produce a reference book addressing the care and services needed by older adults, major disabling physical and mental health conditions, and how social workers might better serve the aging population (Berkman 2006). The text, *Handbook of Social Work in Health and Aging*, provides a pathway to enhance knowledge across disciplines regarding older adults with needs related to poverty, abuse and neglect, immigration, prison, and cultural diversity. Her career includes time as Director of the Ruth D. and Archie A. Abrams Interdisciplinary Research Program at Massachusetts General Hospital, and Professor of Health and Mental Health at Columbia University School of Social Work.

Universities across the country have established centers or institutes of aging excellence. One such Institute on Aging is at the University of North Carolina, and is led by Peggye Dilworth-Anderson, Interim co-Director and co-Lead of the Institute's Aging and Diversity Program. Dr. Dilworth-Anderson is an expert in issues related to health disparities and the care of older adults (http://www.aging.unc.edu/bio/dilworthanderson.html). Research in the Aging and Diversity Program is aimed at reducing or eliminating health disparities in dementia care, cancer conditions, and stroke, among other conditions. The work of the Aging and Diversity Program at the Institute on Aging demonstrates the importance of researcher-community collaboration, which benefits local and national communities.

Another example is the University Center on Aging, established at San Diego State University, where E. Percil Stanford is professor emeritus and Charter Director of the Center. Dr. Stanford developed the National Institute on Minority Aging at the Center. A former AARP Senior Vice President and Chief Diversity Officer, he founded Folding Voice to influence the development of policy, business, and agency outcomes related to minority aging. He supported the inclusion of all adults over age 50-years in the AARP business and political agenda, including gay, lesbian, bisexual or transgendered (LGBT) individuals and groups, and other minority communities (http://www.aarp.org/politics-society/rights/info-06-2009/wisdom_of_the_elders_aarp_and_sage.html). Dr. Stanford demonstrates commitment to work in gerontology, recognizing the experience and life wisdom of older adults. He has contributed to or served on numerous governmental programs, including work in the Department of Health, Education and Welfare with the Administration on Aging, and Social Rehabilitation Service Programs.

Sandra Crewe, Interim Dean at Howard University School of Social Work, is Director of the Multidisciplinary Center on Gerontology at Howard University (HU-MCG). The HU-MCG began with a $900,000 Health and Human Services award in 1992. The HU-MCG mission is to conduct and support professional preparation in geriatrics and gerontology through education and training, collaboration and research, and faculty development. Under the leadership of Dr. Crewe, the HU-MCG supports a variety of seminars featuring academics and practitioners, and boasts partnerships with nationally recognized organizations including the National Caucus & Center on the Black Aged, Inc. (NCBA) focusing on housing, employment, health promotion, and disease prevention among minority and low-income seniors (http://www.ncba-aged.org/about.php).

The work of Thomas LaVeist, author and educator on issues of public health, helps to identify health disparities, the causes that influence disparities, and problems directly linked to health disparities (http://thomaslaveist.com/). His work at the Johns Hopkins Bloomberg School of Public Health includes Director of the Hopkins Center for Health Disparities solutions, and the William C. and Nancy F. Richardson Professor in Health Policy, which afford him opportunity and influence to address major health disparities issues. Dr. LaVeist has produced textbooks relevant to the exploration of health disparities among minorities (LaVeist 2005; LaVeist and Isaac 2013). In the second edition of a textbook used across the nation, he and co-editor Lydia A. Isaac, and a group of talented experts, examined the impact of aging, ethnicity, and healthcare within the United States, and the ensuing minority majority demographic transition (LaVeist and Isaac 2013). *Minority Populations and Health: An Introduction to Health Disparities in the United States* (LaVeist 2005) remains relevant as a primer on contributors to the disparities of the health of minorities in America, including health issues specific to the most prevalent minority groups in America, and recommendations to address identified disparities.

Important contributions to aging and gerontology have occurred through individual participation in crafting guidelines and policy recommendations. Fernando Torres-Gil has a multitude of contributions to aging and gerontology through academic positions in social welfare, public policy, gerontology, and public affairs at UCLA (http://luskin.ucla.edu/fernando-torres-gil). Dr. Torres-Gil is recognized for his research related to aging and diversity issues. His laudable public service includes national appointments under three United States Presidents: 1978 Federal Council on Aging under President Carter; U.S. Assistant Secretary on Aging under President Clinton; Vice Chair of the National Council on Disability under President Obama. Additionally, Dr. Torres-Gil has held positions through which he advocated on behalf of the citizens of Los Angeles, and Salinas, California and California Veterans.

The National Association of Social Workers Foundation recognizes member contributions to policy development on aging (www.naswfoundation.org/pioneers). Among them, Elaine Brody, DSc (Honorary), MSW, is known for voicing concerns about women who provide care for aging parents amid responsibilities for families, jobs, and careers (http://www.gswi.org/careers/Brody_profile.htm). Further, Dr. Brody's contributions to gerontology and aging studies include mental health, caregiving to aging parents, and Alzheimer's Disease. Her past, present, and future perspectives on aging and gerontology are cleverly documented as she describes the importance of listening to what older adults have to say as a strategy for gaining information about the concerns or what affects life for older adults (Brody 2010). Once President of the Gerontological Society of America, Dr. Brody gained early recognition as one of the first social workers in favor of aging as an area of central focus, a specialization.

Claire M. Fagin, a Fellow of the American Academy of Nursing, former Dean of the University of Pennsylvania, School of Nursing worked with the National Hartford Centers of Gerontological Nursing Excellence (formerly known as BAGNC) and garnered the honor of a fellowship named for her: the Claire M. Fagin Fellow Award Program. The purpose of the program is to build academic geriatric nursing capacity (hence, the former title BAGNC) by providing 2-year fellowships in advanced research training for doctorally prepared nurses who focus on gerontological nursing research. The program, coordinated in cooperation with the Gerontological Society of America, prepares nurses committed to academic research careers (http://www.geriatricnursing.org/applications/cmf-fellowship.asp) and promotes advanced training through its nine academic centers.

Nancy R. Hooyman, Hooyman Endowed Professor in Gerontology and Dean Emeritus, University of Washington School of Social Work (http://socialwork.uw.edu/faculty/nancy-hooyman), is a prominent leader and pioneer whose works advance social work contributions in gerontology and the care of older adults. Dr. Hooyman is a mentor to the next generation of social work scholars, faculty, and researchers. Her prolific writing in journals, texts and reference books chronicle and enhance the foundation that builds and transforms engagement of social workers in social gerontology.

Letha Chadiha, professor and principal investigator, is Co-Director of the Community Liaison Core in the Michigan Center on Urban African American Aging Research (http://ssw.umich.edu/faculty/profiles/tenure-track/lethac). Her contributions to aging include research-based knowledge about differences in caregiving among older adults in rural and urban settings. Dr. Chadiha also researches issues related to recruitment of ethnically diverse older adults in research. She serves as a mentor for many social work scholars in aging.

Linda M. Chatters is a researcher, professor, and proliferative writer in the School of Public Health and School of Social Work at the University of Michigan. Dr. Chatters focuses on adult and older adult African Americans, spirituality and religious involvement in well-being, and intergenerational issues that affect families, relations, and health (http://ssw.umich.edu/faculty/profiles/tenure-track/chatters). Her book contributions provide insight into the lives of Black and African American families in America regarding issues of aging, religion, psychology, sociology, and health (Jackson et al. 1993; Taylor et al. 1997, 2003).

Dorothy Smith-Ruiz is a Fulbright Scholar and Associate Professor of Africana Studies at the University of North Carolina, Charlotte. Her major research interests are custodial grandmothers in intergenerational families, the impact on family of the incarceration of American women, and the cultural differences among women of African American, Afro-Caribbean, and African immigrant heritage. Her books focus on the role of African American grandmothers who provide care to family members, and how traditional values are conveyed, and the mental health of persons in the African American community (Ruiz 1990, 2004).

Innovation in social work education with a focus on gerontology is represented in the research expertise of Andrew Scharlach, the Eugene and Rose Kleiner Professor of Aging and professor of Social Welfare at University of California-Berkeley. Dr. Scharlach's interests include aging and aging-friendly communities, families and the issues that affect them, and policies related to informal caregivers and long-term care. He is the Director of the Center for the Advanced Study of Aging Services, which is leading the way in research on innovations within communities to become better places to live and to age.

Joanne Damron-Rodriguez, Professor, the University of California, Los Angeles (UCLA), has participated in numerous advisory capacities to influence geriatric education for social work and nursing students (http://www.geronet.ucla.edu/component/content/article/96-general/104-academic-geriatric-resource-program). Dr. Damron-Rodriguez's publications and participation on the Council of Social Work Education helped shape curriculum for social work careers with a focus in gerontology (http://luskin.ucla.edu/joann-damron-rodriguez).

The leaders and pioneers named above are only a few of the experts whose dedication, commitment, and accomplishments in aging and aging issues have helped to improve the lives of older adults. These experts and allies from the culturally diverse families and communities of the elders helped to shape and implement policies and practices that benefit older adults as well as educate professionals to practice and conduct research in aging and gerontology. Milestones in aging policy, gerontology, education and research mark the contributions and commitment of pioneers to the advancement of this field.

Select Milestones in Aging and Gerontology

National interest in aging and gerontology resulted in major milestones and contributions to older adults through a proliferation of research and dissemination of findings. Such milestones include establishment of the Social Security Administration in 1935 (see Timeline, Table 1) and the National Institute on Aging (NIA) in 1974 as a part of the National Institutes of Health. The Code of Federal Regulations, [statute] 416.110 describes the purpose of the social security program: "to assure a minimum level of income for people who are age 65 or over, or who are blind or disabled…" (http://www.socialsecurity.gov/OP_Home/cfr20/416/416-0110.htm) without resources that meet a federally defined standard of living or minimum income. Few officially recognized federal agencies on aging were in existence before the Social Security Administration (originally named the Economic Security Act by the Committee on Economic Security) was enacted. The age 65 was chosen as a reflection of the age used by retirement, or pension, systems that were in place: private industry and state systems. The notion of income security, however, was described in 1796 by Thomas Paine who recognized the need for a method by which to assure a reasonable income for older adults (http://socialsecurity.gov/history/tpaine3.html). Several aging-related organizations developed over the decade and a half after the development of the Social Security Administration, and more were established after the Older Americans Act of 1965. This section presents an overview of select organizations and milestones established over the last three-fourths of a century with a focus on the well-being of older adults.

Interestingly, the Gerontological Society of America (GSA) had its inception among a group of scientists and physicians in 1939, just 4 years after the signing of the Social Security Act, but was not formally established until 1945, after which a journal dedicated to scientific and applied research related to aging was launched (http://www.geron.org/About%20Us/history). Later, the GSA divided its membership into two primary sections, basic sciences and applied sciences, each of which was further divided into two sections, making four sections of members. The mission of GSA is to support the science of research related to aging, to disseminate new knowledge generated by aging research, and to influence higher education and public policy to support programs in gerontology (http://www.geron.org/About%20Us/Our%20Mission). The GSA supports research funding through its affiliation with the Hartford Geriatric Social Work Initiative and the National Hartford Centers of Gerontological Nursing Excellence, and maintains a policy institute, the National Academy on an Aging Society, to achieve its mission.

The Brookdale Foundation Group, founded in 1950, is a combination of two foundations and a trust, and offers seed grants for its major initiatives related to aging, geriatrics, and gerontology (http://www.brookdalefoundation.org/aboutus.html). Based on endowments from the Schwartz family, The Brookdale Foundation Group supports caregiving across generations through its National Group Respite Program, disseminating information for raising the children of relatives through the Relatives as Parents Program (RAPP), and for developing the next generation of leaders in geriatrics and gerontology through the Brookdale Leadership in Aging Fellowship Program.

The National Council on Aging (NCOA) was formed in 1950 by the National Social Welfare Assembly (NSWA; name includes the 1967 change to the National Assembly for Social Policy and Development and in 2005 became the National Human Services Assembly—NHSA) and was renamed the National Council on Aging in 1960 (http://www.ncoa.org/about-ncoa/ncoa-history.html). The founding organization, NSWA, began in 1923 as the National Social Work Council which existed to support organizations aimed at addressing social needs, with specific attention to issues that surfaced after the Great Depression (http://www.nationalassembly.org/About/History.aspx). The focus of the National Council on Aging is to improve the lives of older adults through education and program initiatives that support healthy aging, helping older adults gain access to and learn more about Medicare benefits including the BenefitsCheckUp program, and helping older adults to plan for and make decisions about remaining independent in a way that is right for each individual. The NCOA has helped to make a difference in the lives of older adults since 1950, more than 60 years.

The American Society on Aging (ASA) was started in 1950 and was originally named the Western Gerontological Society (http://www.asaging.org/about-asa). The purpose of the organization is to work with its members to improve the lives of older adults by supporting individual and collective commitment, knowledge, and skills. ASA provides education for professionals across disciplines through print material, webinar technology, and conferences. The organization boasts a membership that includes clinical practitioners, educators and researchers, agency and educational administrators, people involved in developing policy statements, and the next generation of students.

The American Association of Retired Persons (AARP) was founded in 1958 by a retired high school principal, Dr. Ethel Percy Andrus, who also started the National Retired Teachers Association about a decade earlier (http://www.aarp.org/about-aarp/info-2009/History.html). The aim of both organizations was to promote healthy aging because of the lack of health insurance for retired teachers and to build the ability to negotiate with insurance companies to cover older adults. Although her initial concerns were for retired teachers, she learned over the course of the first decade that there were similar concerns about health benefits after retirement among older adults all across America and in other countries. Not only does AARP promote healthy aging, it helps to disseminate research findings that support informed decision-making for management of chronic conditions, economic security, and physical ability. AARP also aims to serve as the voice for persons aged 50-years and over for policy issues related to Social Security, Medicare, age discrimination, gay rights, and livable communities, to name a few (http://www.aarp.org/politics-society/).

An organization was developed in 1970 when a group of religious and social work retirees were invited into a conversation by founder Maggie Kuhn (http://www.graypanthers.org/). The purpose of the organization was to become advocates for issues of importance to retirees and the general public, including issues of financial or economic security, health care as a right, and opposition to the Vietnam War. After several national presentations, the organization was nicknamed the Gray Panthers in 1972 and the name was retained. The Gray Panthers is formed for younger and older adults willing to advocate for justice, peace, and multiculturalism, recognizing and respecting the strengths in the differences among peoples.

Established in 1974, NIA supported research on aging for decades before its official establishment as an Institute (http://www.nia.nih.gov/about/nia-timeline). The founding director of NIA, Robert Butler, physician and Pulitzer Prize winning author (for his book Why Survive?: Being Old in America) coined the term ageism and advocated for establishing geriatrics as a medical specialty. The current mission of NIA is to support research to develop and disseminate knowledge about the influences and processes of healthy aging as well as the diseases and conditions that contribute to disabilities among older adults (http://www.nia.nih.gov/about/mission).

Area Agencies on Aging (AAAs) and its national constituents aim to represent the collective voice of local offices and Title VI Grants for Indian Tribal Organizations. AAAs were written into law in the Comprehensive Services Amendments (1973) to the Older Americans Act (1965) that established Medicare and Medicaid as part of the Social Security Act (http://www.aoa.gov/AoARoot/

AoA_Programs/OAA/resources/History.aspx). Title VI of the Older Americans Act supports the provision of comparable services to older adults who are American Indians, Alaska Natives, and Native Hawaiians (http://www.n4a.org/about-n4a/).

The Resnick Gerontology Center was established in 1980 with an endowment to the Albert Einstein College of Medicine in partnership with Montefiore, the University Hospital, in the early years of increasing concerns about the rising numbers of older adults in the population (http://www.montefiore.org/research). The program promotes collaboration across health disciplines to prepare medical practitioners in the study of aging and geriatric care. The continuing goal is to strengthen education and research among medicine disciplines: geriatric psychiatry, bioethics, geriatric medicine, and neurology.

Geriatric Education Centers (GECs) are awarded through the U.S. Department of Health and Human Services, Health Resources and Services Administration, and began in 1983 to support education and training in geriatrics for health professionals across disciplines (http://www.hrsa.gov/advisorycommittees/bhpradvisory/acicbl/reports/thirdreport.pdf). GECs are awarded to health professions training programs including but not limited to health administration, clinical psychology and social work, professional counseling, and marriage and family therapy (http://bhpr.hrsa.gov/grants/geriatricsalliedhealth/gec.html). GECs also support continuing education for professors, researchers, and practitioners, and supports clinical training in the places where older adults obtain health care: senior centers, clinics, and long term care.

Established in 1986, Generations United (GU) is an organization focused on quality of life across generations from children to older adults (http://www.gu.org/ABOUTUS.aspx). GU is a coalition of organizations that collaborate to achieve goals to benefit all the individuals the organizations represent. The first two entities to collaborate in forming this organization were the National Council on Aging and the Child Welfare League of America; however participant organizations grew to include AARP and the Children's Defense Fund, among others (http://www2.gu.org/PRESS/PressKit/History.aspx). Simply stated, GU encourages an intergenerational approach and intergenerational collaboration to achieve goals in public policy and application of resources across all levels of aging; across generations.

The Substance Abuse and Mental Health Services Administration (SAMHSA) was established by the United States Congress in 1992 (http://beta.samhsa.gov/about-us/who-we-are). The agency is part of the U.S. Department of Health and Human Services (HHS) and was developed to increase public access to and availability of research related to substance use and information about mental health issues and services to treat related disorders. The overall focus is on improving the mental health of Americans, and reducing drug and alcohol abuse and misuse through appropriately targeted education and treatment programs. The age groups targeted include youth who engage in risky behavior, veterans of all ages, families of military personnel, and members of tribal communities. Research documenting under-use of mental health services by older adults highlights agency efforts to reduce stigma and negative attitudes toward mental health and substance abuse issues by community members, families, health professionals, and persons who suffer with mental illness. Consequently, SAMHSA established the Resource Center to Promote Acceptance, Dignity and Social Inclusion Associated with Mental Health to focus on older adults where they reside or gather, in the home, senior centers, healthcare settings, and in nursing homes (http://promote acceptance.samhsa.gov/audience/adults/default.aspx) with the goal of addressing ageism, perceptions of older adults with mental health or substance use issues, recovery, and access to treatment programs.

The National Institutes of Health, National Institute on Aging and National Institute of Nursing Research issued a request for applications (RFA) in January 1997 (http://www.grants.nih.gov/grants/guide/rfa-files/RFA-AG-97-002.html) to develop Resource Centers for Minority Aging Research (RCMARs). The National Institute on Aging (NIA) supports efforts to not only address older adult health throughout the aging process, but to also reduce health disparities by supporting the dissemination of information, education of researchers, scholars, and health professionals, with particular attention to under-represented minorities (http://www.rcmar.ucla.edu/RCMAR_EVal_Final_Report_4-9-08.pdf). Seven centers are funded for 2012 through 2017 and are located at universities in Alabama, California, Colorado, and Michigan. The mission of the RCMARs is focused on minority older adults: developing and improving knowledge by preparing researchers in minority older adult health research; improving recruitment and retention of older minority older adults in research studies; encouraging the development of instruments and measures that are culturally sensitive to the health status of minority older adults; and investigating interventions designed to improve health status to learn of ways to improve effectiveness among older adults in minority populations. The RCMARs produce a vast array of data-based research reports that cover statistical approaches to measurement review, educational presentations in association with GSA conferences, and papers that focus on race and cross-cultural measurement in research (Sood and Stahl 2011; Stahl and Vasquez 2004; Teresi et al. 2012).

The American Geriatrics Society (AGS) supported the 2008 establishment of an interprofessional workgroup to develop geriatric competencies for inclusion in educational programs that prepare healthcare professionals to work with an aging population (http://www.americangeriatrics.org/files/documents/health_care_pros?PHA_Multidisc_Competencies.pdf). Organized into six domains, the competencies drive advanced education and training in geriatrics and gerontology across disciplines, and are aimed at doctors, nurses, pharmacists, psychologists, and social workers. The six domains are: Health Promotion and Safety, Evaluation and Assessment, Care Planning and Coordination across the Care Spectrum (including End-of-Life Care), Interdisciplinary and Team Care, Caregiver Support, and Healthcare Systems and Benefits. Health disciplines and regulatory bodies have established education program guidelines for training professionals in gerontology by developing competencies that reflect these and additional domains, including Caring for the Elderly Patient, System-Based Care for Elder Patients, or Geriatric Syndromes (http://adgap.americangeriatrics.org/

Table 1 Notable events related to aging and gerontology in the United States

Date	Notable events related to aging and gerontology
1920	The Civil Service Retirement Act provided a retirement system for many governmental employees
1930	The first concept of a National Institute of Health is developed
1935	The Social Security Act passed; provides for old age assistance and old age survivors insurance
1937	The first Institute (Cancer) is established
	Railroad Retirement Act provided pensions for retired railroad employees and spouses
1940	A Unit on Aging is established in the NIH division of chemotherapy. Head: Edward J. Stieglitz
1941	The Unit on Aging (evidently, later the gerontology branch) moves to Baltimore City Hospital under the direction of Nathan Shock
	Surgeon General Thomas Parran forms the National Advisory Committee on Gerontology
1948	The gerontology branch is moved to the National Heart Institute. (Note: There had been an effort to establish an Institute of Aging with the Heart Institute as a subsidiary! This failed when a physician to the Senate opined "We don't need research on Aging. All we need to do is go into the library and read what has been published.")
1950	President Truman convenes the first national conference on aging, sponsored by the Federal Security Agency
	The American Society on Aging (ASA) and the National Council on Aging (NCOA) are established
1951	In his trends in gerontology, Dr. Shock outlines his recommendations for a National Institute of Gerontology
1952	First federal funds appropriated for social service programs for older persons under the Social Security Act
1956	Federal Council on Aging is convened by President Eisenhower. Department of Health, Education, and Welfare established a special staff on aging to coordinate aging details; Center for Aging Research exists within the National Institute of Mental Health
1958	Baltimore Longitudinal Study of Aging established; legislation introduced in Congress calling for a white house conference on aging
1959	A Section on Aging established within NIMH (James E. Birren, head)
	Aging research conducted through NIH intramural programs (Gerontology Branch of the National Heart Institute and NIMH Section on Aging)
	Almost 600 research and training grants on aging are under way through the Center for Aging Research, "the focal point for information on the NIH activities in gerontology"
	Housing Act of 1937 authorized a direct loan program for non-profit rental projects, for the elderly at low interests rates, and lowered eligibility ages for public-low-rent housing, for low-income women to age 62
1960	Social Security Administration eliminated age 50 as minimum for qualifying for disability benefits, and liberalized the retirement test and the requirement for fully insured status

(continued)

Table 1 (continued)

Date	Notable events related to aging and gerontology
1961	First White House Conference on Aging recommends creation of an Aging Institute. Their report reads, in part, "a National Institute of Gerontology should be set up within the National Institutes of Health to conduct research on aging; Federal financial support should be increased for biomedical research in governmental agencies, universities, hospitals, research centers and for building necessary facilities; human population laboratories should be established to study problems associated with aging"
	Social Security Amendments lowered the retirement age for men from 65 to 62, liberalized the retirement test, and increased minimum benefits and benefits to aged widows
1962	The Gerontology Research Center Nathan Shock Laboratory facility is donated to NIH by the City of Baltimore (deed dated 12/6/62; recorded 2/1/63)
	Legislation introduced in Congress, to establish an independent and permanent Commission on Aging
1963	The National Institute of Child Health and Human Development is established to focus on health issues across the life course, including in old age. President John F. Kennedy remarks, "For the first time, we will have an Institute to promote studies directed at the entire life process rather than toward specific diseases or illnesses"
1965	The Older Americans Act establishes the Administration on Aging as "the Federal focal point for activities in aging." It established the Administration on Aging within the Department of Health, Education and Welfare, and called for the creation of State Units on Aging
	Medicare, Title XVIII, a health insurance program for the elderly was established as part of the Social Security Act
	Medicaid, Title XIX, a health insurance program for low-income persons, was added to the Social Security Act
1967	The Gerontology Research Center in Baltimore, long a part of the National Heart Institute, is transferred to NICHD
	Older Americans Act extended for 2 years, and provisions made for the Administration on Aging to study the personnel needs in the aging field
	Age Discrimination Act signed into law
	Administration on Aging moved from the Office of the Secretary of HEW and placed in the newly created Social and Rehabilitative Service Agency within the Department
1968	The Division of General Medical Sciences has responsibility for research grant projects in aging. Aging-related research is conducted and supported throughout NIH—primarily NICHD, but also NCIA, NIAID, the National Heart Institute, and the National Institute of Arthritis and Metabolic Diseases
	Construction of the Gerontology Research Center (GRC) in Baltimore is completed
1969	Older Americans Act Amendments provided grants for model demonstration projects, foster grandparents, and retired senior volunteer programs
1970	The Gray Panthers are started by Maggie Kuhn with a group of five friends
1971	The second white house conference on aging once again recommends creation of a separate National Institute on Aging
1972	A new Title VII is created under the Older Americans Act authorizing funds for a national nutrition program for the elderly

(continued)

Table 1 (continued)

Date	Notable events related to aging and gerontology
1973	President Richard Nixon vetoes a bill for the creation of a National Institute on Aging
	Older Americans Act Comprehensive Services Amendments established area agencies on aging. The amendments added a new Title V, which authorized grants to local community agencies for multi-purpose senior centers, and created the Community Service Employment grant program for low-income persons age 55 and older, administered by the Department of Labor
	Comprehensive Employment and Training Act was enacted; included older persons
1974	May 31, 1974—Public Law 93-296 authorizes the establishment of a National Institute on Aging and requires that the institute develop a national comprehensive plan to coordinate the HEW (Health, Education, Welfare) agencies involved in aging research. October 7, 1974—The National Institute on Aging is established
	Title XX of the Social Security Amendments authorized grants to states for social services. These programs included protective services, homemaker services, transportation services, adult day care services, training for employment, information and referral, nutrition assistance, and health support
	Older Americans Act amendments added transportation under Title III model projects
	Housing and Community Development Act enacted; provided for low-income housing for the elderly and handicapped, pursuant to the Housing Act of 1937
	Title V of the Farm and Rural Housing Program of 1949 expanded to include the rural elderly as a target group
1975	April 23, 1975—First meeting of the National Advisory Council on Aging
	Older Americans Act Amendments authorized grants under Title IV to Indian tribal organizations. Transportation, home care, legal services, and home renovation/repair were mandated as priority services
	July 1, 1975—The Adult Development and Aging Branch and Gerontology Research Center are separated from their parent institute (NICHD) to become the core of the National Institute on Aging
1976	May 1, 1976—Dr. Robert N. Butler appointed first NIA director
	December 8, 1976—The research plan required by P.L. 93-296 goes to Congress
1977	Older Americans Act Amendments required changes in Title VII nutrition program, primarily related to the availability of surplus commodities through the Department of Agriculture
1978	Older Americans Act Amendments consolidated the Title III area agency on aging administration and social services, the Title VII nutrition services, and the Title V multi-purpose senior centers, into a new Title III and added a new Title VI for grants to Indian Tribal Organizations. The old Title V became the Community Service Employment grant program for low-income persons, age 55 and older (created under the 1978 amendments as Title IX)
	Congregate Housing Services Act authorized contracts with local public housing agencies and non-profit corporations, to provide congregate independent living service programs
	OAA amendments required each state to establish a long-term care ombudsman program to cover nursing homes
1981	Third White House Conference on Aging held in Washington, D.C.
	Older Americans Act reauthorized; emphasized supportive services to help older persons remain independent in the community
	Act expanded ombudsman coverage to board and care homes

(continued)

Table 1 (continued)

Date	Notable events related to aging and gerontology
1982	September 20, 1982—NIA Laboratory of Neurosciences Clinical Program admits the first inpatient to a new unit at the NIH Clinical Center
1984	The first Alzheimer's Disease Centers are established
1986	November 14, 1986—P.L. 99-660, section 951-952, authorizes the NIA's Alzheimer's Disease Education and Referral (ADEAR) Center as a part of a broad program to conduct research and distribute information about Alzheimer's disease to health professionals, patients and their families, and the general public
	Reauthorization of the Older Americans Act clarified and reaffirmed the roles of state and area agencies on aging in coordinating community-based services, and in maintaining accountability for the funding of national priority services (legal, access, and in-home)
1987	The first annual Florence S. Mahoney Lecture on aging is held. This series was created in honor of Mrs. Florence Stephenson Mahoney (1899–2002), a woman who tirelessly campaigned for increased Federal spending for medical research and steadfastly championed for the creation of the NIA
	NIA holds its first Summer Institute, a 1-week event to provide junior investigators, particularly those from underrepresented groups, an opportunity to learn about the substance and methodology of aging research from recognized experts in the field
	Omnibus Budget Reconciliation Act provides for nursing home reform in the areas of nurse aide training, survey and certification procedures, pre-admission screening and annual reviews for persons with mental illness
	Reauthorization of the Older Americans Act added six additional distinct authorization of appropriations for services: in-home services for the frail elderly; long-term care ombudsman; assistance for special needs; health education and promotion; prevention of elder abuse, neglect, and exploitation; and outreach activities for persons who may be eligible for benefits under supplemental security income (SSI), Medicaid, and food stamps. Additional emphasis was given to serving those in the greatest economic and social need, including low-income minorities
	The Nursing Home Reform Act (Omnibus Budget Reconciliation Act) mandated that nursing facility residents have "direct and immediate access to ombudspersons when protection and advocacy services become necessary." Simultaneously, the OAA reauthorization charged states to guarantee ombudsman access to facilities and patient records, provided important legal protections, authorized state ombudsmen to designate local ombudsman programs and required that ombudsman programs have adequate legal counsel
1988	November 4, 1988—P.L. 100-607 establishes the Geriatric Research and Training Centers (GRTC)
1990	The GRTCs are expanded and renamed the Claude D. Pepper Older American Independence Centers and charged with conducting research in diseases that threaten independent living
	Americans with Disabilities Act extended protection from discrimination in employment and public accommodations to persons with disabilities
	Cranston-Gonzalez National Affordable Housing Act reauthorized the HUD Section, 202 Elderly Housing program, and provided for supportive service demonstration programs
	Age Discrimination in Employment Act made it illegal, in most circumstances, for companies to discriminate against older workers in employee benefits

(continued)

Table 1 (continued)

Date	Notable events related to aging and gerontology
1992	The Health and Retirement Study, the leading source of combined data on health and financial circumstances of Americans over age 50 and a valuable resource to follow and predict trends and help inform policies for an aging America, is established
	The Substance Abuse and Mental Health Services Administration (SAMHSA) was established by the United States Congress as part of the Department of Health and Human Services (HHS)
	Reauthorization of the Older Americans Act places increased focus on caregivers, intergenerational programs, protection of elder rights and calls for a 1995 white house conference on aging
	The elevation of Commissioner on Aging to Assistant Secretary for Aging
	OAA amendments added a new Title VII "Vulnerable Elder Rights Activities" which included the long-term care ombudsman; prevention of elder abuse, neglect and exploitation; elder rights and legal assistance development; and benefits outreach, counseling and assistance programs. The legislation emphasized the value of the four programs coordinating their efforts. The amendments highlighted the role of local ombudsman programs and the state ombudsman's role as leader of the statewide program and advocate and agent for system-wide change
1993	Six Edward Roybal Centers for Research on Applied Gerontology are authorized to convert research findings into programs that improve the lives of older people and their families
	NIA funds six Exploratory Centers for Minority Aging and Health Promotion in collaboration with the NIH Office of Research on Minority Health
1994	Nine demography of aging centers are funded by NIA to provide research on health, economics, and aging to make more effective use of data from several national surveys of health, retirement, and long-term care
1995	Three Nathan Shock Centers of Excellence in Basic Biology of Aging are established to further the study of the basic processes of aging
	White house conference on aging convened May 2–5, 1995 in Washington, D.C.
	Operation restore trust initiated
1996	NIA introduces its Exercise: A Guide from the National Institute on Aging, providing encouragement and evidence-based guidance for older adults to engage in exercise
1997	The Resource Centers for Minority Aging Research (RCMARs) program is established through a call for applications by the National Institutes of Health, National Institute on Aging, and National Institute of Nursing Research to investigate the variability of health differences experienced across racial and ethnic groups, as well as the mentoring of new scholars in health disparities research
1999	As part of NIA's 25th anniversary celebration, a strategic plan is formulated and made available for public comment. The plan addressees scientific topics holding the greatest promise for advancing knowledge in areas such as the basic biology of aging, geriatrics, and social and behavioral functioning
	International year of older persons: a society for all ages
2000	Older Americans Act Amendments of 2000 signed into law (P.L. 106-501), establishing the new National Family Caregiver Support Program, and reauthorizing the OAA for 5 years on November 13, 2000

(continued)

Table 1 (continued)

Date	Notable events related to aging and gerontology
2001	In a unique private–public partnership, NIA joins the osteoarthritis initiative to bring together resources and commitment to the search for biological markers of osteoarthritis
	NIA and the Icelandic Heart Association announce collaboration on a vast study on the interactions of age, genes, and the environment. The collaboration extends 34 years of data on the health of 23,000 Icelandic residents into the new millennium
	HHS Secretary Tommy G. Thompson released $113 million for first National Family Caregiver Support Programs grants to states on February 15, 2001
2002	NIA begins to offer a selection of its health communications materials in Spanish
	The SardiNIA project, a study to determine the genetic bases for a number of age-related traits and conditions, begins. This 5-year project is a collaboration between the NIA Intramural Research Program and the Italian National Research Council
2003	Enactment of the Medicare Prescription Drug, Improvement and Modernization Act (MMA)
	NIA and the National Library of Medicine (NLM) launch NIHSeniorhealth.gov, a web site designed to encourage older people to use the Internet
	The NIA, joined by the Alzheimer's Association, expands the Alzheimer's Disease Genetics Initiative to create a large bank of genetic materials and cell lines for study to speed up the discovery of risk-factor genes for late-onset Alzheimer's disease
	NIA and the American Federation for Aging Research—in collaboration with the John A. Hartford Foundation, the Atlantic Philanthropies, and the Staff Foundation—establish the Paul B. Beeson Career Development Awards in Aging Research program, a public–private partnership to support clinically trained junior faculty to pursue careers in aging research.
	The NIA IRP's Advanced Studies in Translational Research on Aging (ASTRA) unit, a state-of-the-art facility located at Baltimore's Harbor Hospital, opens in January and becomes the new home of the BLSA
2004	NIA launches the Longevity Consortium, a network of investigators from several large-scale human cohort studies working in collaboration with individual basic biological aging researchers to facilitate the discovery, confirmation, and understanding of genetic determinants of healthy human longevity
	NIA begins the Long Life Family Study, an international multicenter research project to examine families with high numbers of long-lived individuals to better understand the genetic and environmental contributions to exceptional long life in families
	NIA, in conjunction with other Federal agencies and private companies and organizations through the Foundation for the National Institutes of Health, leads the Alzheimer's Disease Neuroimaging Initiative
	NIA launches Healthy Aging in Neighborhoods of Diversity across the Life Span (HANDLS), a multidisciplinary community based, longitudinal, epidemiologic study examining the influences and interaction of race and socioeconomic status on the development of age-associated health disparities among socioeconomically diverse African Americans and whites in Baltimore

(continued)

Table 1 (continued)

Date	Notable events related to aging and gerontology
2005	NIA's Alzheimer's Disease Preclinical Drug Development program is established. Fourth white house conference on aging
2006	NIA leads the NIH conference "AD: Setting the Research Agenda a Century after Auguste D," to assess the state of current Alzheimer's disease research and the most promising routes to progress
	Medicare Part D Prescription Drug program (part of MMA) went into effect
	Enactment of the Lifespan Respite Care Act (administered by AoA)
	Older Americans Act Amendments of 2006 signed into law (P.L. 109-365), embedding the principles of consumer information for long-term care planning, evidence based prevention programs, and self-directed community based services to older individuals at risk of institutionalization. OAA was reauthorized for 5 years on October 17, 2006
2007	"Living Long & Well in the 21st Century: Strategic Directions for Research on Aging," which outlines broad goals and objectives for aging research, is published
	U.S. Secretary of State Condoleezza Rice sponsors the summit on global aging in collaboration with NIA to call attention to challenges and opportunities worldwide from population aging
	A Spanish-language version of the NIA web site (http://www.nia.nih.gov/Espanol) goes live
2008	A biology of aging summit is convened to review NIA's research portfolio, identify areas of opportunity, and facilitate the formulation of cohesive and comprehensive plans for the future
	NIA celebrates the 50th anniversary of the Baltimore longitudinal study of aging
	NIA's five extramural programs (aging biology, geriatrics and clinical gerontology, neuroscience, behavioral and social research, and extramural activities) are reorganized as divisions
2009	An updated version of NIA's award-winning exercise guide for older Americans is published
2010	Enactment of the Affordable Care Act
2011	NIA launches the Go4Life campaign, to promote exercise and physical activity nationwide for people 50 and older, with public and private partners from a variety of aging, fitness, and provider organizations
	NIA and the Alzheimer's Association lead an effort to update diagnostic guidelines for Alzheimer's disease to reflect the full spectrum of the disease, marking the first time in 27 years clinical and research criteria are changed
	First of the nation's baby boomers turn 65
	CLASS (community living assistance and supports) program, part of the Affordable Care Act, is designated to be administered by the Administration on Aging (the CLASS Act was repealed in January 2013)
	The National Alzheimer's Project Act is signed into law. Dubbed NAPA, it requires a coordinated national effort to find ways to treat or prevent Alzheimer's disease and related dementias and to improve care and services. NIH, represented by NIA, participates in the federal Advisory Council on Alzheimer's Research, Care, and Services
	The Trans-NIH GeroScience Interest Group is formed, with leadership from the NIA. The group promotes discussion, sharing of ideas and coordination of activities within the NIH research community working on mechanisms underlying age-related changes, including those which could lead to increased disease susceptibility

(continued)

Table 1 (continued)

Date	Notable events related to aging and gerontology
2012	HHS Secretary Kathleen Sebelius announces the NAPA-required national plan to address Alzheimer's Disease. NIA plays a critical role in developing the first goal of the plan—to effectively treat or prevent Alzheimer's by 2025
	The NIA Intramural Research Program (IRP) was reorganized to recognize new paradigms in the field of aging research. The program now integrates labs and resources bringing together people who share a similar research interest, but are coming at it from different vantage points
	NIA organizes the Alzheimer's disease research summit 2012: path to treatment and prevention. Some 500 researchers and advocates attend the meeting, which results in recommendations aimed at advancing Alzheimer's disease research
	The international Alzheimer's disease research portfolio is launched. Built in collaboration with the Alzheimer's Association, the database captures the full spectrum of research investment and resources and enables public and private funders of Alzheimer's research to share and review funding data
	NIA leads development of the NIH toolbox for neurological and behavioral function. Unveiled in 2012, the toolbox offers researchers a free set of brief tests to assess cognitive, sensory, motor and emotional function in people from toddlers to older adults
	The Administration for Community Living (ACL) was established, bringing together the Administration on Aging, the Office on Disability and the Administration on Developmental Disabilities to focus on reducing fragmentation in community living service and supporting needs of aging and disability populations; enhancing access to quality health care and long-term services and supports; and promoting consistency in community living policy across the federal government

The National Institute on Aging was established in 1974. This timeline provides a snapshot of important dates in the history of aging and policy work in the United States through the National Institute on Aging and later, the Administration on Aging (NIA Timeline, http://www.nia.nih.gov/about/nia-timeline; Historical Evolution of Programs for Older Americans, http://www.aoa.gov/AoA_programs/OAA/resources/History.aspx)

Adapted from http://www.nih.gov/about/almanac/archive/1998/property/acquisition.html and from http://www.aoa.gov/AoA_programs/OAA/resources/History.aspx

Sources Administration for Community Living, U.S. Department of Health and Human Services. http://acl.gov/About_ACL/Organization/Index.aspx

Freeman (1980) and Morley (2004)

National Institute on Aging: Important Events in NIA History. NIH Almanac, 1998. http://www.nih.gov/about/almanac/archive/1998/organization/nia/history.html

National Institute on Aging. NIH Almanac, 2008–2009. http://www.nih.gov/about/almanac/archive/2008-2009/Almanac_2008_2009.pdf

NIA Congressional Justification narratives for FY 2009 and FY 2010 academic-resources/competencies/geriatric-fellowship-curriculum-milestones-december-2012/#top). These domains include competencies in gerontology, medication management, functional impairment and rehabilitation, or diseases in older adults.

Conclusion

Leaders and pioneers in the field of aging and gerontology continually achieve phenomenal milestones in improving the economic, physical, and environmental living conditions of older adults in America. The research in aging and gerontology, however, remains sparse in terms of addressing the diversity of cultures and populations that make up the United States. A search for research specific to African American and Hispanic older adults reveals the work of academicians, practitioners, and government funded programs aimed at developing new knowledge. Titles show that research is focused on comparisons among Hispanic and non-Hispanic populations, leaving questions regarding how race and ethnicity best add value to current knowledge. This book seeks to encourage recognition and celebration of what is unique and different about individuals as aging occurs, and the experiences in the aging process that unite. The collective spirit of aging transcends race, ethnicity, economic position, political affiliations, physical ability, and culture.

Table 1 presents a timeline of notable events related to aging and gerontology in the U.S. This table is offered as a foundation upon which to continue to build on the legacy and rich histories passed on from leaders and pioneers in the field and particularly, the elders, their families, and communities. It is the hope of this work for continued collective contributions and achievements in aging and gerontology for future generations.

References

Alley, D., Putney, N., Rice, M., & Bengston, V. (2010). The increasing use of theory in social gerontology: 1990–2004. *Journal of Gerontology: Social Sciences, 65B*(5), 583–590.

Berkman, B. (Ed.). (2006). *Handbook of social work in health and aging*. New York: Oxford University Press Inc.

Brody, E. M. (2010). On being very, very old: An insider's perspective. *The Gerontologist, 50*(1), 2–10.

Freeman, J. T. (1980). Some notes on the history of the National Institute on Aging. *Gerontologist, 20*, 610–614.

Fry, P. S. (1992). Major social theories of aging and their implications for counseling concepts and practice: A critical review. *The Counseling Psychologist, 20*(2), 246–329. doi:10.1177/0011000092202002.

Hendricks, J., Applebaum, R., & Kunkel, S. (2010). A world apart? Bridging the gap between theory and applied social gerontology. *The Gerontologist, 50*, 284–293.

Jackson, J. S., Chatters, L. M., & Taylor, R. J. (Eds.). (1993). *Aging in black America*. Thousand Oaks, CA: Sage.

LaVeist, T. A. (Ed.). (2005). *Minority populations and health: An introduction to health disparities in the United States*. San Francisco, CA: Jossey-Bass.

LaVeist, T. A., & Isaac, L. A. (Eds.). (2013). *Race, ethnicity, and health: A public health reader* (2nd ed.). San Francisco, CA: Jossey-Bass, A Wiley Impring.

Ljubuncic, P. (2009). The evolutionary theories of aging revisited—a mini-review. *Gerontology, 55*(2), 205.

Morley, J. E. (2004). A brief history of geriatrics. *Journals of Gerontology, Medical Sciences, 59*, 1132–1152.

Qualls, S. H. (1992). Social gerontology theory is not enough: Strategies and resources for counselors. *Counseling Psychologist, 20*(2), 341–345.

Ruiz, D. S. (2004). *Amazing grace: African American grandmothers as caregivers and conveyors of traditional values*. Westport, CT: Praeger.

Ruiz, D. S. (Ed.). (1990). *Handbook of mental health and mental disorder among black Americans*. Westport, CT: Greenwood Publishing Group.

Sood, J. R., & Stahl, S. M. (2011). Community engagement and the resource centers for minority aging research. *The Gerontologist, 51*(suppl_1), S5–7.

Stahl, S. M., & Vasquez, L. (2004). Approaches to improving recruitment and retention of minority elders participating in research: Examples from selected research groups including the national institute on aging's resource centers for minority aging research. *Journal Aging and Health, 16*(5_suppl), 9S-17.

Taylor, R. J., Chatters, L. M., & Levin, J. (2003). *Religion in the lives of African Americans: Social, psychological, and health perspectives*. Thousand Oaks, CA: Sage.

Taylor, R. J., Jackson, J. S., & Chatters, L. M. (Eds.). (1997). *Family life in black America*. Thousand Oaks, CA: Sage.

Teresi, J. A., Stewart, A. L., & Stahl, S. M. (2012). Fifteen years of progress in measurement and methods at the resource centers for minority aging research. *Journal of Aging and Health, 24*(6), 985–991.

Yeo, G. (2003). The ethnogeriatric imperative. *Care Management Journals, 4*(1), 37–45.

Yeo, G. (2009). How will the U.S. healthcare system meet the challenge of the ethnogeriatric imperative? *Journal of the American Geriatrics Society, 57*(7), 1278–1285.

The Fourth Hill of Life: American Indian/Alaska Native Elders

Priscilla A. Day

Abstract The experience of American Indian/Alaska Native Elders is explored through an Indigenous lens from early childhood through the end of life. As the transmitters of culture and knowledge, Elders gain wisdom from a lifetime of experiences embedded in traditional culture and community. Intergenerational trauma has impacted Indigenous communities with 22 % of the population experiencing Post Traumatic Stress Disorder. In spite of this trauma, the role of Elders in Indigenous communities has remained strong. Elders play a critical role in passing on traditions to children and insuring that tribal communities retain their history and identity. The Anishinaabe Grandfather teachings of wisdom, love, respect, bravery, honesty, humility, and truth are shared as examples of the lessons that Elders learn throughout their lives and pass on to future generations.

> I have been fortunate to have known my grandmother and great-grandmother. I can trace the women in my family back seven generations. I know their Indian names. There are no words that can adequately explain what knowing this history means to me. It allows me to keep them alive as now that I am entering elderhood, I pass these names on to my grandchildren.

Many years ago, I was at a gathering listening to one of my elders talk about Anishinaabe culture. He described that, as indigenous people, we are unique in the ways in which we view the world, both the physical and spiritual world. He described how we "look through the eyes of elders" when we see an old tree or the waves on a lake and it evokes deep emotions in us. He described how in these times, we may be feeling the exact feelings that our ancestors felt when they looked upon the same scene. We are connected to our ancestors in moments like this. Throughout my life, I have had these moments many times when I am sitting quietly in nature without the distractions of the modern world. I found it interesting

P. A. Day (✉)
University of Minnesota Duluth, Duluth, USA
e-mail: pday@d.umn.edu

H. F. O. Vakalahi et al. (eds.), *The Collective Spirit of Aging Across Cultures*, International Perspectives on Aging 9, DOI: 10.1007/978-94-017-8594-5_3,

to read the thoughts and feelings of another Anishinaabe writer who described something very similar.

> I sat on a bench my ancestor had built many, many years ago, And I was overcome with a sense of awe and wonderment at the collective spirit of that place…I could feel the spirits of my father and great-grandparents sitting on the beach with me that day, and it was as though I saw the world with the same sense of wonder that children, like my granddaughters, see the world with. The past, present, and future, all were part of the collective spirit of that place (Peacock and Wisuri 2002, p. 45).

To me, this is collective wisdom. This is that sense of belonging to something greater than oneself. It is about cultural identity and cultural continuity despite all the efforts to destroy that link. This chapter will attempt to describe that connection in an effort to retain it for those who come behind.

Most non-native people have little understanding of American Indians, sovereignty, and indigenous family systems. In order to provide a deeper understanding of American Indian/Alaska Native elders, it is important to understand the roles that elders continue to play in their communities and how historical trauma and intergenerational trauma have impacted these roles. Tribal groups are referred to by different names including First Nations, American Indian, Native American, Alaska Natives, indigenous peoples and hundreds of different tribal names from tribal languages (Day 2007). In 1977, the National Congress of American Indians (NCAI) and the National Tribal Chairman's Association issued a joint resolution proclaiming that if a tribe did not specifically use their tribal name that NCAI would support using the general term American Indian and/or Alaska Native (U.S. Department of Health and Human Services 1999). This is the term most often used in treaty language and by federal agencies that deal with American Indians/Alaska Natives and is the term that most American Indian/Alaska Natives preferred to be called according to the Census (Brunner 2013). When working with a member of a tribal group you should always give deference to the way in which people refer to themselves. For the purposes of this chapter, the term American Indian/Alaska Native will be used.

The important role of elders in American Indian/Alaska Native families and communities is well documented (Day 2007; Red Horse et al. 2000). Throughout time, American Indian/Alaska Native elders have been considered reservoirs of traditional knowledge. This wisdom was passed on to others through stories and teachings. A primary role of the elders was to keep communities strong through this knowledge transmission (Cross et al. 2010; Day 2007; Rheault 1999). The presence of elders helped to connect American Indian/Alaska Native people to the past as they taught the next generations. It is believed that elders keep us grounded in the past while children pull us into the future. Both ends of this continuum are an important part of the circle of life.

Who Are We and Where Do We Live?

American Indian/Alaska Native people lived in the Americas long before the arrival of non-natives on this land. As such, tribes have many different origin stories; some tribes believe they came from the land, whereas other tribes have stories of migration. American Indian/Alaska Native currently live throughout the United States with approximately 43 % living on reservations or trust lands while over 53 % reside in urban areas (Indian Health Service 2012).

There are 569 recognized tribes in the U.S. Unlike other groups of color, tribes have a political status not only a racial identity because they retained their sovereignty as the original peoples of this land. Given this political status, tribes have an inherent right to run their internal affairs. Cobb writes "sovereignty is a nation's power to self-govern, to determine its own way of life, and to live that life—to whatever extent possible—free from interference" (2005–2006, p. 118). This political status makes tribes unique from other people of color.

American Indian/Alaska Native elders, defined as those who are age 55 and over, are a growing population (UCLA Center for Health Policy Research 2010). According to the U.S. Census Bureau, American Community Survey (2011), 17.8 % of the American Indian/Alaska Native population is over 55 years old. This percentage has grown since the days when American Indian/Alaska Native life expectancy was less than 50 years old. In 1976, Dr. Ted Marrs, writing in support of the Indian Health Care Life Improvement Act stated, "In 1974 the average age of death of Indians and Alaska Natives was 48.3 years. For white U.S. citizens the average age of death was 72.3 years. For others, the average age was 62.7 years" (Trahant, Para 5, 2010). Today, life expectancy for whites is at a record high, 78.7 years (Murphy et al. 2013). For American Indian/Alaska Natives life expectancy today is 72.3 years (Truhant, Para 5, 2010); however, while American Indians/Alaska Natives have come a long way since 1974, disparities continue to exist.

Intergenerational Trauma

We have only recently come to understand the impact that policies and practices have on individuals, families and tribes as a whole. "Collective trauma unfolds as an omnipresent, community wide phenomenon, while individual trauma unfolds as a personal psychological phenomenon unrelated to a collective common experience" (Red Horse et al. 2000, p. 14). Understanding historical and contemporary individual and community trauma is critical for understanding what is happening today with American Indians (Cross et al. 2010; Peacock and Wisuri 2002; Ziibiwing Center of Anishinaabe Culture and Life Ways 2011).

The history of oppression of American Indian/Alaska Natives across the Americas is similar across tribal groups and the legacy of this oppression continues to resonate. This phenomenon is exacerbated for elders (Cross et al. 2010; Day 2007; Wasserman 2005; Weaver and Yellow Horse Brave Heart 1999). According to the Office of the U.S. Surgeon General, American Indian/Alaska Natives experience Post Traumatic Stress Disorder (PTSD) at almost three times the rate of the general population–22 % prevalence rate for American Indian/Alaska Natives versus 8 % for the general population (Wasserman 2005, p. 5). Wasserman writes:

> We do not know how the experiences of American Indian and Alaska Native people throughout the past generations may have permanently impacted physiological development. If the experience of childhood abuse and neglect can change the structure of the brain, what impact did colonization, massacres, and forced confinement have on the physiological and psychological development of the indigenous people of the past? (Wasserman 2005, p. 21).

Unfortunately it is hard to know the true impact of these traumas on individuals, families, communities and tribes. We can look at the many disparities that exist in tribal communities and speculate about the role of trauma but no one really knows the true impact, except that it is extensive. Therefore, it is important that health and human service providers, gerontologists and other helping professionals understand historic trauma in order to consider the impact of it, particularly on American Indian/Alaska Native elders (Cross et al. 2010). Many elders lived through traumatic events but were never able to talk about their experiences. Those elders are currently in their 70s or older. It is estimated that during the 1930s and 1940s, half of all American Indians attended boarding schools (Ziibiwing Center of Anishinaabe Culture and Life ways 2011). Researchers have measured the impact of historic trauma on American Indian/Alaska Natives and found high levels of emotional distress including "anger, avoidance, anxiety, and depression" (Whitbeck et al. 2004, p. 127).

Those who experienced trauma as children have difficultly reconnecting with family or their tribe. They often have lost their language and cultural lifeways as a result of being at boarding school and away from traditions. Many boarding school survivors learned inappropriate or abusive parenting. They often internalized negative cultural identity. This trauma became embedded in families and tribes affecting generation after generation. Today many elders still carry these wounds (Cross et al. 2010; Day 2007; Ziibiwing Center of Anishinaabe Culture and Life Ways 2011).

Another impact of historic trauma is how it impacted the role of the elders as viewed by some in the American Indian/Alaska Native community. Because of cultural disruption, some elders are not held in high regard by some of their family. This has resulted in abuse of elders. The National Center on Elder Abuse funded the National Indian Council on Aging (NICOA) to conduct one of the few research studies about elder abuse in tribal communities. They found that poverty and isolation of rural tribal members contributed to greater vulnerability of tribal elders (NICOA 2004). However, cultural practices provided a protective factor. Families that believed in traditional values, maintained strong extended family networks and engaged in cultural practices exhibited less elder abuse (NICOA 2004). In other words, those families that retained traditional beliefs about the elders tended to treat those elders with care and respect.

Walking on the Path of Life

The diversity in the tribes has led to different stories and teachings about what is called the path of life. Due to this diversity, I have chosen to describe my tribe, the Anishinaabe, as a case example to assist in guiding the helping professionals who work with American Indian/Alaska Native elders. This tribal group can primarily be found in Canada, Minnesota, North Dakota, and Wisconsin. For Anishinaabe people, walking the path of life is known as "mino-bimaadiziwin" or living the good life. It means following a path in life that is harmonious with the will of the Creator. We believe that as humans, we all have "free will" to choose which path we will walk. When someone decides to choose a path different than what is desired by the Creator, his/her life may be filled with many bumps along the way. According to interviews with Anishinaabe elders by Rheault (1999), there are seven stages of life with tasks that should be learned and mastered. Each stage is developmental building upon the one that came before. In order to live a "good life" and progress through the stages, one must learn the teachings and responsibilities that accompany each phase.

Stages of Life

The first stage starts when the child is conceived and born into the physical world. His/her spirit enters the "good life". All babies and young children are considered sacred and as such are taken care of by their family and community. The next stage, approximately ages 7–13, are the years in which the child/adolescent is learning more about the world. During this time of growing, the child may act without giving his or her actions much thought or reflection. Up until approximately age 20, the two main tasks of the child is to develop a strong sense of who he/she is and his/her life purpose. During these years a child starts to fulfill the purpose he/she was given to pursue. If the child follows this path, he/she becomes and is seen as a "good person". As one moves into late 20s and 30s, life is filled with raising a family and assuming community responsibilities. Gradually people transition into the "doing" stage in their mid-thirties to early forties. In these years, adults generally have more time to further expand on personal and community interests as their children become more independent and interdependent.

Each stage is preparation for becoming an elder, thus as one ages, one begins to assume greater responsibility and understanding of indigenous teachings. As one ages and develops a deeper understanding of the interconnectedness of life, the learner becomes the teacher, and the teacher moves into elderhood. Traditionally elders were treated with a great deal of respect as they had earned being called "elder" through years of experience and learning. Although all older people should be treated with respect, there are some people who are elders by age but not looked to as teachers or role models because they were not able to master all of the tasks in the different stages of the path. The final stage on the path of life is passing over into the next world when we leave our physical existence behind (Rheault 1999).

Peacock and Wisuri (2006) also described the Anishinaabe stages of life. In their description, each stage of life is called a "hill of life". They used the metaphor of the seasons with the journey beginning in springtime. Spring is about birth and infancy. Similarly, Rheault describes this stage as the "good life", as children are born sacred. The second hill of life, summer, occurs as the infant grows into a child and adolescent. The tasks for summer involve helping the child to learn how to behave in ways that help them to develop skills needed to be successful later in life. In this season, the child learns to self-regulate behavior and learn to be compassionate through caring for younger children and elders. The child learns to be responsible for hunting, gardening, and making other contributions to the family and tribe. He/she learns about the importance of interdependency.

The autumn of life, the third hill, encompasses early fall through winter. Similarly, Rheault describes this time as filled with raising families and the work of adults. As this season continues, some adults will take on different roles, some will become healers, and some will become community leaders. These roles are found through dreams, experiences, ceremonies, and circumstance. People step into positions as there is need in the family or tribe. As discussed by Rheault, as people age, they begin to move from one who is a participant in ceremonies to one who is responsible for conducting them.

The final season is winter, the fourth and final hill. During this time, all things including people begin to slow down and eventually we die. For Anishinaabe, this final season is described as the hill of wisdom. This wisdom is acquired throughout one's life. It is hoped that during this season, all that has been learned can be passed on to others who are eager to be taught. Both Peacock and Rheault acknowledged that reaching this stage takes a lifetime of learning, surviving, and developing wisdom through lived and spiritual experiences. No wonder elders are treasured in American Indian/Alaska Native communities.

Paths of Knowledge

In Anishinaabe culture, the four paths of knowledge build on one another. The paths of knowledge are: feeling, watching, reflection, and doing (Rheault 1999). In in the first path, feeling requires one to learn how to balance one's emotions. "The Elders teach that emotional balance-through trusting one's intuition and one's reason-allows one to hear and see with open ears, clear eyes and a good heart" (Rheault 1999, p. 124). We all know what these moments of clarity and peace feels like and how fleeting they can be in modern society. Our elders remind us of the need to have balance in our lives.

The second path, watching and learning, helps us to learn the value of silence. Watching without speaking allows us to "hear" what is meant when people speak and act. It involves being able to read between the lines which allows us to better able to act in an appropriate way. Being in balance with our own emotions allows us to watch and learn more accurately. There are people who ask too many questions when they could learn so much more by listening and watching what is occurring around them. "At times, asking too many questions obscures the obvious knowledge available in the physical-spiritual world. Observation without questioning also teaches patience and humility since one is forced to quiet one's natural inclination to want to know everything all at once" (Rheault 1999, p. 124). Our elders also teach us that we will learn what we need to know when the time is right and that we cannot know everything at once because we are not ready to have this knowledge. Being comfortable with silence helps us to accurately see the world around us and our place in the world. Elders learn this through a lifetime of experience and repeating these teachings many times. They learn about patience and if we are lucky, we can learn from them about how to walk in silence.

The third path teaches us about reflection. "It is here that one uses the power of one's mind and spirit to reflect upon the knowledge that one has gained, understanding it for what it means, and for what it can do in one's life" (Rheault 1999, p. 127). Reflection is how we integrate life's teachings over time allowing us to gain wisdom.

The fourth and final path finally leads to doing. Our elders teach us that only after one has felt, watched and learned, and reflected, is one ready to take action (Rheault 1999). Elders who by definition have lived through many experiences and hopefully have developed mastery in the stages can then become the teachers of a new generation. This process of knowledge transmission has sustained American Indian/Alaska Native since time immemorial.

Overall, when a child is born into the physical world, Anishinaabe elders believe that the child is already complete. The task for a child is to learn about one's role in life. In turn, elders help children learn about their role and how to behave properly in order to live a good life. "Without these Teachers, how else can a person know how to be good? They are oral reference libraries that account for stories, legends, prophecies, ceremonies, songs, dances, language and the philosophy of the people" (Rheault 1999, p. 138). Elders have always played a central role in family and community life. The impact of boarding schools and other kinds of family and cultural disruptions have caused so much damage in tribal families and communities that so many American Indian/Alaska Native elders were not able to learn about their roles. As a result, many of the teachings about family and tribal life were lost. Fortunately, there have always been those who were able to hold onto the traditions.

The Role of the Elders

American Indian/Alaska Native elders' role and status within their families and communities has been well documented (Cross et al. 2010; Day 2007; Henderson 2002; Red Horse et al. 2000). American Indian/Alaska Native elders, as they age, are expected to assume "increased kinship responsibilities" (Red Horse 1997, p. 245). Traditionally, American Indian/Alaska Native children learn about the important role of elders through observation and teachings (Day 2007; Dubray and Sanders 1999).

As elders age, work outside the home often slows or ends. This affords elders to have more time with grandchildren and great grandchildren. This is a time to sing, garden, hunt, fish, play and tell stories to the little ones. Grandparents are often more patient than parents. Elders are treated respectfully at ceremonies and feasts. They are often served food or allowed to go first at meals. Observed at ceremonies is the tendency for those cooking the meals to often make special efforts to cook "old time" foods that appeal to and are comforting to the elders. When the elders speak, people listen to without interrupting. These are some of the rights that the elders have earned.

Understanding American Indian/Alaska Native culture is important in order to make culturally appropriate practice decisions (Cross et al. 2010; Day 2007). Service providers should understand that American Indian/Alaska Native elders differ in the degree to which they identify with and practice their traditional culture. It is not easy to know to what extent someone practices his or her culture. The best way is to observe and to ask. Most elders are probably best described as bi-cultural; they likely have some experience and different levels of comfort in both white and tribal settings. Most elders have learned to respect both traditional and white practices (Day 2007; Lum 1996; Dubray and Sanders 1999).

Unlike whites, most American Indian/Alaska Native elders expect to be cared for within their families (Day 2007; Henderson 2002; Jervis et al. 2010; Redford 2002). This expectation is often met because of the status elders have in their family and community. This can create a burden, both financial and emotional for families, as most tribes do not provide long-term care. Even fewer have the resources to provide tribal nursing homes (Day 2007). Families whose elders are in need of nursing home care are often forced to resort to placing elders in non-Indian nursing homes, often far from their families, or to try to care for elders themselves.

One study found that there are cultural differences in how caregiving is perceived in American Indian/Alaska Native families. Even though poverty is often high and cultural trauma was found, caregivers in this study reported high levels of positive aspects related to providing care to their elders. It should be noted that in this study, the elders did not require full time care. This study found that one of the reasons that caregiving for an elder was not seen has a burden was because of the respect American Indian/Alaska Native have for elders (Jervis et al. 2010). Other reasons included strong beliefs about not placing elders in nursing homes, the need to give back to elders who give so much to others, and that caregivers often receive considerable assistance from other family members (Jervis et al. 2010).

As the population of American Indian/Alaska Native elders continues to grow, long-term care will need to be addressed (Day 2007). Research is needed about the experiences of tribal families and elders who need long term care. Tribes need to develop their own continuum of care for those elders who need assistance outside of the home. If we value our elders, we must provide this care in our communities.

The Wisdom of the Elders

For Anishinaabe elders, wisdom is about living the seven grandfather teachings:

- Wisdom-to value knowledge and learning throughout one's life
- Love-to embrace peace and have unconditional regard for others
- Respect-to treat others in a caring manner and with dignity
- Bravery-to act with integrity even when fearful
- Honesty-to keep one's word and behave in an honorable way
- Humility-to know your place in the larger picture and not put yourself above the Great Spirit, animals, plants, or other life
- Truth-to be honest with yourself and others (Ontario Native Literacy Coalition 2012).

Elders should live these values during their lifetime. As elders, they role model these behaviors and values through how they live their lives. Those who work with the elders need to understand how indigenous people have been impacted by external forces as well as the expectations placed on the elders to keep traditions strong and to be the teachers of the young. "It is the wisdom of everyday things-of humor when it is most needed, of an encouraging hug when something is done right, of singing to children when they have bad dreams" (Peacock and Wisuri 2006, p. 102).

Call to Action

It is imperative to document indigenous ways of knowing across disciplines. It is through the preservation of these ways that our indigenous lifeways will help us to reclaim a quality of life. We need to develop the capacity of our own peoples with traditional ways of healing, of caring for children and the elders, and to have these resources within our communities. We need to teach our young people about the roles of elders. We need to know and teach others about the impact of intergenerational trauma and how it plays out in our families and communities. When we use western ways, we need to frame them within our indigenous values. If we are able to do these things, then our elders will have a better quality of life. When our elders are well cared for, our communities will remain strong.

The Role of Helpers

In order to be of assistance in indigenous communities, one must understand the history of the community as well as the individual and family history (Day 2007; Hart 1999). In addition, understanding cultural beliefs in paramount. In

Anishinaabe culture, knowing about the seven traditional teachings is important. Relationships are seen as the vehicle through which healing can occur (Hart 1999). Healing is about restoring balance and helping one to live a "good life" as defined through the seven teachings. These teachings serve as a guide for helpers. An effective helper comes from a place of wisdom and is someone who approaches others with love and respect. They act with integrity, honesty and truthfulness. They practice humility, understanding that they play a role just like many others. They recognize that they are not better than anyone or anything, no matter how much "education" they have. Methods for helping may include storytelling, understanding the role of spirituality and ceremonies, using humor, being patient and comfortable with silence, utilizing the wisdom of elders, and role modeling appropriate ways of living (Hart 1999). To be an effective helper takes many years of observing and developing meaningful relationships with indigenous people. It means knowing who the indigenous helpers are within communities and being able to utilize them when appropriate.

> As I move into elderhood, I am privileged to reflect upon my experience with my own grandmother-my mother's mother. We were always close because I was born the day before her birthday-49 years apart. Every year of my life, until she died, we celebrated our birthdays together. My grandmother's spirit name was Shawonosakwe- Lady Going South. She was given this name by her uncle who named her after the headwaters of the Mississippi River in northern Minnesota. Her name reflects the path of this powerful river as it flows from northern Minnesota to the Gulf of Mexico. Like the river, she lived through many things but preserved. She nurtured life and provided sustenance to many along her path. She lived to be 3 months shy of 100 years old. Even at the end of her life she taught those of us gathered by her side. She lay in a hospice room unconscious as my brothers and I gathered for the inevitable passing of our family matriarch. She suddenly awoke from her unconscious state and began talking to us. She smiled and started telling stories about each of us. We went from being somber to laughing and joking with one another. She began to describe places she had been in her life as though she were flying over different parts of the country. Finally she got a peaceful look on her face, "I see the lake (the place she was born and lived her whole life). I am home. Thank you." She closed her eyes and never regained consciousness. As she had done throughout my life, at the end of her life she was teaching me about death- about leaving the physical world and entering the spiritual one-with joy and grace. In our beliefs, I am certain she will be waiting for me when I cross over. Until then, she watches over me, as she has done since my birth.

References

Brunner, B. (2013). *American Indian versus Native American: A once-heated issue has sorted itself out*, Para 9. http://www.infoplease.com/spot/aihmterms.html. Retrieved on June 4, 2013.

Cobb, A.J. (Fall 2005/Spring 2006). Understanding tribal sovereignty: Definitions,conceptualizations, and interpretations. *Indigenous Studies Today*, *46*(3/4), 115–132.

Cross, S. L., Day, A. G., & Byers, L. G. (2010). American Indian grand families: A qualitative study conducted with grandmothers and grandfathers who provide sole care for their grandchildren. *Journal of Cross Cultural Gerontology, 25*, 371–383.

Day, P. A. (2007). American Indian elders. In P. J. Kolb (Ed.), *Social work practice with ethnically and racially diverse nursing home residents and their families* (pp. 41–71). New York, NY: Columbia University Press.

Dubray, W. & Sanders, A. 1999. Interactions between American Indian ethnicity and health care. In P.A. Day & H.N. Weaver (Eds.), *Health and the American Indian*, Co-published simultaneously as *Journal of Health and Social Policy, 10*(4), 67–84.

Hart, M. A. (1999). Seeking Mino-pimatasiwin (the good life): An aboriginal approach to social work practice. *Native Social Work Journal, 2*(1), 91–112.

Henderson, J. N. (2002). *How do we understand and incorporate elder's teaching and Tribal values in planning a long term care system? American Indian and Alaska Native roundtable on long term care: Final report 2002*. Albuquerque, NM: National Indian Council on Aging.

Indian Health Service. (2012), Para 3. http://www.ihs.gov/PublicAffairs/IHSBrochure/QuickLook.asp. Retrieved December 8, 2012.

Jervis, L. L., Boland, M. E., & Fickenscher, A. (2010). American Indian family caregivers' experiences and helping elders. *Journal of Cross Cultural Gerontology, 25*, 355–369.

Lum, D. (1996). *Social work practice and people of color: A process-stage approach*. Pacific Grove, GA: Brooks/Cole Publishing Company.

Murphy, S. L., Xu, J., & Kochanek, K. D. (2013). Deaths: Final data for 2010. *Division of Vital Statistics, 61*(4), 1–168.

National Indian Council on Aging. (2004). *A review of the literature: Elder abuse in Indian country: Research, policy, and practice* (pp. 1–29). Washington, DC: National Center on Elder Abuse.

Ontario Native Literacy Coalition Social Community Enterprise (2012). *"Seven grandfather teachings"*, Para 3. http://www.7grandfatherteachings.ca/grandteachings.html. Retrieved December 12, 2012.

Peacock, T., & Wisuri, M. (2002). *Ojibwe waasa inaabidaa: We look in all directions*. Afton, MN: Afton Historical Society Press.

Peacock, T., & Wisuri, M. (2006). *The four hills of life: Ojibwe wisdom*. Afton, MN: Afton Historical Society Press.

Red Horse, J. (1997). Traditional American Indian family systems. *Families, Systems, and Health, 15*(3), 243–250.

Red Horse, J., Martinez, C., Day, P., Day, D., Poupart, J., & Scharnberg, D. (2000). Family preservation concepts in American Indian communities. *National Indian Child Welfare Association*. http://www.nicwa.org/research/01.family%20pres01.rpt.pdf. Retrieved May 25, 2012.

Redford, L. (2002). *Long term care in Indian country: Important considerations in developing long term care services. American Indian and Alaska Native roundtable on long term care: Final report 2002*. Albuquerque, NM: National Indian Council on Aging.

Rheault, D. (1999). *Anishinaabe Mino-Bimaadiziwin: The way of a good life*. Peterborough, Ontario: Debwewin Press.

Trahant, M. (2010) Indigenous people's literature blog. http://tempxyp2440ym0wn.blogspot.com/2010/05/life-expectancy-of-native-americans.html Retrieved June 5, 2013.

UCLA Center for Health Policy Research. (2010). *American Indian elder health care: Critical information for researchers and policymakers*. CA: Los Angeles.

U.S. Census Bureau, American Community Survey (2011). http://www.census.gov/newsroom/releases/archives/facts_for_features_special_editions/cb11-ff22.html. Retrieved December 8, 2012.

U.S. Department of Health and Human Services. (1999). *Center for disease control and prevention*. National Health Statistics Reports: Division of Health Interview Statistics.

Wasserman, E. (2005). *Understanding the effects of childhood trauma on brain development in Native children*. West Hollywood, CA: Tribal Law and Policy Institute.

Weaver, H. N., & Yellow Horse BraveHeart, M. (1999). Examining two facets of American Indian identity: Exposure to other cultures and the influence of historic trauma. *Journal of Human Behavior in the Social Environment, 2*(1/2), 19–33.

Whitbeck, L. B., Adams, G. W., Hoyt, D. R., & Chen, X. (2004). Conceptualizing and measuring historical trauma among American Indian people. *American Journal of Community Psychology, 33*(3/4), 127.

Ziibiwing Center of Anishinaabe Culture and Life Ways. (2011). *American Indian boarding schools: An exploration of global ethnic and cultural cleansing* (pp. 1–28). MI: Mount Pleasant.

Aging and Mistreatment: Victimization of Older Adults in the United States

Pamela B. Teaster, Debra A. Harley and Amani Kettaneh

Abstract Elder mistreatment is a single or repeated act or lack of appropriate action occurring within any trust relationship that causes harm to an older person. Elder mistreatment has cultural implications across the world because some societies value elders while others devalue them. Approaches to define, detect, and address elder mistreatment must be placed within a culturally sensitive context and considered alongside culturally specific risk factors. Examining Caucasians, African Americans, Native Americans, Israelis, Filipinos, and lesbian, gay, bisexual, and transgender (LGBT) elders reveals that it is a critical and growing public health and societal problem: approximately 1–2 million Americans, age 65 or older have been abused or neglected by family members and people entrusted with their care and protection. Four of every five cases of elder abuse, neglect, exploitation, and/or self-neglect go unreported to authorities. Affecting both sexes, elder mistreatment is highly correlated with disability. Elder mistreatment occurs across different cultures and ethnic groups and other facets of identities, in various settings, and most commonly at the hands of family members. Elder mistreatment most immediately affects individual elders, but prevention and intervention efforts must also include macro-level systemic factors. Appropriate funding and training for professionals is critical. Elder mistreatment is a problem that entails societal values that impede problem solutions: ageism, sexism, individualism, victim-blaming, and familial norms. Without a deep understanding of the complexities involved, elders will increasingly be mistreated.

The National Center on Elder Abuse (NCEB) and the World Health Organization (WHO) recognize elder mistreatment as a continually increasing and serious problem in the United States and worldwide. Concern over the mistreatment of older people has amplified because of increased recognition of the problem and that older adults are living longer such that by the year 2025, in developed and in

P. B. Teaster (✉) · D. A. Harley · A. Kettaneh
University of Kentucky, Lexington, USA
e-mail: pamela.teaster@uky.edu

H. F. O. Vakalahi et al. (eds.), *The Collective Spirit of Aging Across Cultures*,
International Perspectives on Aging 9, DOI: 10.1007/978-94-017-8594-5_4,

developing countries, the global population aged 60 years and older will more than double from the year 2020 (Krug et al. 2002; Phelan 2013). Elder mistreatment, elder abuse, senior abuse, abuse in later life, and abuse of older adults, is defined as "a single, or repeated act, or lack of appropriate action, occurring within any relationship where there is an expectation of trust, which causes harm or distress to an older person" (WHO 2012).

Elder mistreatment can be the result of intentional or unintentional neglect. In addition, mistreatment has cultural implications in societies across the world as some societies value their elders while others devalue them. Some cultures (e.g., China, Japan) that have long revered their elders are shifting in their attitudes due to factors such as rapid technological advancements and changing market economies. According to WHO, approaches to define, detect, and address elder mistreatment must be placed within a cultural context and considered within culturally specific risk factors. In this chapter we provide a discussion of elder mistreatment in relation to certain cultural groups, including African Americans, Native Americans, Israelis, Filipinos, and lesbian, gay, bisexual, and transgender (LGBT) elders as an initial contribution to the conversation on this critical issue. Too often, elder mistreatment is a silent crime.

Elder mistreatment is a critical and growing public health and societal problem. Approximately 1–2 million Americans, age 65 or older have been abused or neglected by family members and people entrusted with their care and protection (National Center on Elder Abuse 2005). Four of every five cases of elder abuse, neglect, exploitation, and/or self-neglect will go unreported to authorities (UK Justice Center for Elders and Vulnerable Adults 2005). Overall, prevalence studies indicate that between six and eight percent of older persons in the United States are likely to have experienced significant abuse in the last month (Acierno et al. 2010; Cooper et al. 2008). In addition, research suggests that domestic elder abuse is primarily a matter of family abuse. A significant portion of elder abuse cases reported in the United States involve spouse or partner violence, and older women are more likely than are younger women to experience violence for a longer period of time, to be in current violent relationships, and to have health and mental health problems as a result (National Center on Elder Abuse 2005). Although women are disproportionately at risk and are victims of domestic violence and mistreatment, intimate partner violence and abuse in later life occurs frequently among both women and men (Wisconsin Coalition Against Domestic Violence 2009). Elder abuse and domestic violence are highly correlated with disability (i.e., presence of a disability or acquisition of a disability as result of abuse and neglect). As the baby boom generation ages, it is highly likely that more victims of late life violence and abuse will seek out or be referred to domestic violence programs (Aravanis 2006).

To prevent and intervene in elder mistreatment, it is critical that education and training be trans-disciplinary in nature, including gerontology, public health, social work, family studies, rehabilitation counseling and disability studies, criminal justice, pharmacy, psychology, and law. In addition, policymakers need to understand elder abuse from a trans-disciplinary approach, which can best inform policy.

A multidisciplinary understanding of elder mistreatment is presented in this chapter in order to help service providers recognize signs of elder mistreatment and to be knowledgeable about the array of resources available to treat and prevent this critical issue. It is crucial for service providers across disciplines to be able to recognize signs of elder mistreatment and to be knowledgeable about an array of available resources and safety and support services. The chapter will first describe types of mistreatment, with a focus on abuse and neglect, prevalence, and risk factors, including a theoretical model of causation. Next, the signs of mistreatment in later life, cultural issues affecting informal and formal care of older persons, and barriers to intervention, including legal, forensic, and legislative issues, will be discussed. Finally, a trans-disciplinary approach to intervention will be presented with a selection of resources available.

Types, Prevalence, and Risk Factors

Elder mistreatment can take the form of physical, emotional, psychological and sexual abuse, financial exploitation, passive and active neglect by a caregiver, domestic violence, and healthcare fraud (see Table 1 for definitions of types of elder abuse). Every year, an estimated 2.1 million older Americans are victims of physical, psychological, and other forms of abuse and neglect (American Psychological Association 2005). As an example, victims of elder sexual abuse are most often females over age 70 who are totally dependent on another for care or who are functioning at a low level. Older victims are less likely to report sexual assault than are younger victims (Ramsey-Klawsnik and Teaster 2012; Roberto and Teaster 2005; Teaster and Roberto 2003). Although a variety of circumstances are regarded as elder mistreatment, definitions vary, for example, those by the National Center on Elder Abuse (2013) and Bonnie and Wallace (2003) that indicate mistreatment does not include general criminal activity such as home break-ins, muggings or distraction burglary (where a stranger distracts an older person at the door while another person enters the property to steal).

Settings of Abuse

Elder mistreatment tends to take place where the elder lives, most often in the elder's home, in a relative's home, or in a facility responsible for care, such as a long-term care facility. In the majority of cases, family members are the perpetrators (Acierno et al. 2010; Teaster et al. 2006a). Perpetrators, particularly adult children, are often dependent on their victims for financial assistance, housing, and other forms of support (Nerenberg 2008). In addition, abusers who live with the elder have opportunities to isolate him or her from the larger community so that the abuse is not discovered (Acierno et al. 2010; National Center on Elder Abuse 2005).

Table 1 Types of elder abuse

Physical abuse: Physical force that results in bodily injury, pain, or impairment. It includes assault, battery, and inappropriate restraint
Sexual abuse: Non-consensual sexual contact of nay king with an elderly person. It includes forcing them to participate in conversation of a sexual nature against their will. In addition, it may include situations where a person is no longer able to give consent (e.g., dementia)
Emotional/Psychological abuse: Infliction of anguish, pain, or distress through verbal or non-verbal actions. It includes humiliation, threats, coercion, name-calling, ridicule, constant criticism, accusations, blaming, ignoring, silence, or shunning
Domestic violence: An escalating pattern of violence by an intimate partner where the violence is used to exercise power and control
Financial/material exploitation: Illegal, improper, or unauthorized use of an elder's funds, property, assets, or other valuables. It includes changing the person's will to name the abuser as heir, fraudulently obtaining or use of a power of attorney, deprivation of money or other property, theft, undue influence, deception, coercion, eviction from their home, or misrepresentation
Neglect: Refusal, or failure, to fulfill any part of a person's obligations or duties to an elderly person. It includes depriving a person of food, heat, clothing or comfort or essential medication, or depriving a person of needed services to force certain kinds of actions, financial and otherwise. Neglect may be intentional (active neglect) or happen out of lack of knowledge or resources (passive neglect)
Self-neglect: Behaviors of an elderly person that threaten the elder's health or safety. Self-neglect (harm by self) is treated as conceptually different than abuse (harm by others)
Abandonment: Desertion of an elderly person by an individual who has physical custody of the elder or by a person who has assumed responsibility for providing care to the elder
Rights abuse: Denying the civil and constitutional rights to a person who is older, but not declared by the court to be mentally incapacitated
Scam by strangers: Sweepstakes that elderly persons must pay in order to collect winnings, fraudulent investment schemes, predatory lending, and lottery scams

Adapted from National Center on Elder Abuse (2005), National Committee for the Prevention of Elder Abuse (2008) and Shilling (2010)

Although studies vary, in a nationally representative sample, .2 % of 3,005 community-dwelling older adults (aged 57–85) reported experiencing physical abuse by a family member (Laumann et al. 2008). A study by Acierno et al. (2010) of 5,777 community dwelling elders found that a similar number reported being physically abuse (1.6 %). Other theories about mistreatment include victim vulnerability, dependency or impairment of the older person, personal characteristics of the elder (e.g., dementia, disruptive behaviors, trusting behaviors, high levels of independence), significant needs for assistance, external stress (e.g., financial problems, job stress, additional family stress), intergenerational transmission of violence, intra-individual dynamics or personal problems of the abuser (e.g., alcoholism, drug addiction, emotional disorder), and stress on the part of the caregiver (Johannesen and LoGiudice 2013; National Center on Elder Abuse 2005; Woolf 1998).

Although the majority of abuse and neglect happens in community settings, abuse and neglect can also occur in facilities (e.g., congregate living, assisted living, nursing homes) including those that appear to provide high-quality care (Krug et al. 2002). Many elders residing in long-term care settings have a dementing illness, are frail and vulnerable, and/or are dependent on their caregivers for food, housing, health needs, and activities of daily living. Physical functioning problems, activities of daily living (ADL) limitations, and behavioral problems are significant risk factors for at least half of the types of abuse and for multiple abuse types. Among older adults who have experienced at least one type of abuse, more than half (51.4 %) has experienced another type of abuse (Post et al. 2010).

Paid caregivers or residents in long-term care settings can perpetrate physical, caretaking, verbal, emotional, neglect, and material mistreatment. Causes of abuse and neglect in care settings may be attributed to inadequate staffing levels, poorly trained staff, low pay for employees, and inadequate medical and nursing care. Some facilities do not have enough direct care providers to meet levels recommended by federal officials and experts (Payne and Fletcher 2005). Employees in care settings may not receive enough training regarding techniques for working with patients with disabilities and appropriately documenting their conditions. A combination of inadequate pay, job burnout, responsibility for too many residents on a shift, poor working conditions, and stress (both at work and at home) may result in an environment ripe for elder mistreatment. In addition, some physicians treating residents in long-term care settings may know little about them and may be pressed for time to diagnose and treat (Purcell 2012; Robinson et al. 2012).

When residents abuse, some may have dementing conditions that make them unaware or only moderately aware of what they are doing. Other residents have mental capacity and are in the facility with the express intent to abuse. These residents may find less cognitively intact and more physically frail residents an easy mark for mistreatment. Furthermore, senior bullying, which in many ways looks similar to bullying among younger age groups (e.g., name calling, insults, threatening, pointed jokes), is on the rise. Victims of senior bullying often experience considerable emotional distress and engender health-related problems as a result (Bonifas and Frankel 2012).

A form of exploitation, healthcare fraud and abuse may be carried out by a variety of professional providers of healthcare, including but not limited to healthcare administrators, doctors, nurses, nurses' aides, hospital personnel, and others. It may include over-and under medicating; not providing appropriate or timely care, but charging for it; overcharging or double-billing for medical care services; receiving kickbacks for referrals to other providers or for prescribing certain drugs; recommending fraudulent remedies for an illness or condition; and Medicaid fraud (Robinson et al. 2012).

Theoretical Constructs

In an effort to understand and explain risk factors associated with elder mistreatment, several researchers on this topic (Krug et al. 2002; Schiamberg and Gans 2000; Schiamberg et al. 2011; Teaster et al. 2006) have applied the Social-Ecological Model, which takes into account interactions that occur across a number of nested systems. The Social-Ecological Model is a comprehensive public health approach that addresses an individual's risk factors and the norms, beliefs, and social and economic systems that create the conditions for mistreatment to occur. The Centers for Disease Control and Prevention employs the Social Ecological Model as a framework to better understand violence and the effect of potential prevention strategies. The Social-Ecological Model considers the complex interplay between individual, relationship, community, and societal factors. This model allows for addressing the multiple factors that put people at risk for experiencing or perpetrating violence (Dahlberg and Krug 2002). Krug et al. (2002) used the model to describe the four factors of the environment in elder abuse.

Reilly and Gravdal (2012) applied the ecological model for family violence prevention across the life cycle to examine violence and its interconnectedness to family, community, and society and demonstrated family violence in multiple arenas of human activity. The authors concluded that educational initiatives are needed at many levels, for all disciplines and in many venues and underscore the imperative to develop collaborative educational programs for health care professionals, educators, community workers, the legal system, public safety, and governmental agencies and officials. In addition, Reilly and Gravdal stress that this model offers a holistic approach for developing comprehensive intervention programs. When applied to service providers, an overarching premise of the Social-Ecological Model is that their work is neither solitary nor rests solely on any one person.

Another model that is used to examine elder mistreatment is the Power and Control Theory (sociocultural theory, which grew out of the work of Lev Vygotsky on cognitive development) (Rieber and Hall 1997). Power and Control Theory was developed in the intimate partner violence literature and focused on the power differential between men and women [similar to Feminist Theory (Walker 1990; Burnight and Mosqueda 2011)]. There is also a power differential in the United States between older adults and younger adults that typically manifests itself as ageism. According to Burnight and Mosqueda, stemming in part from ageism, in Power and Control Theory, perpetrators often believe they have exclusive rights entitling them to mistreat, abuse, and/or neglect an older adult.

A number of other theoretical approaches have been applied to the etiology of elder mistreatment. The reader is referred to these theories for further description and detail: (a) interpersonal theory [e.g., caregiver stress theory (Wolf 2000), social exchange theory (Homans 1958), dyadic discord theory (Riggs and O'Leary 1996)], and (b) intrapersonal [e.g., social learning theory (Bandura 1978)], and (c) multisystemic theory [e.g., ecological (Bronfenbrenner 1977)].

Risk Factors for Abuse

Mistreatment of older adults affects both men and women (Acierno et al. 2010; Lifespan 2011; Teaster et al. 2006). Older women appear to be victims more often than older men, a fact that may be attributable to the longer lifespan of women

Table 2 Risk factors for elder abuse

Individual factors
Age (very old)
Severity/intensity of impairment
Cognitive and physical impairments
Financial difficulty on the part of the abuser
Resentment by family members at having to spend money on the care of the older person
Relationship factors
Depression in the caregiver
Substance abuse by the caregivers
Lack of support from other caregivers
A history of domestic violence in the home
Level of stress of caregivers/lack of resilience
Dependency of the care recipient on the abuser
Strong emotional attachment between the abused and the abuser
The elder's role, at an earlier time, as an abusive parent or spouse
Existence of disruptive behavior and aggression by the care recipient
Living arrangements, particularly overcrowded conditions and a lack of privacy
Nature of the relationship between the caregiver and the care recipient before abuse began
Dependency of the abuser on the older person, usually adult children for housing and financial assistance
Community and societal factors
Poverty
Social isolation
Reduced opportunities for social interaction
Cultural norms and traditions (e.g., sexism, ageism, cultural violence)
Perception (i.e., older people are depicted as frail, weak, dependent) and less worthy of government investment or family care than other groups

Adapted from Krug et al. (2002) and Robinson et al. (2012)

than men and the growing numbers of older adults generally. Table 2 provides a list of risk factors for elder mistreatment. Though mistreatment can occur at any age, older women who are abused differ from their younger counterparts in a number of ways. First, some older women may have been socialized with specific attitudes and values, particularly relating to gender roles, marriage, and family. Second, financial barriers have greater implications for older women. Many older women did not engage in paid employment when they were younger and, even those in their preretirement years may be unemployable because of lack of work experience as well as unfounded ageist beliefs concerning an older workforce. Third, older women are more likely than younger women to have health or functional problems that may make them dependent on someone for care, and consequently, make it more difficult for them to seek help or to leave an abusive relationship. Fourth, because of a strong care ethic and a commitment of typically remaining true to their marriage vows to stay together in sickness and health, it can be extremely difficult for an older woman to leave a dependent, yet abusive partner (Seaver 1996; Vinton 1991). Finally, older women may have become resigned to living in situations of long-standing abuse and may be unable to realize that there are choices available to them (Straka and Montminy 2006; Teaster et al. 2006). Indeed, older women may remain in abusive relationships because of cohort, period, and aging effects (Zink et al. 2003).

Elders, Intimate Partner Violence, and Risk Factors

One phenomenon of elder mistreatment is that of domestic abuse in later life. Too often, older women who experience domestic violence are an invisible group who fall into the gap between elder abuse and domestic violence Teaster et al. (Teaster et al. 2006; Straka and Montminy 2006). Mentioned above, many older women have endured years of abuse from their husbands or male partners. In fact, spouses make up a large percentage of elder abusers, and a substantial proportion of these cases are domestic violence grown old (National Center on Elder Abuse 2005; Teaster et al. 2006). Risk factors for this form of abuse may be moderated by cultural values and family dynamics. For example, in Latino (the general term used in this chapter to be inclusive of various populations) families, married Latina elders are at a higher risk for being victims of abuse or neglect than are Latina elders who are not in marital relationships because of gender role expectations that "mandate" women to tolerate abuse and focus on serving others, economic dependence, and societal or religious expectations disapproving of marital dissolutions (Parra-Cardona et al. 2007). Similarly, many older African American women are only willing to report more serious maltreatment, particularly if they are used to the abuse from their husbands as a behavior that has occurred for a long period of time (Brown 2000). Women in rural areas who have experienced mistreatment are mistrustful of local law enforcement and other local agencies because of the tight networks that bind small communities (Teaster et al. 2006).

Indicators of Mistreatment in Later Life

Tactics used by abusers are remarkably similar, no matter the age of the victim. Abusers frequently scour the environment for elders they can dominate, people believed to be weak, and/or people unlikely or unable to retaliate (Aravanis 2006). A desire for power and control is present in many types of abusive relationships of elders particularly those that involve intimate partner violence [National Clearinghouse on Abuse in Later Life (NCALL) 2013]. The exercise of power and control is revealed in both physical and emotional manifestations.

Generally, signs of mistreatment include frequent arguments or tension between the perpetrator and the elder as well as changes in personality or behavior in the elder (Robinson et al. 2012). The most obvious signs of abuse are due to physical abuse, which includes battery, assault, and inappropriate restraint. Indicators include broken bones, fractures, sprains, choke marks, and dislocations; burns from cigarettes, hot water, and appliances; abrasions on arms, legs, or torso resembling rope or restraints; and bruises, including bilateral and "wrap around" bruises. There may also be unexplained injuries or explanations that do not fit the pattern of injury, as well as a history of similar injuries and numerous or suspicious hospital visits A victim may be brought to many different medical centers to prevent medical personnel from noticing a pattern of abuse, or medical care may not be sought at all. In addition, there may be unexplained delays between onset of injury and seeking medical attention for which it is obvious that injuries have not received proper care. For bedridden adults, neglect injuries (e.g., bed sores) may be present. Other potential indicators of physical abuse are dehydration, signs of malnourishment (e.g., sunken eyes, loss of weight), hypothermia, and pain upon being touched (National Coalition Against Domestic Violence 2007; Woolf 1998).

Signs of sexual abuse, such as trauma to the breasts and genital area, are frequently overlooked in investigations of physical abuse among older people (Ramsey-Klawsnik and Teaster 2012). Other signs include venereal disease, genital infections or unusual discharge or smell, unexplained vaginal or anal bleeding, torn, stained, or bloody underclothing, and indicators common to psychological abuse, which may be concomitant with sexual abuse (Robinson et al. 2012; Woolf 1998; Women of Color Network 2008). The most common types of sexual abuse of older women living in facilities involved instances of sexualized kissing and fondling and unwelcomed sexual interest in the women's body by older male residents in the nursing home (Teaster and Roberto 2003, 2004). In national work by Acierno et al. (2010) sexual abuse was reported by fewer than 1 % of individuals.

Signs of psychological or emotional abuse include ambivalence, deference, passivity, shame, mild to severe anxiety, confusion, and disorientation. Behavioral signs of psychological abuse may manifest themselves as trembling, clinging, cowering, lack of eye contact, hypervigilance, agitation, and evasiveness. Family members are the most frequent perpetrators, accounting for 57 % of emotional abuse events (Acierno et al. 2010). In many cases, older adults know when they are being abused; unfortunately, they are often powerless to advocate on their own behalf. Some are culturally bound not to do so. Cultural implications related to elder mistreatment are discussed in the following section.

In addition to "power and control" issues as victim indicators of physical, psychological, emotional, and sexual abuse, NCALL (2013) describes suspect indicators and environmental indicators of abuse and neglect. Power and control in the mistreatment of elders result from deviance and dependence of abusers (Pillemer and Finkelhor 1989). Examples of suspect indicators include the following: providing inconsistent and conflicting explanations of the older adult's injuries; belittling, threatening, or insulting the older adult; handling the older adult roughly; ignoring the older adult's need for assistance; or demonstrating reluctance to help the older adult. Suspect indicators may also include a past history of being abusive; not speaking to or providing companionship to the older adult; and isolating him or her from the outside world, friends, or relatives. The abuser shows behaviors of controlling and dominating the older adult, always being present when anyone talks with the older adult, speaking for him or her, and being overly protective or defensive. The abuser portrays him or herself as a victim or as the only caring person in the older adult's life. Often, the abuser may be charming and helpful toward professionals and the victim while others are present, abuses the older adult's pets, controls and dominates the older adult's life or activities, and justifies and minimizes his or her own actions.

Environmental indicators include an unclean or unsanitary living environment, strong orders of urine and/or feces, lack of food, lack of medication or assistive devices if needed by the older adult, lack of heat or air cooling systems, electricity, or running water, safety issues, abused or neglected pets, damage to home caused by abusive behavior (e.g., broken door frames, holes punched in walls, and broken items), evidence of hoarding, house, roof, yard in disrepair, infestation of insects or rodents, and dangerous environment due to basic safety and health standards not being met.

In general, signs that someone may be harming an older adult include a physical assault ranging from a pinch to a punch, intimidation to verbal threats, limited social contact to isolation, stealing to theft by deception or embezzlement, and inappropriate enmeshed relationship to sexual assault. The signs of elder abuse in later life are not mutually exclusive in classifications of abuse, violence, or neglect. These classifications overlap and represent similar symptomology and description. Suspect indicators may be present in a spouse or partner, family member, or caregiver.

The Intersection of Cultural Issues and Elder Mistreatment

It is critical for service providers working with the elders to understand differences in perceptions about elder mistreatment as a function of culture. According to Tauriac and Scruggs (2006), "because abuse is a culturally defined phenomenon, an emic definition (specific to a cultural group) is crucial for enriching the understanding of what abuse means" to ethnic minority groups (p. 38). For ethnic minority American elders, cultural identity is a characteristic too important to lose to the aging process, especially when a nursing home or other facility mistakenly defines all older persons in the same way, thus creating a melting pot rather than individual identities (Orlovsky 2004). Culture includes values, beliefs, and traditions that significantly affect family life. Cultural values dictate the roles of family members, their interactions with one another, how decisions are made within families, how family resources are distributed, and how problems are defined. Further, culture influences how families cope with stress and their propensity to initiate help-seeking behaviors (National Committee for the Prevention of Elder Abuse 2003). With regard to elder care in culturally diverse families, several questions should be explored (see Table 3).

Table 3 Question to ask of diverse cultures

What conduct is considered abusive?
To whom will families turn for help?
Who make decisions about family resources?
What role do elders play in the family? In the community?
Under what circumstances will families seek help from outsiders?
Is it considered abusive to use elder's resources for the benefit of other family members?
When did immigrant elders come to the United States and under what circumstances?
Who within the family is expected to provide care to elderly members? What happens when they fail to do so?
What religious beliefs, past experiences, attitudes about social service agencies or law enforcement, or social stigmas may affect community members' decisions to accept or refuse help from outsiders?

Adapted from the National Committee for the Prevention of Elder Abuse (2003)

Podnieks et al. (2010) conducted the first large-scale global survey of elder abuse in order to understand the nature and response to the problem in individual countries worldwide. A total of 53 countries from the six World Health Organization regions responded to the study, with a total of 190 surveys representing 362 respondents. The researchers found that elder abuse was recognized as a problem that exists in all responding countries and that no single definition of elder abuse applied globally. However, definitions most often used were those of the International Network for the Prevention of Elder Abuse and the National Research Council (Bonnie and Wallace 2003). Poverty and lack of social support were considered leading causes of elder abuse in developing countries while living alone and lack of social support were regarded as elder abuse contributors in developed countries. Most professionals and the general public seemed to lack an understanding or awareness of laws that protect older adults. There is little specialized training and research on elder abuse outside of a few select countries, such as the United States and Canada.

Although "there is a scarcity of theoretical frameworks that describe precursors and dynamics associated with elder abuse and neglect" in ethnic minority cultures (Parra-Cardona et al. 2007, p. 451; Women of Color Network 2008), Bronfenbrenner's (1977) ecological model has been employed in studies to explain elder abuse and the role that culture plays (Horsford et al. 2011; Parra-Cardona et al. 2007). When applied within a sociocultural context, ecological theory helps to identify variables for consideration in the study of ethnic minority families affected by elder mistreatment (Burnight and Mosqueda 2011). Schiamberg et al. (2011) proposed the ecological model as framework for understanding elder abuse in nursing homes. In addition, family systems theory has been used in mental health issues and interventions with Muslims to explain broad patterns of family structure and process of influences of interactional and intergenerational behaviors. In fact, family systems theory is considered a practical model with Muslim families because it examines the interactions and roles of each member across generations as a potentially reciprocal influence and affect on the emotional makeup and behaviors of all members and their relationship patterns within the entire system (Daneshpour 2012).

In the presentation of cultural issues and abuse of ethnic minority elders, it is important to stress that cultural groups are hardly monolithic; they involve both intracultural and generational differences. For example, in many Latino populations, the elder may be foreign-born and the caregiver U.S.-born, which affects family dynamics and values. Elders in ethnic minority families may have cultural values and beliefs that discourage them from reporting abuse. These values include collectivism (putting the family's or group's needs before self), avoiding bringing shame (*pena*) to the family, family harmony, gender role expectations, hierarchal family structure, extended family, kinship ties and interdependence, strong religious ties, and reverence towards elders. These characteristics explain ways in which multigenerational, immediate and extended families, and fictive kin live together and share financial and familial responsibilities (Tauriac and Scruggs 2006). However, reverence towards elders does not necessarily prevent mistreatment.

The remainder of this section presents information on specific cultures. Due to space limitations, all cultures cannot be addressed. The intent is not to exclude any particular culture but rather to highlight some information that does not typically appear in the literature.

Cultural Considerations for Native Americans

According to customs and traditions of American Indian and Alaska Native communities, native elders are regarded as repositories of knowledge and as invaluable community resources. Traditionally, elders held positions of power and were cherished for their experience and wisdom. Although many tribes, villages, and pueblos continue to hold elders in high esteem, elder abuse is on the rise. Among many tribal people, the recent trend of elder abuse and neglect is viewed as a symptom of the loss of tribal cultural values and as an indicator of serious imbalance within their communities (White 2004). Less abuse occurred on wealthier reservations, with reports of abuse secondary (not deliberate) neglect. Tribal elders and other members regarded abuse as a health problem caused primarily by community, rather than individual or family dysfunction (Maxwell and Maxwell 1992). Tribal elders in rural areas may be isolated and living in poverty. For Native American victims in urban settings, gender was found to be the most significant risk factor, with women 9.4 times more likely than men to be physically abused, and in 88 % of the cases, the abuse concerned a male harming a female (Buchwald et al. 2000).

Cultural Considerations for African Americans

Cultural differences in the definition of elder abuse also exist between African Americans and other ethnic groups. Hudson et al. (1999) examined African Americans' (aged 40 and over) attitudes toward elder abuse and found that they did not view a single incidence of yelling or swearing at, slapping, or hitting an elder as sufficient to warrant the label of elder abuse. Similarly, Tauriac and Scruggs (2006) examined perceptions of extreme, moderate, and mild forms of elder mistreatment among African Americans and found that they emphasized physical abuse when giving examples of extremely abusive behavior. Along with physical abuse, verbal abuse was the most frequently identified form and was significantly related to age in complex ways. They found that the older the person, the more he or she perceived verbal abuse as an extreme form of mistreatment, while the younger the person, the more he or she perceived verbal abuse as a mild form of mistreatment.

Cultural Considerations for Groups Receiving Little Attention: Israelis, Filipinos, and LGBT Elders

Israeli Elders

In a study on Israeli (i.e., Ashkenazi and Sephardic) perspectives on elder abuse, Rabi (2006) found that psychological neglect, neglect/abandonment, and physical aggression were considered the most prominent exemplars of elder abuse within Israeli populations and across gender and ethnicities. The Ashkenazi group regarded physical and economic abuse as forms of moderate abuse, and neglect/abandonment and disrespect as forms of mild abuse. Israeli women place more emphasis on disrespect than their male counterpart.

Filipino Elders

An ethnic group that is rarely separated from their families and found living in nursing homes is Filipinos. Similar to other Asian societies, Filipinos embrace the value of *filial piety* (duty to care for one's parents). For example, placement of Filipino elders in nursing homes is considered by their families to be abandonment (de Guzman et al. 2012). However, Guzman et al. point out that the concept of family loyalty has changed over time and that family responsibilities are understood differently across generations. In general, ethnic minority elders are often poorer, less educated, unemployed, more prone to illness than are nonminority elders, and tend to visit medical specialists less frequently (Filippo et al. 2007).

LGBT Elders

Also included in any classification of cultural diversity is the population of sexual minorities [i.e., lesbian, gay, bisexual, transgender (LGBT)]. Sexual minorities cut across ethnic groups, gender, socioeconomic status, geographic location, religion, and a host of other intersecting identities and social positions. Sexual minorities have rarely been the focus of research in the literature on abuse beyond intimate partner violence in younger populations and are particularly scarce in the literature on elder mistreatment. Due to the lack of focus on older LGBT persons, one may speculate that researchers believe that as LGBT populations age they somehow morph into another existence, or into nonexistence. That is, many scholars omit older LGBT populations from curricula and research to the extent that it would appear that sexual minorities do not exist (Hall and Fine 2005; Harley et al., in this book). Failure to include explicit research of sexual orientation and gender identity among elders limit the ability of service providers and healthcare professionals to address the complete needs and issues of LGBT older adults. A research focus and collecting data on older LGBT populations is an important principle of person-centered care because as service providers know about the individual client, the better the service they will be able to provide (National Resource Center on LGBT Aging 2013).

Not surprisingly, LGBT elders face the same types of abuse, violence and neglectful situations as their heterosexual counterparts; however, they encounter additional exploitation because of their sexual identity. For example, homophobic mistreatment of LGBT older adults may manifest in such ways as nursing home staff refusing to bathe an elder because he or she does not want to touch anyone who is lesbian or gay, or an elderly transgender person with Alzheimer's disease may be forced by healthcare facility staff to dress in ways abhorrent to the elder, thus disrespecting his or her gender identity. In a study of victimization of older LGBT adults, Balsam and D'Augelli (2006) found that two-thirds had experienced some form of verbal abuse because of their sexual orientation and 29 % had been threatened with violence. The LGBT elders also faced threats of being "*outed*" (others disclosing their sexual orientation) if they reported abuse or neglect. The threat of being "*outed*" is a particularly frightening possibility for older adults who have lived much of their lives during a time when being LGBT was either illegal or classified as a psychiatric disorder in the *Diagnostic and Statistical Manual of Mental Disorders*.

It is imperative that the role of culture as it concerns abuse and neglect be well understood. Elderly populations represent perhaps the most diverse of age groups because of their rich array of life experiences, interactions, and length of time lived (Tornstam 2005). Unique cultural experiences of older adults and their beliefs and attitudes about abuse and cultural expectations must be considered for any type of intervention to be truly successful. Failure to consider cultural attributes may serve to only create additional barriers to addressing the needs of a vulnerable population, such as confounding intervention and legislative strategies discussed below.

Considerations for Intervention Strategies

A multiplicity of factors and complex issues arise in the prevention, identification, and intervention of elder abuse and neglect. In addition to cultural beliefs and practices described earlier, many ethnic minority elders may be reluctant to seek help because of the possible risk of being separated and isolated from family, friends, and community (Women of Color Network 2008). For some ethnic minorities, immigration status, fear of deportation, or jeopardizing citizenship status may prevent them from reporting elder abuse and neglect. Other barriers to intervention are mistrust toward social services and agencies, law enforcement, and criminal justice systems; histories of oppression; social exclusion, discrimination, structural segregation and racism; and linguistic and communication challenges. Many of these experiences can increase stress levels in the family, blur the lines of what constitutes abusive behavior, and interfere with help-seeking (Abugideiri 2012).

As an example, among tribal communities living on reservations, three major challenges exist: (a) a lack of protocols for addressing elder abuse issues, (b) a lack of policies and procedures for tribal agencies handling elder abuse and neglect issues, and (c) a need for increased training on elder abuse and neglect (White 2004). These barriers are compounded by major gaps in knowledge regarding definitions of abuse and neglect; the incidence, prevalence, and types of elder mistreatment; and current preferred treatment strategies. In addition, very limited statistics on American Indian elder mistreatment exist, and state mandatory reporting structures do not apply to Tribal groups living on reservations due to Tribal sovereignty (Holkup et al. 2007). Barriers to providing services to American Indians range from the reluctance of some Indian victims, their families, or others to acknowledge that abuse is occurring at all to such a paucity of resources in the social service system that that it effectively negates any services that are offered (Baldridge et al. 2004).

Preexisting Trauma

A related and frequently overlooked factor in elder mistreatment is that of preexisting trauma (e.g., inner city crimes, torture and maltreatment in refugee camps, war or politically motivated abuse, homophobia) experienced by elders (Acierno et al. 2010; Davis and Waites 2008; Schnurr et al. 2002). The way in which the elder viewed and resolved preexisting trauma may either facilitate or hinder intervention. Unresolved trauma that occurred early in life may present barriers to help seeking for older adults. Early life trauma may include physical abuse, sexual abuse, neglect, or discrimination. Those elders who experienced trauma across an intersection of identities (e.g., race, gender, sexual orientation), as they aged and became reliant on others for care, especially forced vulnerability to those who may otherwise discriminate against them, may experience substantial stress. For sexual minorities, in particular, homophobia serves to keep victims from seeking help. Legal discrimination against them may make it extremely difficult for them to leave an abusive relationship, and internalized homophobia and social prejudice may lead to self-neglect and isolation (Balsam and D'Augelli 2006).

Issues related to intimate partner violence cut across all races and ethnicities and sexualities in old age. For example, in a study of domestic violence against older women, eleven concepts appear to be connected to barriers to help seeking. Of these concepts, seven are connected to both domestic abuse and barriers to help-seeking and include isolation, jealousy, intimidation, protecting the family, self-blame, powerlessness, and spirituality. In addition, isolation, jealousy, and intimidation are abuser driven. Those concepts related more directly to barriers to help seeking only include hopelessness, secrecy, concern for the abuser, and the justice system response. The concepts of protecting family, self-blame, powerlessness, secrecy, hopelessness, and concern for the abuser are victim driven. The justice system response concept does not appear to fit into either of these categories (Dunlop et al. 2005). Dunlop et al. indicate that the preponderance of participants in the study, both non-victims and victims, view the justice system as the correct and appropriate cluster of institutions to address individual incidents of domestic abuse. In addition, both groups of participants overwhelmingly agreed that emotional abuse is as bad as, or worse than, physical abuse.

A major barrier in the implementation of services for older adults experiencing intimate partner violence is that the responses and interventions of many women's shelters are designed for younger victims of abuse and domestic violence. Often, shelter staff members do not have knowledge about issues of aging, disability, or health and mental health issues that elders may experience. For older victims, restrictions on the length of time women can stay in shelters may create further barriers, especially when the victim has no other recourse. Moreover, expectations for what victims will be able to do, such as enter the workforce, may be unreasonable for older victims given their age, support networks, education, health and other life experiences (Reingold 2006; Vinton 1992).

Finally, geographical location may impose additional barriers to help-seeking and increase risk for continued abuse and neglect for older adults. For example, people in rural areas tend to have fewer available services, a lack of access to public transportation, are more isolated, and have decreased access to family support because of the family's relocation. Older adults living in rural areas may have inadequate healthcare and limited access to aging network services. In a study of elder abuse and neglect among rural and urban women, more rural women were represented in physical abuse, emotional abuse, and deprivation categories (Dimah and Dimah 2004). The factors of poverty, remoteness, and higher levels of need for medical and mental health care among older adults in rural areas further exacerbate an elder's risk potential for abuse and neglect. In addition, issues related to culture can either enhance or impede intervention and prevention efforts.

Legal, Forensic, and Legislative Issues

Although federal and state laws address elder abuse, neglect, and exploitation, state law is the primary source of sanctions, remedies and protections related to elder abuse. According to Straka and Montminy (2006), responses to elder abuse are largely shaped by legislative context. Several federal laws relate specifically to elder abuse and neglect; however, none of these laws provides broad regulatory mechanisms for state or local programs established specifically to support services for victims of elder abuse (Center for Elders and the Courts, nd). Key federal laws include these:

1. *Elder Justice Act of 2009*, which was enacted in March 2010 as part of the Patient Protection and Affordable Care Act (P.L. 111–148). The Act, passed but as yet unfunded, coordinates elder abuse detection and prevention programs within the Office of the Secretary of Health and Human Service (HHS). It established the *Elder Abuse Coordinating Council* and an *Advisory Board on Elder Abuse, Neglect, and Exploitation*. The Act also requires owners, operators and employees of long-term care facilities to report suspected crimes committed in their facilities and to provide 60 days written notice to the HHS Secretary and the state of a facility's impending closure. The notice must include a plan for transfer and adequate relocation of all residents. See Table 4 for components authorized by the Act.
2. *Older Americans Act* (under Title 7) contains definitions of elder abuse and authorizes federal funding for the National Center on Elder Abuse, which collaborates with several organizations to promote and support elder abuse awareness initiatives, multidisciplinary responses to elder mistreatment, and professional training and education.
3. *Violence Against Women Act*, which established federal domestic violence that may be applied in cases of elder abuse. These federal crimes are limited to abuse that occurs in the federal territories or involves crossing state, federal, or tribal boundaries in the commission or attempted commission of a crime of violence against an intimate partner, to stalk or harass, or to violate a qualifying protection order.
4. *Elder Victims Act of 2009* is designed to authorize grant funding for training state and local prosecutors, courts, and law enforcement personnel handling elder justice-related matters. This act also established the Elder Serve Victim program to facilitate and coordinate emergency services to victims of elder abuse (Center for Elders and the Court, nd.).

Other avenues of protection against elder abuse include state laws, adult protective services, criminal law, civil remedies, probate codes or trusts and estates statutes, the Long-Term Care Ombudsman Program, and institutional abuse laws. One area in which the law and forensics have fallen short is their failure to

Table 4 Components authorized by the Elder Justice Act[a]

Stationary and mobile elder abuse forensic centers to develop forensic markers, methodologies for intervention, forensic expertise, and capacity to collect forensic evidence
Enhancement of long-term care through programs to recruit, train and retain long-term care staff; programs to improve management practices; and adoption of standards for electronic exchange of clinical data
Grants to enhance the provision of adult protective services by state and local agencies and to conduct demonstration programs to test training in and methods to detect and prevent elder abuse and financial exploitation
Grants to support long-term care ombudsman programs
Evaluations of grant funded activities
A national institute for training, technical assistance, and development of best practices to improve investigations of elder abuse reported in long-term care facilities

[a] As noted in the text, the Elder Justice Act has been passed, but no money has been devoted to its implementation

Adapted from the Center for Elders and the Courts

consider the link between intimate partner violence and elder abuse, which in turn limits knowledge about the most effective response strategies for this group of older victims (Payne 2008).

Trans-Disciplinary Approach to Intervention

Elder mistreatment permeates every segment of society. Informed citizens and professionals who work with the elder population must understand the magnitude of violence against older people, as it affects them as well as their families and future generations. Indeed, professionals need better training around intervening with perpetrators, planning for the safety of elders and effective response strategies, understanding the dynamics that distinguish mistreatment involving older adults from that involving younger victims, recognizing legal and forensic procedures, and recognizing age, gender, and culturally specific service needs (Payne 2008).

Thus, as suggested above, the provision of services to elders requires a trans-disciplinary approach in order to develop effective interventions and relevant social policies in response to abuse and neglect. Payne (2008) offers several policy recommendations for human and social service programs: (1) current elder abuse training content should be reviewed to make sure that the trainings include a focus on interventions with abusers, communicating lethality, and worker safety; (2) development of specific training courses devoted to elder abuse cases involving domestic violence; (3) development of policies mandating or requiring workers to participate in domestic violence training; (4) development of elder abuse/domestic violence Web-learning to enhance current training techniques; and (5) collaborative training between domestic violence professionals, adult protective services, and other human services professionals.

Similar to trans-disciplinary approaches, in the 1950s, Madeleine Leininger introduced transcultural nursing to the healthcare field to improve the cultural needs of patients (Orlovsky 2004). Other disciplines in human services and social sciences have followed suit with the inclusion of cultural competence as a component of their codes of ethics (e.g., American Psychological Association, Certified Rehabilitation Counselor Commission, National Association of Social Workers) to accommodate cultural differences.

A trans-disciplinary approach to both prevention and intervention can lead the way for health and human service providers, law enforcement, and community partners as first responders to maximize care and safety of elders who are victims of mistreatment. In addition, a continuum of trans-disciplinarily driven prevention and intervention initiatives may assist in a better understanding of human development and a lifespan perspective of elders. As service providers from many disciplines come together in response to elder mistreatment, the potential for an effective, holistic approach may well be the outcome.

Conclusion

Elder mistreatment is an expanding problem because of an older adult population that is living longer than ever before in history. A variety of factors discussed earlier in this chapter put elders at risk for mistreatment. Elder mistreatment occurs across different cultures and ethnic groups and other facets of identities, in various settings, and most commonly at the hands of family members.

The magnitude of the problem is gaining visibility at local, state, and national levels. Elder mistreatment most immediately affects the individual elder, but efforts for prevention and intervention must take other systemic factors into account if any efforts are to be sustained and successful. It is crucial that agencies and entities that provide services to elders and address the problem of elder mistreatment are equipped with adequate and appropriate resources to do so. This imprimatur means appropriate funding and training for professionals. At this point, many curricula in many disciplines barely address the problem and too many do not at all. Elder mistreatment should be recognized as a problem that is reflective of societally held values, some of which impede efforts to address the problem: ageism, sexism, individualism, victim-blaming, and familial norms. Initiatives that fail to take into account challenges and opportunities at all levels will inevitably prove unsuccessful, despite the best of intentions or for that matter, a misdirected and unaccountable infusion of dollars.

Finally, it is critical to understand those who perpetrate mistreatment as well as the systems that are designed to intervene to address the problem. Such efforts will be best approached when researchers and educators enter into efforts with practitioners and elders themselves so that evidence-based approaches to the problem are developed and implemented. Without a deep understanding of the complexities involved in elder mistreatment, abuse and neglect of elders will continue to increase rather than be abated by a holistic, comprehensive approach to decrease and, eventually eliminate, the problem.

References

Acierno, R., Hernandez, M. A., Amstadter, A. B., Resnick, H. S., Steve, K., Muzzy, W., et al. (2010). Prevalence and correlates of emotional, physical, sexual, and financial abuse and potential neglect in the United States: The national elder mistreatment study. *American Journal of Public Health, 100*, 292–297. doi:10.2105/ajph.2009.163089.

American Psychological Association. (2005). *Elder abuse and neglect In search of solutions*. Washington, DC: Author. Available at http://www.apa.org/pi/aging/eldabuse.html

Aravanis, S. (2006). Domestic violence in later life: A guide to the aging network for domestic violence and victim service programs. *Issue Brief. National Center on Elder Abuse*. Washington, DC: National Center on Elder Abuse.

Baldridge, D., Nerenberg, L., & Benson, W. F. (2004). *A review of the literature. Elder abuse in Indian country: Research, policy, and practice*. Washington, DC: National Indian Council on Aging.

Balsam, K. F., & D'Augelli, A. R. (2006). The victimization of older LGBT adults: Patterns, impact and implications for intervention. In D. Kimmel, T. Rose, & S. David (Eds.), *Lesbian, gay, bisexual, and transgendered aging: Research and clinical perspectives* (pp. 110–130). New York: Columbia University Press.

Bandura, A. (1978). Social learning theory of aggression. *Journal of Communication, 28*(3), 12–29.

Bonifas, R., & Frankel, M. (2012). *The senior bullying series. Part one: What is bullying*. Retrieved May 21, 2013 from http://www.mybetternursinghome.com/senior-bullying-part-1-what-is-bullying/

Bonnie, R. J., & Wallace, R. B. (Eds.). (2003). *Elder mistreatment: Abuse, neglect, and exploitation in an aging America*. Washington, DC: National Academies Press.

Bronfenbrenner, U. (1977). Toward an experimental ecology of human development. *American Psychologist, 32*, 513–531.

Brown, E. A. (2000). Elder mistreatment in the African American community. *Research Perspective, 7*, 105–114.

Buchwald, D., Tomita, S., Hartman, S., Furman, R., Dudden, M., & Manson, S. M. (2000). Physical abuse of urban native Americans. *Journal of General Internal Medicine, 15*, 562–564.

Burnight, K., & Mosqueda, L. (2011). *Theoretical model development in elder mistreatment*. Washington, DC: U.S. Department of Justice.

Center for Elders and the Courts (n.d.) http://www.eldersandcourts.org/

Cooper, C., Selwood, A., & Livingston, G. (2008). The prevalence of elder abuse and neglect: A systematic review. *Age and Aging, 37*, 151–160.

Dahlberg, L. L., & Krug, E. G. (2002). Violence: A global public health problem. In E. Krug, L. L. Dahlberg, J. A. Mercy, A. B. Zwi, & R. Lazano (Eds.), *World report on violence and health* (pp. 1–56). Geneva: World Health Organization.

Daneshpour, M. (2012). Family systems therapy and postmodern approaches. In A. Ahmed & M. M. Amer (Eds.), *Counseling Muslims: Handbook of mental health issues and interventions* (pp. 119–134). New York: Routledge.

Davis, M. E., & Waites, C. (2008). Intergenerational caregiving: Family and long-term care. In C. Waites (Ed.), *Social work practice with African-American families: An intergenerational perspective* (pp. 143–168). New York: Routledge.

de Guzman, A. B., Lacorte, J. C., Lacsamana, A. K. G., Lagac, M. L. M., Laguador, J. M., Lapid, J. J. R., et al. (2012). Who says there is no life after abandoned Filipino elderly in nursing homes. *Educational Gerontology, 38*, 890–901.
Dimah, K. P., & Dimah, A. (2004). Elder abuse and neglect among rural and urban women. *Journal of Elder Abuse and Neglect, 15*(1), 75–93.
Dunlop, B. D., Beaulaurier, R. L., Seff, L. R., Newman, F. L., Malik, N., & Fuster, M. (2005). *Domestic violence against older women: Final technical report*. North Miami: The Center on Aging of Florida International University.
Filippo, S. M., Reiboldt, W., White, B., & Hails, J. (2007). Perceptions of elderly self-neglect: A look at culture and cohort. *Family and Consumer Sciences Research Journal, 35*(3), 215–231.
Hall, R. L., & Fine, M. (2005). The stories we tell: The lives and friendship of two older black lesbians. *Psychology of Women Quarterly, 29*, 177–187.
Harley, D.A., Stansbury, K., Nelson, M., & Espinosa, C. T. (in press). A profile of rural elderly African American lesbians: Meeting their needs. In H. F. O. Vakalahi, G. M. Simpson & N. Giunta (Eds.), *The collective spirit of aging across cultures*. New York: Springer.
Holkup, P. A., Salois, E. M., Tripp-Reimer, T., & Weinert, C. (2007). Drawing on wisdom from the past: An elder abuse intervention with tribal communities. *Gerontologist, 47*(2), 248–254.
Homans, G. C. (1958). Social behavior as exchange. *The American Journal of Sociology, 63*(5), 597–606.
Horsford, S. R., Parra-Cardona, J. R., Post, L. A., & Schiamberg, L. (2011). Elder abuse and neglect in African American families: Informing practice based on ecological and cultural frameworks. *Journal of Elder Abuse and Neglect, 23*, 75–88.
Hudson, M., Beasley, C., Benedict, R., Carlson, J., Craig, B., & Mason, S. (1999). Elder abuse: Some African American views. *Journal of Interpersonal Violence, 14*, 915–939.
Johannesen, M., & LoGiudice, D. (2013). Elder abuse: A systematic review of risk factors in community-dwelling elders. *Age and Aging, 42*, 292–298.
Krug, E. G., Dahlberg, L. L., Mercy, J. A., Zwi, A. B., & Lozano, R. (Eds.). (2002). *Abuse of the elderly. World report on violence and health* (pp. 124–145). Geneva: World Health Organization.
Laumann, E. O., Leitsch, S. A., & Waite, L. J. (2008). Elder mistreatment in the United States: Prevalence estimates from a nationally representative study. *The Journals of Gerontology Series B: Psychological Sciences and Social Sciences, 63*, S248–S254.
Lifespan. (2011). *Elder Abuse: Under the Radar*. Lifespan of Greater Rochester Inc. Retrieved from http://www.lifespanroch.org/documents/ElderAbusePrevalenceStudyRelease.pdf
Maxwell, K. E., & Maxwell, R. J. (1992). Insults to the body civil: Mistreatment of elderly in two Plains Indian tribes. *Journal of Cross-Cultural Gerontology, 7*(11), 3–23.
National Center on Elder Abuse. (2005). *Fact sheet. Domestic violence: Older women can be victims too*. Washington, DC: Author.
National Center on Elder Abuse (2013). What is elder abuse? Retrieved from http://www.ncea.aoa.gov/faq/index.aspx
National Clearinghouse on Abuse in Later Life (NCALL). (2013). Retrieved March 11, 2013 from http://www.ncall.us/information/landing
National Coalition Against Domestic Violence. (2007). *Abuse in later life*. Washington, DC: Author.
National Committee for the Prevention of Elder Abuse. (2003). The role of culture in elder abuse. Retrieved January 23, 2013 from http://www.preventelderabuse.org/issues/culture.html
National Resource Center on LGBT Aging. (2013). *Inclusive questions for older adults: A practical guide to collecting data on sexual orientation and gender identity*. New York: Services and Advocacy for LGBT Elders.
Nerenberg, L. (2008). *Elder abuse prevention: Emerging trends and promising strategies*. New York: Springer Publishing Company.
Orlovsky, C. (2004). Transcultural nursing: Providing cultural care for the elderly. Retrieved January 23, 2013 from http://www.nursezone.com/printArticle.aspx?articleID=21956

Parra-Cardona, J. R., Meyer, E., Schiamberg, L., & Post, L. (2007). Elder abuse and neglect in Latino families: An ecological and culturally relevant theoretical framework for clinical practice. *Family Process, 46*(4), 451–470.
Payne, B. K. (2008). Training adult protective services workers about domestic violence: Training needs and strategies. *Violence Against Women, 14*(10), 1199–1213.
Payne, B. K., & Fletcher, L. B. (2005). Elder abuse in nursing homes: Prevention and resolution strategies and barriers. *Journal of Criminal Justice, 33*(2), 119–125.
Phelan, A. (Ed.). (2013). *International perspectives on elder abuse*. New York: Routledge.
Pillemer, K., & Finkelhor, D. (1989). Causes of elder abuse: Caregiver stress versus problem relatives. *American Journal of Orthopsychiatry, 59*(2), 179–187.
Podnieks, E., Anetzberger, G. J., Wilson, S. J., Teaster, P. B., & Wangmo, T. (2010). WorldView environmental scan on elder abuse. *Journal of Elder Abuse and Neglect, 22*, 161–179.
Post, L., Page, C., Conner, T., Prokhorov, A., Fang, Y., & Biroscak, B. J. (2010). Elder abuse in long-term care: Types, patterns, and risk factors. *Research on Aging, 32*(3), 323–348.
Purcell, B. (2012). Elder abuse information. Retrieved January 20, 2013 from http://www.elder-abuse-information.com/abuse/abuse_causes.htmm
Rabi, K. (2006). Israeli perspectives on elder abuse. *Educational Gerontology, 32*, 49–62.
Ramsey-Klawsnik, H., & Teaster, P. B. (2012). Sexual abuse of health care facility residents: Adult Protective Services and facility policy and practice implications. *Generations, 36*(3), 53–59.
Reilly, M., & Gravdal, J. A. (2012). An ecological model for family violence prevention across the life cycle. *Family Medicine, 44*(5), 332–335.
Rieber, W. (Vol. Ed), & Hall, M. J. (Trans). (1997). *The collected works of L. S. Vygotsky, Vol. 4: The history of the development of higher functions*. New York: Plenum Press (Original work published in 1941).
Riggs, D. S., & O'Leary, K. D. (1996). Aggression between heterosexual dating partners: An examination of a causal model of courtship aggression. *Journal of Interpersonal Violence, 11*, 519–540.
Reingold, D. A. (2006). An elder abuse shelter program: Build it and they will come, a long term care based program to address elder abuse in the community. *Journal of Gerontological Social Work, 46*(3–4), 123–135.
Roberto, K. A., & Teaster, P. (2005). Sexual abuse of vulnerable young and old women. *Violence Against Women, 11*(4), 473–504.
Robinson, L., de Benedictis, T., & Segal, J. (2012, December). *Elder abuse and neglect: Warning signs, risk factors, prevention, and help*. Retrieved January 20, 2013 from http://www.helpguide.org/mental/elder_abuse_physical_emotional_sexual_neglect.htm
Schiamberg, L. B., Barboza, G. G., Oehmke, J., Zhang, Z., Griffore, R. J., Weatherill, R. P., et al. (2011). Elder abuse in nursing homes: An ecological perspective. *Journal of Elder Abuse and Neglect, 23*, 190–211.
Schiamberg, L. B., & Gans, D. (2000). Elder abuse by adult children: An applied ecological framework for understanding contextual risk factors and the intergenerational character of quality of life. *International Journal of Aging and Human Development, 50*, 329–359. doi:10.2190/DXAX-8TJ9-RG5K-MPU5.
Schnurr, P. P., Avron Spiro, A., Vielhauer, M. J., Findler, M. N., & Hamblen, J. L. (2002). Trauma in the lives of older men: Findings from the normative aging study. *Journal of Clinical Geropsychology, 8*(3), 175–187.
Seaver, C. (1996). Muted lives: Older battered women. *Journal of Elder Abuse and Neglect, 8*(2), 3–21.
Shilling, D. (2010). Improving the court system's response to elder abuse. *Victimization of the Elderly and Disabled, 11*, 49, 51–52, 62–63
Straka, S. M., & Montminy, L. (2006). Responding to the needs of older women experiencing domestic violence. *Violence Against Women, 12*(3), 251–267.
Tauriac, J. J., & Scruggs, N. (2006). Elder abuse among African Americans. *Educational Gerontology, 32*, 37–48.

Teaster, P., Otto, J., Dugar, T., Mendiondo, M. S., Abner, E., & Cecil, K. (2006a). *The 2004 Survey of State Adult Protective Services: Abuse of adult 60 years of age and older*. Washington, DC: National Center on Elder Abuse.
Teaster, P. B., & Roberto, K. A. (2003). Sexual abuse of older women living in nursing homes. *Journal of Gerontological Social Work, 40*(4), 105–119.
Teaster, P. B., & Roberto, K. A. (2004). The sexual abuse of older adults: APS cases and outcomes. *The Gerontologist, 44*, 788–796.
Teaster, P. B., Roberto, K. A., & Dugar, T. A. (2006b). Intimate partner violence of rural aging women. *Family Relations, 55*, 636–648. doi:10.1111/j.1741-3729.2006.00432.x.
Tornstam, L. (2005). *Gerotranscendence: A developmental theory of positive aging*. New York: Springer.
UK Justice Center for Elders and Vulnerable Adults. (2005). Retrieved November 28, 2012 from http://www.mc.uky.edu/gerontology/jceva.html.
Vinton, L. (1991). Abused older women: Battered women or abused elders. *Journal of Women and Aging, 3*(3), 5–19.
Vinton, L. (1992). Battered women's shelters and older women: The Florida experience. *Journal of Family Violence, 7*(1), 63–72.
Walker, C. (1990). Feminist literary criticism and the author. *Critical Inquiry, 16*(3), 551–571.
White, H. B. (2004). *Elder abuse in Tribal communities*. Southwest Center for Law and Policy and Office on Violence Against Women.
Wisconsin Coalition Against Domestic Violence. (2009). *Elder abuse, neglect, and family violence: A guide for health care professionals*. Madison: Wisconsin Coalition Against Domestic Violence and Wisconsin Bureau of Aging and Disability Resources.
Wolf, R. S. (2000). Elder abuse: An introduction. *Generations, 20*(4), 301–305.
Women of Color Network. (2008). *Fact and stats: Elder abuse in communities of color*. Available from http://womenofcolornetwork.org
Woolf, L. M. (1998). *Elder abuse and neglect*. Retrieved January 20, 2013 from http://www2.webster.edu/~woolflm/abuse.html
World Health Organization (WHO). (2012). *Aging and life course*. Retrieved November 28, 2012 from http://www.who.int/ageing/projects/elder_abuse/en
Zink, T., Regan, S., Jacobson, C. J., & Pabst, S. (2003). Cohort, period, and aging effects. *Violence Against Women, 9*(12), 1429–1441.

Resources: Websites

Adult Protective Service: (may be found in each state as a link to the state government website)
Administration on Aging: http://www.aoa.gov/AoARoot/index.aspx
American Psychological Association—*Elder Abuse and Neglect In Search of Solutions*: http://www.apa.org/pi/aging/resources/guides/elder-abuse.aspx
International Network for the Prevention of Elder Abuse: http://www.inpea.net
National Abuse in Later Life Resource Directory (2011): www.ncall.us
National Domestic Violence Hotline: www.ndvh.org
National Center on Elder Abuse: www.ncea.aoa.gov
National Committee for the Prevention of Elder Abuse: http://www.preventelderabuse.org
National Long Term Care Ombudsman Resource Center: www.ltcombudsman.org
Senior Service Association: http://www.seniorserviceassoc.org/elder_abuse.html
Survivor Project: LGBT Elder Abuse and Neglect Issues: http://www.survivorproject.org/elderabuse.html
WomensHealth.gov: http://womenshealth.gov/aging/safety-abuse/elder-abuse.cfm
World Health Organization: http://www.who.int/aging/projects/elder_abuse/en/

Resources: Training Modules

Help Guide—*Reporting Elder Abuse*: http://www.helpguide.org/mental/reporting_elder_abuse.htm

National Center on Elder Abuse: http://www.ncea.aoa.gov/training/index.aspz

LATIN@ ELDERS: Securing Healthy Aging Inspite of Health and Mental Health Disparities

Catherine K. Medina and Lirio K. Negroni

Abstract The elder population is the fastest growing subgroup of the U.S. This demographic reality has been referred as the "graying of America" and is the result of improved and longer life expectancy of elders because of the medical advances, technology, lifestyle choices and the "baby boomer" generation. The "graying of America" is also ethnically and racially diverse, and Latin@s are the fastest growing segment of the elder population. This chapter discusses the diversity among Latin@ elders and addresses their perceptions of aging and well-being given their diverse realities and existing health and mental health disparities. Using a social justice frame, the authors examine Latin@ health and mental health disparities and the intersectionality of particular social determinants of health: acculturation and health status, chronic illness and the Hispanic mortality paradox, poverty, adjustment to daily living and health literacy. In spite of health disparities that result from structural inequalities, the authors emphasize three positive influences that affect healthy aging by Latin@ elders: (1) family relationships, (2) social support networks, and (3) health/mental health service delivery system within a community context. The authors illustrate community programs as best practice models that include cultural competence and interprofessional collaborations that challenge the structural inequities to reduce health and mental health disparities within the Latin@ elderly population.

Since the twentieth century, the United States (U.S.) and the global world have experienced an unprecedented growth in the number and proportion of older adults (Vincent and Velkoff 2010). The elder population is the fastest growing subgroup

---

Latin@s is used in this chapter. The @ sign is the deference to the quite recent determination to develop and use nonsexist language and to identify with Latin America (Wallerstein 2005).

C. K. Medina (✉) · L. K. Negroni
University of Connecticut, Connecticut, USA
e-mail: catherine.medina@uconn.edu

H. F. O. Vakalahi et al. (eds.), *The Collective Spirit of Aging Across Cultures*,
International Perspectives on Aging 9, DOI: 10.1007/978-94-017-8594-5_5,

of the U.S. This demographic reality has been referred to as the "graying of America" and is the result of improved and longer life expectancy of elders because of medical advances, technology, and lifestyle choices. Another factor influencing the number of elders is the "baby boomer" generation, consisting of individuals born from 1946 to 1964, when our nation experienced an unusually high post-World War II increase in birthrates. As the austerity of the Great Depression and the war years was replaced by economic security, the baby boom was a temporary aberration in the long-term trend of declining fertility in the U.S. (Barusch 2009). Since the 1970s, the U.S. has experienced a decline in fertility and an increase in longevity. Therefore, increases in the elder proportion of the population are attributed to a decline in both death and birth rates. By 2030, it is forecasted that one in five Americans will be over the age of 65 and this will put unprecedented demands on social programs serving elders.

The "graying of America" is also ethnically and racially diverse, and Latin@s are the fastest growing segment of the elder population. Minority populations have increased from 16.3 % of the elder population in 2000 to 20 % of elders in 2010, and are projected to increase to 24 % of elders in 2020. Between 2010 and 2030, the White population 65+ is projected to increase by 59 % compared with 160 % for older minorities, with Latin@ growth projected at 202 % (U.S. Census Bureau Decennial Census 2010). The Latin@ elder population 60 and over is projected to quadruple over the next quarter century, and by 2030 it will be the largest population of minority elders (Tucker et al. 2010). Currently, Latin@ elders constitute a large group (8 %) and by 2019, the Latin@ population 65 and older is projected to be the largest racial/ethnic minority in this age group [Administration on Aging (AOA) 2010]. This Latin@ boom is driven by the fact that Latin@s live longer than their White counterpart. Contrary to expectations based on Latin@s' lower socio-economic profile and living conditions, national studies have documented lower-age adjusted and income-adjusted mortality rates due to heart disease, stroke, and cancer among Latin@s relative to Whites (Falcon et al. 2009; Stratton et al. 2009). This phenomenon of longevity has been termed the *Latin@ or Hispanic epidemiological paradox* and will be further explained in the acculturation and health status section.

Despite longer life expectancies, Latin@ elders face mental health disparities. There is some evidence that depression among Latin@ elders is slightly higher than for the general population and that depression is strongly associated with physical illness for Latin@ elders (Chavez-Korell et al. 2012; Falcon and Tucker 2000; Areán et al. 2005). According to the Center for Disease Control and Prevention (CDC 2012) Behavioral Risk Factor Surveillance System (BRFSS), Latin@ elders had a higher prevalence (13.2 %) of frequent mental distress (FMD) compared to their White (8.3 %), and Black and non-Hispanic counterparts (11.1 %). This same study indicated that Latin@ elders (14.5 %) were slightly more likely to report a lifetime diagnosis of anxiety disorder compared to White (12.6 %), Black (11 %) and other non-Hispanic older adults (14.2 %). Latin@ elder mental health status can be impacted by many factors such as coping with life changes and loss, impact of acculturation within the family, lack of supports and limited access to services. These factors create conditions for the development of depression and anxiety in Latin@ elders which affects health and well-being.

This chapter discusses the diversity among Latin@ elders and addresses their perceptions of aging and well-being given their diverse realities and existing health and mental health disparities. Aging is a fundamental aspect of the life cycle, and one of continuing development. According to Rowe and Kahn (1998), the goal of healthy aging is to minimize decline by emphasizing the importance of maintaining health, mental functioning and active engagement in life. Yet Cohen (2005) recognizes the huge potential for positive growth in later life. Using a social justice frame, the authors examine Latin@ health and mental health disparities and the intersectionality of particular social determinants of health. Social determinants are defined as the conditions in which people are born, grow, live, work, and age, including the health system [World Health Organization (WHO 2011)]. Social justice promotes a fair and just society that upholds the right to equality of access, opportunity, and outcomes for all individuals (Rawls 1971). This view of social justice refers to egalitarian theories that contend that distributive justice redress existing social and economic inequalities by redistributing resources. As a moral imperative, distributive justice calls for distributing goods to provide the greatest benefits to the least advantaged members in society in a fair and just matter (Rawls 1971; Iatridis 1994). The authors discuss the following social determinants of health in relationship to the well-being of Latin@ elders: acculturation and health status, chronic illness and the Hispanic mortality paradox, poverty, adjustment to daily living and health literacy from a social justice perspective.

The final section of the chapter proposes ways in which Latin@ healthy aging can be promoted with particular consideration of community interventions and interprofessional collaborations. These proposed practices will be discussed from a social justice perspective because the national and global health environment requires social scientists and healthcare providers to expand their knowledge base about health and well-being and to promote health equity for all.

Theoretical Framework: Well-Being and Healthy Aging from a Social Justice Perspective

The U.S. health care delivery system, especially in older life, focuses on health as the absence of disease by curative treatment, and in the presence of disease that directs care to stabilizing acute exacerbation of chronic disease by symptom management (Day 2010). The definition of health as the absence of disease keeps the discourse in the realm of the health care industry, as it is currently structured, with an emphasis on medical and technological efforts to diagnose and cure disease. An alternative system of care views healthy aging as one that centers on the multidimensional human capacity to thrive and flourish from prenatal to end-of-life. For example, healthy aging would be a normal life span that emphasizes well-being as a dynamic state. The elder continues to develop their potential, works productively and creatively, builds strong and positive relationships and contributes to their society. When theorists argue that healthy aging encompasses human thriving, human dignity and self-determination, they are connecting health and social justice (Day 2010; Powers and Faden 2006; Nussbaum 2006). Social justice promotes a fair and just society that upholds the right to equality of access to goods and resources, opportunities, and the probabilities of life chances for actualizing successful outcomes for all societal members. Healthy aging is a function of a just society, allowing all members of society their human capacity to thrive and flourish.

Diversity Within Latin@ Elders

Although the research literature refers to the Latin@ elder population as one group, Latin@ elders are quite diverse and culturally fluid. Latin@ elders represent 21 nations from Central and South America (including Brazil), Mexico and the Spanish speaking Caribbean. Among Latin@ subgroups Mexicans are the largest group in the U.S., followed by Puerto Ricans, Cubans, Salvadorans, and Dominicans (CDC 2012). Increasingly, Latin@ immigration to the U.S. is coming from Central and South America. Within-group differences exist on the basis of sociopolitical and economic histories within their countries of origin, immigration and migration experiences, specific customs, and their mixed racial backgrounds (Negroni-Rodríguez and Morales 2001). Most Latin American populations have White (European), Black (African), and Native Indian heritage. Most Latin@ elders share a history of colonization by Spain, a connection to the Catholic Church and Spanish as the primary official language, except for Brazil, and the diversity of indigenous languages spoken in mainly Mexico and Central America (Náthuatl, Zoque, Chontal, Miskito, and Mayan languages). Their self-identity as members of the same ethnic group is influenced by this history and by the sharing of similar cultural values, beliefs and traditions such as *confianza* (trust in relationships), *familismo* (centrality of family) and *respeto* (respect for elders or individuals with experience) .

Latin@ elders residing in the United States comprise of two groups: those who were born and raised outside the United States and immigrated at some point in their childhood or adulthood (foreign-born) and those who were born in this country (native-born). In regards to Puerto Rican elders, because they are U.S. citizens, they have a different migration history that occurred in two different cohorts. The first cohorts are Puerto Ricans from the mainland (Puerto Rico) who migrated to stateside (U.S) in the 40 and 50s because of the need for economic security. The second cohorts consist of those Puerto Ricans who arrived at stateside in the last decades for education and health reasons or to accompany or reunite with relatives. Latin@ elders are more likely to be foreign-born than native-born (Motel and Patten 2013). This distinction is important because membership to each category presupposes different challenges concerning social determinants of health for healthy aging, such as health status, level of acculturation, access and use of services, health literacy and language proficiency in both languages (Spanish and English).The intersectionality of social determinants and health and mental health disparities will be further discussed.

Also depending on the period of immigration or migration to the U.S., many Latin@s elders may have experienced racism and discrimination resulting in a lack of access to goods and resources. Racism and discrimination were more openly practiced and accepted than in present times (Ramos 2007). Elders with an African heritage may have experienced additional burdens, given the prejudices and racism against Blacks (Aguirre and Turner 1995, cited by Ramos 2007, p. 49). The discrimination might have denied access to employment, health care, housing, insurance and a living wage. The interpersonal (person to person) and institutional (inherent in structural systems) racism presents major challenges to illness prevention, healthy outcomes, and overall well-being in Latin@ elders.

Latin@ Elders Perceptions of Aging and Well-Being

In the Latin@ culture old age is highly valued and considered a time when the elder enjoys the rewards for a life lived according to one's cultural values (Beyene et al. 2002). For many Latin@ elders aging is not just about chronologically maturing into old age, it is a twofold process. Aging is viewed from the perspectives of one's physical conditions (functional role) and one's "spirit" (relational function). For Latin@s there may be a difference between "being old" and "feeling old". "Being old" is associated with the person's inability to carry out daily activities while "feeling old" depends on the person's spirit based on perceived self-esteem derived from their relational status with family members (Beyene et al. 2002). Feelings of well-being may be influenced by the type of family interactions and sense of fulfillment based on culturally defined roles they have in family interactions. The elders' sense of morale is supported by their perceived closeness with their adult children, family and extended family members. Latin@ elders who expressed unsatisfactory family relations defined their health status as poor. These Latin@ elders expressed feeling lonely, and perceived aging as a very sad phase of their life. On the other hand, those who had satisfactory family relationships defined their health status as good, and portrayed a positive perception of aging (Beyene et al. 2002).

In most Latin@ families, respect (respeto) for elders is an important value taught to children very early in life (Sanchez-Ayendez 1988). Because Latin American cultures have norms and values that reflect a high level of positive valuation for the elders, children have social and moral obligations to support elder parents (Ortiz 1995; Vega 1995; Sotomayor 1992; Paz and Applewhite 1998; Beyene et al. 2002). Therefore, it is common for Latin@ elders to live with family members not only as a result of health or economic necessity, but because of culturally bound expectations governed by norms of mutual reciprocity among families (Vega 1995; Angel and Tienda 1987). Latin@ elders expect to be respected and valued. In the Latin@ community there is the cultural expectation that family members take care of the needs and best interests of their older adults (Vélez and Cole 2008), and treat them with respect. Children feel an obligation and responsibility to reciprocate for the care and love they received. Female family members may encounter a higher level of stress in caring for their elders, as they are the ones expected to provide most of the care.

Acculturation aand other factors can influence how a Latin@ elder is cared for and treated. Some families lack awareness of the elder's value orientation about family and aging; others may not realize the challenges faced by their elders as a result of their immigration, migration and acculturation process (Negroni 2012). Sometimes the younger generations do not understand what it means for their elders to grow old in a new country and culture. This can be a source of stress and pain both for the elders—who may feel they are not as important to the family—and for the family. In turn, there is a strain on the relationship and this can trigger feelings of insecurity, fear, incompetence, frustration, anger, grief, and separation for both generations.

Although many Latin@ elders rely on their families for support during this life span, Rosnick and Reynolds (2003) note the importance of control for Latin@ elders in decision making about the course of their life. Other researchers have examined the importance of autonomy and the desire to control their lives (Heyman and Gutheil 2010). Control and autonomy are factors that influence self-determination to promote healthy aging, human thriving, and well-being. The human capacity of Latin@ elders to control bodily health, bodily integrity, and engage in practical decision-making about their health is to maintain dignity and self-determination throughout the aging process (Nussbaum 2006).

Another factor that influences the relational function of "spirit" and a positive aging process is the Latin@ elders' relationship to religion. Religious and spiritual beliefs and faith play an important role in helping Latin@ elders cope with the stressors of old age (Negroni 2007; Krakauer et al. 2002; Phipps et al. 2003). Because of their religious roots they believe that God will help them deal with their problems (Aranda and Knight 1997). They often state "*Está en las manos de Dios*"—meaning the issue is in God's hands. This cultural expression is within the context that the Latin@ elder is in a spiritual relationship with a supreme being and that the locust of control about a situation is restored. It is not in the context of fatalism or helplessness, but rather in the context of a spiritual relationship with a source of power. Aging is regarded as a gift from God even when people's health status is poor (Beyene et al. 2002). According to Heyman and Gutheil (2010), research indicates that older Latin@ adults often turn to their churches and embrace the concept of God's will.

Health and Mental Health Disparities: Intersectionality of Social Determinants of Health

According to the Institute of Medicine (IOM), there has been a long-standing and well-documented pattern of health disparities for certain groups in the U.S. These health disparities refer to differences in the incidences, prevalence, mortality, and burden of diseases and other adverse health conditions that exist among specific population groups in the U.S. (National Institute of Health 2000). Specifically, health disparities refer to those avoidable differences in health that result from cumulative social disadvantage.

The discussion on Latin@ elders' health and mental health requires a multi-dimensional view and analysis of the different social determinants of health that occur within the context of power and oppression and how these determinants are the cause of health disparities (World Health 2011). Social determinants include the circumstances in which people are born, grow up, live, work and age, and the systems put in place to deal with illness. These circumstances are in turn shaped by a wider set of forces: economics, social policies, and politics (World Health 2013). When examining the intersectionality of these determinants among Latin@ elders, it is essential to remember their status as an underrepresented and oppressed group. Different social determinants that impact elders' potential for health and well-being come from the larger societal system both in their countries of origin and the U.S., their families and themselves (Parra-Cardona et al. 2007).

Higher rates of chronic illnesses and longer years of disabling conditions may define the aging experience of many Latin@ elders (Angel et al. 2005; Ruiz and Ransford 2012). Findings from the 2011 National Health Interview Survey and other research indicate that Latin@ elders have higher poverty rates, lower education and greater risk factors related to environmental barriers (physical health, language, access to quality care). In the context of health disparities, social justice refers to the minimization of the social and economic conditions that adversely affect the health and well-being of Latin@ elders. In the next section, relevant social determinants of health that influence Latin@ elders aging and well-being are discussed. Among the factors that intersect and can increase health disparities for Latin@ elders are: acculturation and health and mental health status, language and health literacy, chronic illnesses and the Hispanic mortality paradox, limited adjustment to daily living, poverty, lack of access to quality health care, and social supports (CDC 2012).

Acculturation and Health and Mental Health Status

Franzini et al. (2002) define acculturation as the process by which an individual raised in one culture enters the social structure and institutions of another, and internalize the prevailing attitudes and beliefs of the new culture, while simultaneously retaining aspects of one's culture of origin. Several studies demonstrate that there is a relationship between acculturation, health and mental health, and findings indicate that the more a group acculturates to the dominant culture, their physical and mental health disparities move closer to minority health disparities in the U.S. (Coonrod et al. 2004). The research suggests that Latin@ foreign-born have better health outcomes than U.S. native-born because they are less acculturated and have not adopted U.S. health risk behaviors (smoking, consumption of alcohol or other drugs), nor have the extended exposure to the cumulative social stress resulting from U.S. racism and structural discriminatory practices toward low-socio-economic status minority populations previously mentioned.

Too often the native Latin@ in the U.S. knows the burden of economic insecurity, lack of access to health care, lack of health insurance, and lack of health literacy because of poor language proficiency. The cumulative stress for native Latin@ elders come from years of exerted energy to respond against difficult odds when one does not have the necessary resources to cope when confronted with an external stimulus [(stressor i.e. lower social status, racism, discrimination, lack of resources) (LaVeist 2005)]. This experienced stress can lead to emotional, behavioral and physiological changes that can put the native Latino elder under greater risk for developing mental disorders and physical illnesses. Latin@ adults consistently show that they have poorer health than Whites of the same age group across measures of disease, disability, and self-assessed health (Dilworth et al. 2012).

Existing findings raise questions regarding the role of acculturation and other factors on self-assessed physical well-being. For example, in the study by Beyene et al. (2002) with foreign-born mostly Central American and Mexican elders, most of whom had lived in the U.S. for 20 years or longer, participants rated their health as good to excellent (61.7 %), fair (25 %), and poor (13.6 %). Yet in the study by Negroni-Rodriguez and Bok with mostly Puerto Ricans who had lived in the U.S. an average of 30 years, participants rated their health as fair (65 %), and poor to very poor (17.3 %). Min and Barrio (2009) found a significant difference on self-assessed health status with 60 % of Mexican-Americans rating their health as either "fair" or "poor", compared with only 23.3 % of non-Latino Whites. These findings suggest that more research needs to be conducted with the different Latin@ subgroups because of the varying socio-political factors that influence the relationship between acculturation and self-assessment of physical well-being.

Although there is a dearth of recent data on Latin@ elder's acculturation and mental health status, researchers acknowledge that Latin@ elders' help-seeking patterns and English language limitations, along with the experiences of immigration and adjustment to a new country are stressors that can influence the development of mental health problems. Studies document a slightly higher level of depression among Latin@ elders than among the general older adult population (Arean et al. 2005; Falcon and Tucker 2000; Gonzalez et al. 2001). In the Negroni-Rodriguez and Bok (2004) study, Latina elders reported feeling depressed (65.2 %), having anxiety problems (60.9 %) and having problems with their memory (43.5 %). These women were born and raised outside of the United States (100 %), lived alone (61 %), identified family as their primary source of support (74 %) and primarily spoke and preferred to be addressed in Spanish (100 %).

Most research on Latin@ elders' mental health and acculturation generally focuses on caregivers (Janice et al. 2009; Min and Barrio 2009). The study by Min and Barrio (2009) addressed caregiver's preferences and showed that Latin@ elders (45.4 %) are more likely to prefer informal caregivers such as *promotores (as)* than formal/paid helpers compared to non-Latin@ whites (16.7 %). Regarding cultural values about the role of caregivers, more Latin@ elders than non-Latin@ elders agreed that adult children should be more responsible than the elders parents themselves for making long term care arrangements (74.2 % vs. 33.3 % respectively); and that care should be provided by family members, not by an outsider (73.3 % vs. 32.6 %). Those who agreed to these two statements were three times more likely to select an informal caregiver than those who disagreed (Min and Barrio 2009).

Language and Health Literacy

For many Latin@ elders, language discordance between their medical provider and themselves has been cited as a particular structural barrier to accessing quality health services (Born et al. 2004). Communication with health care providers may not be effective because of the language barrier or sometimes medical information is lost in translation. A recent study to understand cultural competence found that Latin@s experienced language-based discrimination related to their medical encounters (Napoles-Springer et al. 2005).

There is a growing interest in health literacy as a major factor contributing to medical adherence. Health literacy refers to "the degree to which individuals have the capacity to obtain, process, and understand basic health and service needed to make appropriate health decisions" (Hispanic Health Council 2006, p. 14). According to this profile, current research indicates that over 90 million people (including English speaking individuals) in the U.S. struggle to understand basic health information such as reading materials, prescription labels, filling out medical and insurance forms, and communicating with their providers. The disproportionate low level of educational attainment and the large portion of Latin@s elders who are monolingual in Spanish are factors that influence their degree of health literacy and their adherence to medical treatment.

Chronic Illness and the Hispanic Mortality Paradox

Among Latin@ elders there is a prevalence of chronic illness with 85 % reporting at least one chronic illness, while others reported two to four chronic illnesses (Villa and Torres-Gil 2001). In a study of 83 Latin@ elders, Beyene et al. (2002) reported that the most frequent chronic illnesses were arthritis (88 %) followed by heart disease and hypertension (34.2 %), and diabetes (30.1 %). On the favorable side, Latin@s tend to have lower mortality rates. This finding has been referred to as the *Hispanic epidemiological paradox* or the *Hispanic mortality paradox* (Franzini et al. 2002). Franzini et al. offer four different explanations for the Hispanic epidemiological paradox. The first is poor-quality data reporting. Prior to 1978, death rates for Latin@s were not collected by providing a separate category for Latin@s; instead, they were estimated by Latin@ categorized surnames. This resulted in inconsistencies in counts of Latin@-origin decedents that may have a European sounding surname such as one of the authors of this chapter. This type of data collection leads to misclassification and underestimation of deaths that occur among Latin@s. Often the data on death rates among Latin@s is under-reported or not collected at all (LaVeist 2005; Medina 2011).

The second plausible, and most common, explanation given is the "salmon bias" phenomenon. The coining of the "salmon bias" hypothesis is based on salmon swimming upstream to spawn before their death (LaVeist 2005, p. 277). Similarly, it is hypothesized that Latin@s return "home" (to their country of origin) when they become seriously ill or reach mature age. Therefore, their deaths are not recorded in U.S. statistics, producing an artificial death rate for Latin@s. This hypothesis has been refuted by some studies (Abraido-Lanzo et al. 1999). The third explanation is that only the healthiest individuals will migrate to the U.S. from other countries. Studies support that newly arrived migrants have better health status than persons in their country of origin (Marmot et al. 1984). This healthy migrant effect has been argued by Abraido-Lanza et al. who state that all migrants from Europe should have this selective migration advantage. The final explanation of the Hispanic mortality paradox is the distribution of health risks and protective factors among Latin@s. Although many Latin@s have definite health risks through their life span such as lower rates of immunizations, higher prevalence of hypertension, unintentional injuries, and living in environments with poorer quality of air and exposure to biological, chemical and other toxin elements, they also show lower rates of smoking, a higher proportion of women getting prenatal care, and a higher fruit and fiber diet than Whites (LaVeist 2005; Medina 2011).

The resiliency of the Latin@ population through their family and community networks has been a source of strength, and can be a protective factor for cumulative stress in their environments (Stratton et al. 2009). How these factors contribute to lower mortality rates or healthy outcomes for Latin@s is an area for future research. An understanding of the Hispanic epidemiological paradox is worth investigation, because how health risks and protective factors translate into health outcomes may play a role in lowering incidence rates of certain diseases in Latin@s.

Limited Adjustment to Daily Living

Similar to other elders, maintaining functional independence is important for Latin@ elders and their families. The process of dependence can be slowed down if elders are able to successfully adjust to life changes, and if they are cared for appropriately. In 2008, 9.2 % of Latin@ elders aged 65+ needed help from other persons for personal care as compared to 5.7 % for non-Hispanic Whites and 10.3 % for non-Hispanic Blacks (AOA 2010). A study by Villa and Torres-Gil (2001) found that 45 % Latin@ elders reported difficulty with adjustment to living tasks such as: managing hygiene, toileting, dressing, or ambulating around the house. Other research indicated that compared to Whites, twice the number of Latin@ elders reported difficulty in performing instrumental activities of daily living such as: using the phone, shopping, managing their money, or doing light housework (AOA 2012; Hazuda and Espino 1997). The ability to perform these ADLs competently can be threatened by the presence of diseases and disability. Collins et al. (2002) found that Latin@s elders reported their health as fair to poor because of their high level of physical disability and functional limitations. From a social justice perspective, the high level of poverty along with low educational levels, can limit Latin@ elders' access to the resources that can guarantee quality of life. When elders cannot perform their daily activities they will need help. Conditions such as the lack of support systems and limited care giving services resulting from the structural factors discussed in this chapter can make this needed help less available.

Poverty

In 2010, almost 3.5 million elders (9.0 %) in the U.S. were living below the poverty level (AOA 2012). Another 2.3 million (5.8 %) were classified as "near-poor" (calculated as income between the poverty level and 125 % of this level). Among those who fell below the poverty level, 6.8 % were Whites, 18 % were African Americans and 18 % were Latin@s. Higher poverty rates were found in cities (11.2 %) compared to rural areas (10.4 %). Most Latin@ elders live in cities with a majority residing in the following four states: California (27 %), Texas (19 %), Florida (16 %) and New York (9 %). Older women had a higher poverty rate (10.7 %) than older men (6.7 %); and older persons living alone were more likely to be poor (16 %) than older persons living with family members [(5.3 %); (AOA 2012)].

According to the National Hispanic Council on Aging (NHCOA), Latin@ older adults are more likely to be wholly dependent on Social Security than any other racial or ethnic group and without this financial support, 50 % or more Latino elders would fall into the poverty level. When faced with such economic insecurity, elders alter their behaviors to save money. Despite their hard work throughout their life, they are not able to save enough money to support their basic needs in older age. In a national survey of older adults with diabetes in which the majority of respondents (51 %) were of Latin@ descent, severe measures were taken by these elders to meet their economic needs. Nearly one in five (19 %) cut back on the use of medications, 20 % limited food intake and other essentials, 14 % increased credit card debt, and 10 % borrowed money from family and friends for medications (Piette et al. 2004). Lack of economic resources has been shown to contribute to cost-related medication non-adherence (Heisler et al. 2007).

Because many older Latin@s do not live alone or reside in nursing homes, families pool resources to meet basic needs (Fennell et al. 2010). According to the AOA (2010) statistical profile of Hispanic Older Americans, 67 % of Latino older men lived with their spouses, 15 % lived with another relative, 5 % lived with a non-relative, and 13 % lived alone. For Latina women, 41 % lived with their spouses, 32 % with another relative, 1 % with a non-relative and 27 % lived alone. Although, the percent of older Latin@s living alone is lower than that of the general population, the highest poverty rate of any elder group was experienced by Latina women living alone. However, the percent of Latin@ elders living with other relatives is almost twice that of the total general population (AOA 2010). The high number of elders living with relatives may be related to the cultural values of *familismo* and *respeto* which increases adult children's sense of responsibility for caring for their elder parents and the emphasis on the collective in Latin@ families (Fennell et al. 2010).

Lack of Access to Health Care

Latin@s tend to underutilize health and mental health, social services, and other public delivery systems, and drop out early after beginning treatment (Ortega et al. 2006; Sue 2003; Vega and Alegria 2001). Latin@s throughout the life span have substantially lower rates of healthcare insurance compared to other ethnic and racial groups (Medina 2011), and it has been reported that 31 % do not have a usual source of health care (LaVeist 2005). Some of the difficulties with service utilization may stem from the actual helping process itself. For a patient to adhere to medical treatment, the process is dependent on the level of health literacy—"the degree to which individuals have the capacity to obtain, process, and understand basic health and service needed to make appropriate health decisions" (Hispanic Health Council 2006, p. 14). Because the United States has a complex health care delivery system which involves many specializations and coordination of services, the consumer is dependent on their level of health literacy to navigate their healthcare (Medina 2011). For many Latin@ elders this can be a challenge because of their low education attainment and lack of language proficiency in English and sometimes Spanish will influence their level of health literacy and their ability to access and utilize health care.

In Latin@ communities, many elders seek a natural helper—individuals that people naturally turn to for advice, emotional support, and tangible aid (neighbor, *bodega* (grocery) owner, *botanica (herb shop with spiritual meanings)* spiritualist and/or healer) to help access and utilize healthcare services. Because of their skill sets and expertise, natural helpers assist with referrals to external resources (DiClemente et al. 2002). They are local residents that provide informal, spontaneous, culturally relevant healthy advice associated within the context of everyday living. Community residents will refer to them as madrinos/madrinas (godfathers/ godmothers) because of their access to resources based on their networks and ties to local services and leaders that are central to many activities in the community. The health prevention field has realized that the collaboration of formal systems of care with natural helpers can influence health promotion and outcomes for Latin@ communities (DiClemente et al. 2002).

Viewed as a community support system of care not to replace specialized or professional services, but to complement and foster an appropriate and cultural response to clients' and community needs, natural helpers are often trained by health professionals. Medical providers will refer to the natural helpers as promotores (as) who integrate information about health and the health care delivery system into the Latin@ community's culture, language and value system, thus reducing many of the barriers to health access and utilization of services (Migrant Health Promotion 2013). They also help make health care systems more responsive to Latin@ elders. In a collective spirit, promotore (as) with the appropriate resources, training and support influence the meaning of health, prevention, and well-being. They link Latin@ elders to health/mental health care and social services; educate elders about disease and injury prevention, and work to make available services more accessible (Migrant Health Promotion 2013). Promotores (as) mobilize Latin@ communities to create positive change toward well-being.

Latin@ elders generally are undertreated for mental health (Chavez-Korell et al. 2012), and mental health services are usually provided by a primary health care physician (Negroni-Rodríguez and Bok 2004). Understanding Latin@ elders' use of health and mental health services requires an integrated inquiry that incorporates factors such as views about well-being, health and mental health problems, help-seeking patterns, availability of supports, and availability of services. In their review of the literature, Kim et al. (2009) found lower rates of mental health service use and different quality of care from mental health service providers in foreign-born as compared to native-born adults. Their comparison study between Asian and Latin@ elders documented that for Latin@ elders four factors influenced their use of mental health services: being a younger elder, a woman, having a mood disorder and self-rating their mental health as poor.

Social Support Networks

One of the many factors that leads to good health and long life is the amount of social support a person enjoys (Donelly et al. 2001). Research documents that individuals get innumerable benefits when they have adequate social supports (Vega 2010), and social support networks are important social determinants of health particularly for aging adults (Forster and Stoller 1992). Support systems provide emotional, informational and instrumental assistance on an every-day basis as well as at times of crisis. They also serve as a mechanism for helping individuals maintain their cultural heritage. Their support systems can influence

Latin@ elders' view of their needs and problems and influence help-seeking patterns and access to health care services (Negroni-Rodríguez and Bok 2004; Negroni 2007). Negative interactions with those support systems, such as constant criticism and belittlement from other people, can undermine emotional well-being. Older adults who perceive themselves as socially engaged and supported are in better mental and physical health than those socially isolated (Siebert et al. 1999). The types of social support networks that Latin@s use include: nuclear and extended family, friends, close neighbors, and church members. Religious activities, social clubs, merchants (markets, botanical shops, restaurants, and beauty parlors) and senior centers are important sources of social support networks (Delgado 1998; Negroni-Rodríguez and Bok 2004).

Although Latin@ elders depend primarily on their families, the availability of family members to care for them may be limited. Different factors account for the scarcity of such support, for instance, children facing their own stressors may find it difficult or impossible to fulfill their filial responsibility (Moody 1998; Gallagher-Thompson et al. 2003, 2008); social support declines as a result of family members becoming more Americanized and changing their views about aging and caring for their elder; and/or because elders' face the loss of significant supports for instance, the death of a spouse or a significant other (Vega 2010). The stressors and changes faced by caregivers not only can limit their availability but cause negative interactions with the elders, such as constant criticism and belittlement, and can undermine elders' emotional well-being. Thus, the support provided by non-family members is quite important and more now than in the past, Latin@ elders may seek their needed social support through their connections and networking with community resources (Gallagher-Thompson et al. 2008).

Best Practices: Community Efforts that Include Cultural Competence and Inter-professional Collaborations to Increase Latin@ Healthy Aging

Health disparities may be reduced through community participation that addresses social determinants of health. As previously indicated, family, friends, and neighbors represent the primary care givers and advocates of Latin@ elders. Latin@ community organizations represent the next layer of support and help. Throughout the United States local community efforts have been implemented in cities, at the state level and nationally as a result of communities' concerns about the well-being of Latin@ elders. In the following section the authors will provide examples to describe the focus of these efforts.

The National Hispanic Council on Aging (NHCOA) is considered the leading national organization working to improve the lives of Latin@ older adults, their families, and caregivers. NHCOA promotes, educates, and advocates for research, policy, and practice in the areas of economic security, health, and housing. It has been in existence for more than 30 years. The organization has identified four priority areas to work on: Health, Economic Security, Leadership Development and Empowerment, and Housing. Within each area the organization has been developing initiatives, addressing policy issues and conducting research. In the area of health, NHCOA has been active in influencing the national agenda around health and health care by implementing health promotion and education programs; providing cultural competency training to physicians and patients; working on a national level to prevent Medicare fraud within the Hispanic community; and advocating on issues of critical importance to the health of Hispanic older adults, their families, and caregivers, including the equitable implementation of the Affordable Care Act or health care reform on Capitol Hill.

As indicated in 2007 by the U.S, Department of Health and Human Services (DHHS), Latin@ elders may look to the leaders and staff in community organizations for referrals to a wide range of services and for services themselves. In different states there are community organizations focused on helping to educate Latin@ elders on Medicare benefits, helping address language barriers and advocating for optimal health/mental health service delivery. These community organizations may provide medical care, counseling, nutrition programs, exercise classes, or other programs (DHHS 2007).

The Latino Geriatric Mental Health Group of New York City (2007) was organized with the representation from the Association of Hispanic Mental Health Professionals, The Geriatric Mental Health Alliance of New York, the NYC Department for the Aging, and the NYC Department of Health and Mental Hygiene. They convened a meeting of more than 100 individuals to discuss how services for the Latin@ elder population could be improved. Their discussion resulted in seven major areas of focus: (1) language, (2) cultural competence, (3) workforce development, (4) outreach and education, (5) home and community-based services, (6) family support, (6) best practices, and (7) limited service capacity. Together these organizations seek to better serve Latin@ elders. Among their accomplishments are: involvement in the New York Governor's Commission on Hispanic Affairs leading to the development of the first bilingual/bicultural psychiatric impatient and out-patient units in the City of New York; increased awareness of Hispanic mental health issues by disseminating cutting edge Mental Health knowledge through conferences (Association of Hispanic Mental Health Professionals 2013); and advocacy for laws and programs that support older adults (Geriatric Mental Health Alliance of New York 2012). Also, the Department for the Aging (DFTA) has become the lead agency for implementation of the 59 strategies for improving New York's livability for older persons.

In 2007, various federal organizations in the United States joined efforts and created The Hispanic Elders Project, a pilot project that involved eight teams from different metropolitan areas with large numbers of Latin@ elders. These teams were charged with the task of assisting local communities in developing coordinated strategies for improving the health and well-being of elder Latin@s. The major purpose was to link Latin@ elders and their families with service providers, medical care providers, community organizations, and public agencies to promote the use of the new benefits, prevention programs, and other initiatives that would reduce health disparities among Latin@ elders (DHHS 2007).

The Latino Age Wave Initiative in Colorado and California is a project developed by the Hispanics in Philanthropy (HIP) organization and supported by grants under their HIP Hispanic Aging program. The initiative's goals are to expand and strengthen services, advocate on behalf of Latin@ elders and raise awareness of Latin@ aging issues. Another focus of this initiative is to engage Latin@ older adults as advocates for policies that improve their health and well-being (Global Policy Solutions LLC 2011). The Latino Center on Aging in New York, established in 1991, has a similar focus: to improve the lives of Latino seniors through advocacy and education, increase knowledge on the hardships faced by these elders, and assist in the creation of new programs and services. Another organization serving Latin@ elders is the Institute for Puerto Rican/Hispanic Elderly also in New York. In 2006, the Institute was administering 23 programs with services for Latin@ elders (Institute for Puerto Rican/Hispanic Elderly 2013).

Senior centers are a strong support network for Latin@ elders, and often the staff and elder become part of the family (Negroni-Rodríguez and Bok 2004). The centers coordinate different activities and services that address the cultural, social, emotional and physical needs of the Latin@ elders in their geographical locations. For example, The Ethel Macleod Hart Senior Center in Sacramento, California has created the *Manitos* Program. *Manitos (*Spanish term used in Mexico to refer to one's pals, buddy or friends) is a club open to older adults who speak Spanish. The program features games, food, songs, exercise, and educational activities.

In the state of Connecticut where Latin@s represent 14 % of the population, there are several local centers for Latin@ elders. Casa Otoñal, a nonprofit organization, consists of senior housing, a senior center, housing for grandparents who have custody of their grandchildren, companion and homemaker, information and assistance, benefits counseling, translation, and other social and support services to traditionally underserved Latin@ elders in the south side of the state. Senior centers provide for the many needs of older adults (Miltiades et al. 2010).

Another strategy to increase Latin@ healthy aging is for interdisciplinary teams to participate in an Interprofessional Collaborative Practice (IPCP). This is a model of collaboration of various disciplines (medicine, dentistry, pharmacy, physical therapy, social work, and public health) to train students and to improve health outcomes for a geriatric population. Using the various expertise, the threefold goals are: (1) to facilitate access, (2) to promote a continuum of services that insure healthy outcomes, and (3) to provide and coordinate culturally and linguistically appropriate services in primary, dental and mental health care, while training students for the growing geriatric population in global societies. The intersectional framework of the IPCP model offers an opportunity to effectively increase access to care and achieve high quality patient and population-centered outcomes by providing coordinated care from multiple perspectives and disciplines in their homes and communities.

This health promotion strategy will also increase the number of nurses skilled in interprofessional collaborative practice and promote the development of diverse and culturally competent skills needed to practice in existing and emerging health care systems. These partnerships, in doing interventions in the homes of the elders, can improve direct health outcomes for an underserved aging population through meaningful collaborations with other faculty and students from various disciplines. Working collectively, the team can enhance the training outcomes of an emerging healthcare workforce in geriatrics and reduce the inappropriate and high utility of emergency room visits by older adults. This health promotion strategy is inter-professional collaborative practice that promotes the development of diverse and culturally competent skills needed to practice in existing and emerging health care systems.

Conclusion

This chapter focuses on the health and mental health realities of aging for Latin@ elders in low-income communities, and advocates for community intervention strategies that promote equity for Latin@ elders in aging well. The authors examine several social determinants of health and note that social support networks and community strategies influence and secure healthy aging within the Latin@ elder community in spite of health and mental health disparities. The networks of social support lead Latin@ older adults toward well-being while "aging in place". Vega (2010) refers to this process as Latin@ elders' "ability to retain an appropriate level of independent living in one's community and place of residence and have personal control over one's lifestyle" (p. 3). In spite of health disparities that result from structural inequalities, the authors emphasize three positive influences that affect healthy aging among Latin@ elders: (1) family relationships, (2) social support networks, and (3) health/mental health service delivery system within a community context. The authors illustrate community programs as best practice models that include cultural competence and interprofessional collaborations that challenge the structural inequities to reduce health and mental health disparities within the Latin@ elder population. In a collective spirit, community programs respond to ensure conditions for Latin@ elders to lead healthy lives and sustain optimal functioning.

References

Abraido-Lanza, A. F., Dohrenwen, B. P., Ng-Mak, D. S., & Turner, J. B. (1999). The Latino mortality paradox: A test of the "salmon bias" and health migrant hypotheses. *American Journal of Public Health, 89*(10), 1543–1548.

Administration on Aging. (2010, January). *A statistical profile of Hispanic older Americans aged 65*. Retrieved from www.aoa.gov/AoAroot/Aging_Statistics/minority_aging/Facts-on-Hispanic-Elderly-2008.aspx

Administration on Aging. (2012, February 10). *A profile of older Americans: 2011*. Retrieved from http://www.aoa.gov/Aging_Statistics/Profile/2011/10.aspx

Angel, J., Kahlert, R., & Whitfield, K. (2005, September). *Hispanic and Mexican American health and aging in a new century, an overview*. Key issues in Hispanic health and health care policy research, Lyndon B. Johnson School of Public Affairs. The University of Texas at Austin.

Angel, J. L., & Tienda, M. (1987). Determinants of extended household structure: Cultural pattern or economic need?. *American Journal of Sociology, 87*, 1360–1383.

Aranda, M. P., & Knight, B. G. (1997). The influences of ethnicity and culture on the caregiver stress and coping process: A sociocultural review and analysis. *The Gerontologist, 37*, 342–354.

Areán, P., Ayalon, L., Hunkeler, E., Lin, E., Tang, L., Harpole, L., et al. (2005). Improving depression care for older minority patients in primary care. *Medical Care, 43*(4), 381–390.

Association of Hispanic Mental Health Professionals. (2013). *Our accomplishments*. Retrieved on July 19, 2013 from http://www.ahmhp.org/accomplishments/index.php

Barusch, A. S. (2009). *Foundations of social policy: Social justice in human perspective* (4th ed.). Connecticut: Wadsworth Press.

Beyene, Y., Becker, G., & Mayen, N. (2002). Perception of aging and sense of well-being among Latino elderly. *Journal of Cross-Cultural gerontology, 17*, 155–172.

Born, W., Gerniner, K. A., Sylvia, E., Butler, J., & Ahluwalia, J. S. (2004). Knowledge, attitudes and beliefs about end-of-life care among inner-city African Americans and Latinos. *Journal of Pallatiative Medicine, 7*, 247–256.

Center for Disease Control and Prevention. (2012, October 11). *Hispanic or Latino populations*. Retrieved from http://www.cdc.gov/minorityhealth/populations/REMP/hispanic.html

Chavez-Korell, S., Delgado-Rendón, A., Beer, J., Rodríguez, N., Garr, A. D., Pine, C. A., et al. (2012). Improving access and reducing barriers to depression treatment for Latino elders: Un Nuevo Amanecer (a new dawn). *Professional Psychology: Research and Practice, 43*(3), 217–226.

Cohen, G. D. (2005). *The mature mind: The positive power of the aging brain*. New York: Basic Books.

Collins, K. S., Hughes, D. L., Doty, M. M., Ives, B. L., Edwards, J. N., & Tenney, K. (2002). *Diverse communities, common concerns: Assessing health care quality for minority americans* (pp. 1–68). New York: The Commonwealth Fund.

Coonrod, D., Bay, C., & Balcazar, H. (2004). Ethnicity, acculturation and obstetric outcomes: Different risk factor profiles in low-and high acculturation Hispanics and in White Non-Hispanics. *Journal of Reproductive Medicine, 49*(1), 17–22.

Crist, J. D., McEwen, M. M., Herrera, A. P., Kim, S. S., Pasvogel, A., & Hepworth, J. T. (2009). Caregiving burden, acculturation, familism, and Mexican American elders' use of home care services. *Research and Theory for Nursing Practice: An International Journal, 23*(3), 165–180.

Day, L. (2010). Health care reform, health. and social justice. *American Journal of Critical Care, 19*(5), 459–461. doi:10.4037/ajcc2010742.

Delgado, M. (1998). Puerto Rican elders and merchant establishments: Natural caregiving systems or simply businesses? In M. Delgado (Ed.), *Latino elders and the twenty-first century: Issues and challenges for culturally competent research and practice* (pp. 33–46). New York: The Haworth Press Inc.

Department of Health and Human Services. (2007). *Improving Hispanic elders' health: Community partnerships for evidence-based solutions: Project overview*. Retrieved on April 30, 20113 from http://www.academyhealth.org/ahrq/elders/overview.pdf

DiClemente, R. J., Crosby, R. A., & Kegler, M. C. (Eds.). (2002). *Emerging theories in health promotion practice and research: Strategies for improving public health*. San Francisco, CA: Jossey-Bass, John Wiley & Sons.

Dilworth-Anderson, P., Pierre, G., & Hilliard, T. (2012). Social justice, health disparities, and culture in the care of the elderly. *Journal of Law, Medicine, & Ethics, 40*, 26–31.

Donelly, J. W., Eburne, M., & Kittleson, M. (2001). *Mental health: Dimensions of self-esteem and emotional well-being*. Boston: Allyn & Bacon.

Falcon, L. M., & Tucker, K. L. (2000). Prevalence and correlates of depressive symptoms among Hispanic elders in Massachusetts. *Journal of Gerontology: Psychosocial and Social Sciences, 55*(2), S108–S116.

Falcon, L. M., Todorova, I., & Tucker, K. (2009). Social support, life events and psychological distress among the Puerto Rican population in the Boston area of the United States. *Aging and Mental Health, 13*(6), 863–873.

Fennell, M. L., Feng, Z., Clark, M. A., & Mor, V. (2010). Elderly Hispanics more likely to reside in poor-quality nursing homes. *Health Affairs, 29*(1), 65–73. doi:10.1377/hithaff. 2009.0003.

Forster, L., & Stoller, E. P. (1992). The impact of social support on mortality: A seven-year follow-up of older men and women. *Journal of Applied Gerontology, 2*(2), 173–186.

Franzini, L., Ribble, J. C., & Keddie, A. M. (2002). Understanding the Hispanic paradox. In T. A. Laveist (Ed.), *Race, ethnicity, and health: A public health reader* (pp. 280–310). San Francisco: Jossey-Bass.

Gallagher-Thompson, D., Coon, D., Solano, N., Ambler, C., Rabinowitz, Y., & Thompson, L. (2003). Change in indices of distress among Latino and Anglo female caregivers of elderly relatives with dementia: Site-specific results from the REACH national collaborative study. *The Gerontologist, 43*(4), 580–591.

Gallagher-Thompson, D., Gray, H., Dupart, T., Jimenez, D., & Thompson, L. (2008). Effectiveness of cognitive/behavioral small group intervention for reduction of depression and stress in Non Hispanic White and Hispanic/Latino women dementia family caregivers: Outcomes and mediators of change. *Journal of Rational-Emotive & Cognitive-Behavior Therapy, 26*(4), 286–303.

Geriatric Mental Health Alliance of New York. (July, 2012). *Geriatric mental health news*. Retrieved on June 13 from http://archive.constantcontact.com/fs073/1101456716765/archive/1110464920159.html#LETTER.BLOCK7

Global Policy Solutions LLC. (February, 2011). *The Latino age wave: What changing ethnic demographics mean for the future of aging in the United States*. Report by Hispanics in Philanthropy (HIP). Retrieved on May 1, 2013 from http://www.hiponline.org/storage/documents/HIP_LatinoAgeWave_FullReport_Web.pdf.

Gonzalez, H. M., Haan, M. N., & Hinton, L. (2001). Acculturation and the prevalence of depression in older Mexican Americans: Baseline results of the Sacramento area Latino study on aging. *Journal of the American Geriatrics Society, 49*(7), 948–953.

Hazuda, H. P., & Espino, D. V. (1997). Aging, chronic disease, and physical disability in hispanic elderly. In K. S. Markides & M. R. Miranda (Eds.), *Minorities, aging, and health* (pp. 127–148). Thousand Oaks: Sage.

Heisler, M., et al. (2007). Mechanisms for racial and ethnic disparities in Glycemic control in middle-aged and older Americans in the health and retirement study. *Archives of Internal Medicine, 167*(17), 1853–1860.

Heyman, J. C., & Gutheil, I. A. (2010). Older Latino's attitude toward and comfort with end-of-life planning. *Health and Social Work, 35*(1), 17–26.

Hispanic Health Council. (2006). *A profile on Latino Health in Connecticut: The case for change in policy & practice*. Retrieved May 1, 2013, from http://www.hartfordinfo.org/issues/wsd/health/Profile_Latino_Health.pdf

Iatridis, D. S. (1994). *Social policy: Institutional context of social development and human services*. Pacific Grove: Brooks/Cole Publishing.

Institute for Puerto Rican/Hispanic Elderly. (2013). Retrieved July 19, 2013 from http://www.iprhe.org/

Kim, G., Jang, Y., Chiriboga, D. A., Ma, G. X., & Schonfeld, L. (2009). Factors associated with mental health services use in Latino and Asian immigrant elders. *Aging & Mental Health, 14*(5), 535–542. doi:10.1080/13607860903311758.

Krakauer, E. L., Crenner, C., & Fox, K. (2002). Barriers to optimum end of life care for minority patients. *Journal of the American Geriatrics Society, 50*, 182–190.

Latino Geriatric Mental Health Workgroup of New York City. (2007). *Meeting the mental health needs of Latino elders*. Network of Care of Behavioral Health Online Newsletter. Retrieved on A pril 29, 2013 from http://losangeles.networkofcare.org/mh/library/article.aspx?id=2587&cat=42

LaVeist, T. A. (2005). *Minority populations and health an introduction to health disparities in the United states* (1st ed.). San Francisco: Jossey-Bass.

Marmot, M. G., Adelstein, A. M., & Bulusu, L. (1984). Lessons from the study of immigrant mortality. *Lancet, 2*, 1455–1457.

Medina, C. (2011). Creating the Latino agenda for eliminating health disparities in communities: A call to action. In J. H. Schiele (Ed.), *Social welfare policy: Regulation and resistance among people of color* (pp. 271–291). Thousand Oaks: Sage.

Migrant Health Promotion. (2013). Retrieved July 19, 2013, from http://www.migranthealth.org/

Miltiades, H. B., Grove, S. A., & Drenovsky, C. (2010). *Understanding the impact of senior community center participation on elders' health and well-being*. Commonwealth of Pennsylvania—Department of Aging. Latino Center on Aging (nd). Mission. Retrieved on April 25, 2013 from http://www.gerolatino.org/english/5552/5615.html

Min, J. W., & Barrio, C. (2009). Cultural values and caregiver preference for Mexican-American and Non-Latin@ White elders. *Journal of Cross-Cultural Gerontology, 24*, 225–239.

Moody, H. R. (1998). *Aging: Concepts and controversies*. Thousand Oaks: Sage Publications Incorporated.

Motel, S., & Patten, E. (2013). *Statistical portrait of Hispanics in the United States, 2011*. Pew Hispanic Center, Washington, DC. Retrieved on April 12, 2013 from http://www.pewhispanic.org/2013/02/15/statistical-portrait-of-hispanics-in-the-united-states-2011/

Napoles-Springer, A. M., Santoyo, J., Houston, K., Perez-Stable, E. J., & Steward, A. L. (2005). Patients' perceptions of cultural factors affecting the quality of their medical encounters. *Health Expectations, 8*(1), 4–17.

National Institutes of Health. (2000). *NIH strategic research plan to reduce and ultimately eliminate health disparities.*

Negroni, L. K. (2007). [Latin@s stories of growing older in the United States]. Unpublished focus group data. Puerto Rican and Latin@ Studies Project, University of Connecticut School of Social Work.

Negroni, L. K. (2012). The Puerto Rican family. In R. Wright, C. H. Mindel, R. W. Habenstein, & T. V. Tran (Eds.), *Ethnic families in America: Patterns and variations* (5th ed., pp. 129–147). New York: Pearson Education Inc.

Negroni-Rodríguez, L. K., & Bok, M. (2004). *Understanding the impact of natural support networks on the physical and mental health of Latina elderly and the utilization and satisfaction with services*. Unpublished findings report, Puerto Rican and Latin@ Studies Project, University of Connecticut School of Social Work.

Negroni-Rodríguez, L. K., & Morales, J. (2001). Individual and family assessment skills with Latino/Hispanic. In R. Fong & S. Furuto (Eds.), *Culturally competent practice: Skills, interventions and evaluations* (pp. 132–146). Boston: Allyn and Bacon.

Nussbaum, M. C. (2006). *Frontiers of justice: Disability, nationality, species membership*. Cambridge: The Belknap Press of Harvard University Press.

Ortega, A. N., Feldman, J. M., Canino, G., Steinman, K., & Alegŕia, M. (2006). Co-occurrence of mental and physical illness in US Latinos. *Social Psychiatry and Psychiatric Epidemiology, 41*(12), 927–934.

Ortiz, V. (1995). The diversity of Latino families. In R. Zambrana (Ed.), *Understanding Latino families: Scholarship, policy, and practice* (pp. 18–39). Thousand Oaks: Sage.

Parra-Cardona, J. R., Meyer, E., Schiamberg, L., & Post, L. (2007). Elder abuse and neglect in Latino families: An ecological and culturally relevant theoretical framework for clinical practice. *Family Process, 46*(4), 451–470.

Paz, J., & Applewhite, S. L. (1998). Culturally competent practice. *Journal of Gerontological Social Work, 30*, 1–15.

Phipps, E. J., True, G., & Murray, G. F. (2003). Community perspectives on advance care planning: Report from the community ethics program. *Journal of Cultural Diversity, 10*(4), 118–123.

Piette, J. D., Heisler, M., & Wagner, T. H. (2004). Problems paying out-of-pocket medication costs among older adults with diabetes. *Diabetes Care, 27*(2), 384–391.
Powers, M., & Faden, R. (2006). *Social justice*. Oxford: Oxford University Press.
Ramos, B. M. (2007). Housing disparities, caregiving and their impact for older Puerto Ricans. *Journal of Gerontological Social Work, 49*(1/2), 47–64.
Rawls, J. A. (1971). *A theory of justice*. Cambridge: Harvard University Press.
Rosnick, C. B., & Reynolds, S. L. (2003). Thinking ahead: Factors associated with executing advance directives. *Journal of Aging Health, 15*, 409–429. doi:10.1177/0898264303015002005.
Rowe, J. W., & Khan, R. L. (1998). *Successful aging*. New York: Pantheon Books.
Ruiz, M. E., & Ransford, H. E. (2012). Latino elders reframing familismo: Implications for health and caregiving support. *Journal of Cultural Diversity, 19*(2), 50–57.
Sanchez-Ayendez, M. (1988). The Puerto Rican family. In C. H. Mindel, R. W. Habenstein, & R. Wright Jr (Eds.), *Ethnic families in America*. New York: Elsevier.
Siebert, D. C., Mutran, E. J., & Reitzes, D. C. (1999). Friendship and social support: The importance of role identity to aging adults. *Social Work, 44*, 522–533.
Sotomayor, M. (1992). *Empowering Hispanic families: A critical issue for the 90s*. Milwaukee: Family Service of America.
Stratton, A., Hynes, M. M., & Nepaul, A. N. (2009). *The 2009 Connecticut health disparities report*. Hartford: Connecticut Department of Public Health.
Sue, S. (2003). In defense of cultural competency in psychotherapy and treatment. *American Psychologist, 58*, 964–970.
Tucker, K. L., Mattei, J., Noel, S. E., Collado, B. M., Mendez, J., Nelson, J., et al. (2010). The Boston Puerto Rican health study, a longitudinal cohort study on health disparities in Puerto Rican adults: Challenges and opportunities. *BMC Public Health, 10*, 107. doi:10.1186/1471-2458-10-107.
U.S. Census Bureau Decennial Census. (2010). *American fact finder*. Retrieved July 19, 2013, from http://factfinder2.census.gov/faces/nav/jsf/pages/wc_dec.xhtml
Vega, W. A., & Alegria, M. (2001). Latino mental health and treatment in the U.S. In C. W. Molina & R. E. Zambrana (Eds.), *Health issues in the Latino community* (p. 179). San Francisco: Jossey-Bass.
Vega, W. A. (1995). The study of Latino families: A point of departure. In R. E. Zambrana (Ed.), *Understanding Latino families: Scholarship, policy, and practice* (pp. 3–17). Thousand Oaks: Sage Publications.
Vega, M. (2010). *Latinos "aging in place": Issues and potential solutions*. Unpublished paper presentation by the University of Southern California. Retrieved on 4/12/2013 from http://www.utexas.edu/lbj/caa/2010/presentations/VegaW.pdf
Vélez, D., & Cole, S. A. (2008). Culture, place of origin, and service delivery for older adult Latino immigrants: The case of Puerto Rican older adults. *Journal of Gerontological Social Work, 51*, 300–314.
Villa, V., & Torres-Gil, F. (2001). The health of elderly Latinos. In M. Aguirre-Molina, C. W. Molina, & R. E. Zambrana (Eds.), *Health issues in the Latino community* (pp. 157–178). San Fancisco: Jossey-Bass.
Vincent, G. K., & Velkoff, V. A. (2010). *The next four decades, the older population in the United States: 2010–2050*, Current Population Reports, P25-1138, U.S. Census Bureau, Washington, DC.
Wallerstein, I. (2005). Latin@s: What's in a name? In R. Grosfoguel et al. (Eds.), *Latin@s in the world-system: Decolonization struggles in the twenty-first century U.S. empire* (pp. 31–39). Boulder: Paradigm Press.
World Health Organization (WHO). (2013). *Commission on social determinants of health.* Retrieved May, 1, 2013 from http://www.who.int/social_determinants/thecommission/en/index.html
World Health Organization. (2011). *Social determinants of health: Key concepts*. Final Report of the Commission on Social Determinants of Health. Retrieved on May 1, 2013 from http://www.who.int/social_determinants/thecommission/finalreport/key_concepts/en/index.html

Aging, Intimacy, and Sex

Nancy Giunta and Stephanie A. Jacobson

Abstract The chapter brings together a range of current knowledge around intimacy and sexuality among older adults. The authors offer evidence to counter the common misconception that older adults are not sexually active. Sexuality and aging is then discussed within an historical context that includes HIV and AIDS, the medicalization of sexuality, the introduction of the internet and social media, and the movement for equality among the lesbian, gay, bisexual and transgender communities. Next, the physical and mental health implications of sexual activity and aging are examined along with potential social benefits of sex and intimacy. Along those lines, cultural norms around sexuality across diverse communities intersect with physiological changes in sexual functioning that occur with age. Finally, the authors offer implications for research, practice, and policy in this important aspect of aging. A call to professionals is made to continue to learn more and develop cultural sensitivity, thus embracing the normative nature of sexuality and intimacy throughout adulthood.

Many Americans erroneously believe that older adults are not sexually active (Hooyman and Kiyak 2005; Zeiss 1982). This belief that older adults are not interested in sex has contributed to a prevailing stigma about older adults and sexuality (Schiavi 1999), which leads to misinformation and potential risks to physical and mental health. There is undeniable evidence that people continue to be interested in sex throughout the lifespan and that the need for intimacy does not diminish with age (AARP 2005; Lindau et al. 2006, 2007; Zeiss and Kasl-Godley 2001). This chapter brings together a range of current knowledge around intimacy and sexuality among older adults. First, evidence will be provided to counter the

N. Giunta (✉)
Silberman School of Social Work at Hunter College, City University of New York, New York, USA
e-mail: ngiunta@hunter.cuny.edu

S. A. Jacobson
Quinnipiac University, Hamden, CT, USA

H. F. O. Vakalahi et al. (eds.), *The Collective Spirit of Aging Across Cultures*, International Perspectives on Aging 9, DOI: 10.1007/978-94-017-8594-5_6,

assumption that older adults are not sexually active. This will be followed by a presentation of historical events that have helped shape the current context of sexuality among older adults. Next, the role of physical and mental health in sexual activity and aging will be discussed, followed by an examination of the social benefits of sex and intimacy. Finally, cultural norms around sexuality across diverse communities will be discussed and considered along with typical physiological changes in sexual functioning that occur with age. The chapter will close with implications for research, practice, and policy in this important aspect of aging.

One of the authors of this chapter had worked as a social worker in a busy urban hospital well known for its geriatric expertise. Through this experience, she witnessed hesitation of medical professionals to discuss sexual activity and sexual health with older people. It became clear to her one day that action needed to be taken to address this problem. During a social work assessment with a 90-year-old hospitalized woman, the patient raised issues and had questions related to her sexual activity and sexual partners. The social worker explored these issues and counseled the woman about condom use and safe sex practices, unaware that the woman's physician (a geriatrician) was outside the room and had overheard the end of the conversation. Later, the physician commented that he felt the conversation was unnecessary because the patient was not sexually active. Clearly, it is these assumptions on the part of medical practitioners and reinforced by societal norms that inhibit communication about sexual activity and sexual health among older adults.

Vignette:

There is substantial evidence that the need for intimacy and interest in sexual activity continues throughout the life course (AARP 2005; Lindau et al. 2006, 2007; Schwartz 2007; Starr and Weiner 1981; Zeiss and Kasl-Godley 2001). Previous sexual activity remains one of the strongest predictors of sexual activity among older adults. Although frequency of sexual intercourse may generally decline with age, sexual activity tends to remain stable among older adults who have been sexually active in the past (Beutel et al. 2002; Kontula and Haavio-Mannila 2009; Lindau et al. 2006, 2007; Schick et al. 2010; Waite et al. 2009; Zeiss and Kasl-Godley 2001). Biological, medical and psychosocial factors have been shown to control for decreases in intercourse (DeLamater and Moorman 2007). Caution must be taken in the interpretation of sexuality research in the field of aging. Many studies do not explicitly define sexual experience, behavior, and interest. For example, many researchers (Kontula and Haavio-Mannila 2009; Waite et al. 2009) define sexual activity only as sexual intercourse with a partner, while others do not provide a specific definition. Additionally, multicultural perspectives are rarely captured in current research on sexuality and aging (Huang et al. 2009), despite tremendous cultural variation in the role of sexuality throughout the life course. Training on clinical practice with multicultural older adults is also rare (Hinrichsen 2006), especially around areas of sexuality (Huang 2009).

A recent study (Schick et al. 2010) found that 42.0 % of women and 53.4 % of older men age 50 and older reported being sexually active in the past year. Although there is a steady decline in partnered sex that correlates with age (Beutel et al. 2002; Kontula and Haavio-Mannila 2009; Lindau et al. 2006, 2007; Schick et al. 2010; Waite et al. 2009; Zeiss and Kasl-Godley 2001), frequency of sex remains stable for those who have been sexually active (Lindau et al. 2007; Waite et al. 2009). For individuals who remain sexually active, frequency of partnered sex remains stable through age 75 and only declines modestly for adults over age 75 (Lindau et al. 2007; Waite et al. 2009). This maintained frequency with age suggests continued interest in sexuality throughout the lifespan (AARP 2005; Lindau et al. 2006, 2007; Zeiss and Kasl-Godley 2001). Women report decreased urgency for sex, rating it as less important with age, but they retain sexual desire and describe it as an important aspect of their lives both emotionally and psychologically (Lindau et al. 2007).

Older adults are often marginalized as sexual beings based on their age. The intersection of age and sexual orientation further marginalizes older adults who are lesbian, gay, bisexual or transgender (LGBT). Heteronormative assumptions around older adults, particularly among medical and social service providers, often lead LGBT elders to "closet" themselves when seeking long term supportive services. Today's cohort of older adults came of age during a time when the medical profession pathologized homosexuality and subjected it to *cures* including medications, shock therapy, castration, hysterectomy, lobotomy, and aversion therapy (Adam 1995; Katz 1976, 2007). LGBT people face health disparities in certain physical and mental health conditions, lack of healthcare, higher rates of substance abuse, higher rates of STIs, and larger incidences of certain cancers. Additionally, LGBT patients often feel stigmatized in healthcare agencies (Fredriksen-Goldsen and Muraco 2010; Fredriksen-Goldsen et al. 2013; Wight et al. 2012) and those particularly marginalized in these settings are elders of color and elders who are transgendered or have a transgender history (Fredriksen-Goldsen et al. 2013).

A Historical Context of Sexuality and Aging

A rich history has shaped the current context of sexuality among older adults. Four historical developments that will be discussed here include the emergence of HIV/AIDS, the introduction of Viagra to treat erectile dysfunction, the development of the Internet and social media, and the gay rights movement. These historical events are just four of many that have shaped the current context of sexuality and aging. They have been selected for this chapter because the authors believe the elders of today have lived through the emergence of these activities, which makes the landscape of sexuality and aging quite different than that of previous cohorts and quite different when elders of today came of age in the middle of the twentieth century.

HIV and AIDS

In 1981, the CDC published a Morbidity and Mortality Weekly Report (MMWR) on Pneumocystic carinii pneumonia in gay men in Los Angeles (CDC 1996). This MMWR was the first report of what later became known as AIDS. AIDS quickly became the focus of headlines and prominent in the minds of Americans (McLaren 1999; Parascandola 2008). With rising rates of HIV/AIDS and its association with sex, but no treatment for the disease, Americans became more conscious of the potential health consequences of sexual behavior.

Similar to the initial response to syphilis, people viewed AIDS as a moral problem (Giami 2008; Parascandola 2008). Individuals were uncomfortable discussing HIV/AIDS because of its association with sexual activity among socially stigmatized groups, such as men who have sex with men and injection drug users. At the same time, healthcare providers were concerned about the lack of a viable treatment during the early years of the epidemic (Parascandola 2008).

The earliest HIV/AIDS prevention initiatives developed in community-based agencies, often founded by gay men who recognized the need for HIV education and prevention in their community. The fight against the spread of HIV/AIDS targeted gay men, IV drug users, young people, and individuals with multiple sexual partners. It initially ignored lesbians, heterosexuals, older adults, and monogamous or married couples (Giami 2008). Because of targeted sex education, women and older adults did not receive information about how to protect themselves against HIV/AIDS.

Although HIV is spread in a number of ways it is classified as an STI. Recent studies and public health surveys have focused on HIV and AIDS infections in older adults. According to the CDC, a quarter of all people living with HIV in 2006 were aged 50 and older (2008a). The number of adults over age 50 living with HIV or AIDS nearly doubled from 2001 to 2005 (Kuehn 2008). By 2015, more than half of individuals living with HIV will be aged 50 or older (Brick et al. 2009). Adults over the age of 50 account for 15 % of new diagnoses and over a third of AIDS related deaths (CDC 2008b). In a recent report, the New York City Department of Health (NYC-DOH) stated, "persons aged 50 and older play an important but under-recognized part in the HIV/AIDS epidemic" (2010, p. 4). Despite the growing prevalence of HIV among older adults, many heterosexual elders misjudge the threat of HIV (Henderson et al. 2004; Maes and Louis 2003; Tangredi et al. 2008) and believe that their risk of contracting HIV is low.

Healthcare professionals encouraged certain groups that were identified as "at risk" to modify their sexual behaviors (Giami 2008). While making recommendations about HIV prevention, doctors and research scientists also worked to discover therapeutic treatments (Hart and Wellings 2002). Although HIV remains incurable at this time, antiretroviral treatments have extended life expectancy and increased quality of life. Today, people infected with HIV are living into older adulthood.

Viagra and the Medicalization of Sexuality

Male sexual dysfunction became an increasingly popular area of scientific study in the beginning of the 1990s (Castelo-Branco et al. 2008). The first International Conference for Consensus on Erectile Dysfunction took place in 1993. During this conference, the term *impotence* was abandoned and replaced by the term *erectile dysfunction* (ED) defined as "the incapacity to achieve and maintain a penile erection enough to attain satisfactory sexual relations" (Castelo-Branco et al. 2008, pp. 55–56). Following this change in terminology, Viagra, the first pharmaceutical product to treat ED, arrived on the market in March 1998 and there are now three such drugs available (Herzog 2008). Changes in terminology and pharmaceutical marketing efforts directly to consumers rather than to physicians helped make Viagra the most popular pharmaceutical drug ever sold (Hart and Wellings 2002). There are now three such drugs available (Hirsch 2011). Viagra requires good timing and communication with the partner because it creates a small window for erection (Westheimer and Lehu 2005). Cialis was developed to manage these issues and is advertised to last thirty-six hours. Today, about one in five men report using medication for erectile dysfunction (Schick et al. 2010) with the use of these medications increasing with age until age 70, when the usage declines. Perhaps due to side effects such as flushing and headache, few men continue use for over a year (Brick et al. 2009; Westheimer and Lehu 2005).

Until the introduction of Viagra, decline in sexual functioning was assumed a normal part of aging. The availability of an oral medication for erectile dysfunction challenged the stereotype that there was an age limitation on men's sexual performance and interest (Castelo-Branco et al. 2008). Some argue that the introduction of Viagra has medicalized sexuality (Kingsberg 2002; Loe 2004), especially for older adults. Medical specialists (urologists) have displaced psychologists and therapists as the primary clinicians treating sexual dysfunction (Loe 2004).

The medicalization of sexuality has affected older women as well, with the emergence in the late 1990s of the phenomenon known as female sexual dysfunction (Berman et al. 1999; Moynihan 2003). The study of female sexual dysfunction also has primarily fallen under the purview of urologists (Goldstein 2008). In 1998, the American Foundation of Urologic Disease (AFUD) agreed on a definition of female sexual dysfunction (Berman et al. 1999). However, after some debate, Basson and colleagues (2004) conceptualized female sexual dysfunction to include women's sexual interest/desire disorders, subjective sexual arousal disorders, genital sexual arousal disorders, women's orgasmic disorders, and pain disorders.

No definitive treatment is available for female sexual dysfunction, but hormone replacement therapy (HRT) with estrogen alone or in combination with progesterone has been popular for decades to relieve sexual symptoms associated with menopause (Harvard Medical School 2011; Palacios 2008). As HRT has grown in popularity, the risks and benefits have been hotly debated (Palacios 2008). In 2001, research indicated that HRT relieved symptoms of menopause and protect against osteoporosis, heart disease, colon cancer, and Alzheimer's disease (Harvard Medical School 2011). The risks included increased susceptibility to breast cancer and gallbladder disease. More recent research links HRT to increases in those conditions, in addition to increased risk of stroke, blood clots, and dementia. Since HRT is highly controversial (Palacios 2008), other hormonal therapies are being considered for women. Physicians now recommend vaginal creams, inserts, or rings that contain estrogen instead of HRT. Approximately fifteen percent of women over age 60 report taking a hormone pill, while only four percent use a cream and one percent use a patch. Women aged 45 and older decreased their use of hormone pills between 1999 and 2004, while the use of hormone creams and patches remained consistent. This decline may have been due to studies that reported potential risks of HRT (Palacios 2008).

The Internet and Social Media

Nearly half of American older adults are using the Internet, and among older internet users, about one in three uses social networking sites like Facebook, LinkedIn or others (Leist 2013; Madden 2012; Zickuhr and Madden 2012). Substantial evidence has emerged that internet use among elders contributes to increases in social networks and personal well-being (Nimrod 2009). Relatively unexplored phenomena within the context of aging and relationships are the emergence of online dating and the role of social media in older adult sexuality. Studies of internet dating suggest that older adults and women in particular are more selective in choosing a potential date than younger adults (William et al. 2011). There is still very little known about the experiences or outcomes of online dating among older adults, including the perceptions of intimacy and sexuality among elders who use the internet for this purpose.

There is a proliferation of online content offering dating, sex, and intimacy advice for people over age 50. For example, the AARP website has a dating section as well as a section on sex and intimacy. Content in these sections, available in Spanish and English, includes online dating opportunities, articles about a variety of issues related to dating in older adulthood, and regular columns by dating experts (one written by a man and one by a woman). While these sections target mostly heterosexual audiences, the AARP website has a separate section focused on LGBT content. This section, AARP Pride, however, is difficult to find on the website and not integrated into the general sex and intimacy or dating sections described above. Despite this limitation, the content on the AARP Pride site covers a range of issues related to LGBT aging, including some coverage of dating, sex, and intimacy.

There are benefits and risks to using the internet for dating and intimacy that must be considered for older adults. Just as the benefits for younger generations include opportunities to increase social connections and avoid isolation, older adults are at greater risk of abuse and predatory activities on the internet. Older adults need to be informed and educated on how to navigate internet dating and social media with caution and learn how to identify potentially harmful online risks.

LGBT Rights

In the U.S., there are approximately three million adults aged 65 or older who identify as LGBT (National Gay and Lesbian Task Force 2011). Experts estimate that this number will double by 2030. Today's older adults grew up during a period when homosexuality was not widely accepted (Brick et al. 2009). The expected increase by 2030 may reflect increased comfort about being "out" as older adults (Brick et al. 2009). Despite these increasing numbers of self-identified LGBT older adults, in a recent study only eight of the 2,000 older adults who reported being in a relationship stated it was with someone of the same sex (Lindau et al. 2007). The low rates of same sex relationship in this older adult sample are difficult to explain, but the wording of the questions could have prevented respondents from answering reliably.

Homosexuality was pathologized in the medical literature beginning in the late 19th century, followed shortly by civil rights cases for gay men in Germany and England (Adam 1995; Katz 2007). In America, resistance to the medical attack on homosexuality was demonstrated mostly in the arts (Katz 1976), until the middle of the twentieth century when Harry Hay and others founded the Mattachine Society (Katz 1976), a social and advocacy group for gay men's rights. Del Martin and Phyllis Lyon founded a lesbian rights group called Daughters of Bilitis (DOB) in 1955. Gay and lesbian people continued to develop advocacy organization such as these through the 1960s, at which time the Stonewall riots catalyzed gay liberation groups to resist against institutions that were anti-homosexual, including medicine (Adam 1995). Ultimately, advocacy by organized gay liberation groups led to the removal of homosexuality from the APA diagnostic manual in the 1970s (Adam 1995). In 1972, the NASW rejected the medical model that discriminated against homosexuality. In 2009 the Matthew Shepard and James Byrd, Jr., Hate Crimes Prevention Act (2009) resulted in gender identity and sexual orientation being added to the list of categories of hate crimes. As of this writing, 15 countries recognize same sex marriage. Same-sex marriage is legal in 13 states however 31 states restrict marriage to one woman and one man.

Today, elders from the LGBT community continue to serve as pioneers in forging rights for same sex couples. In 2013, the U.S., Supreme Court decided it was unconstitutional to deny federal spousal benefits to same sex married couples. Edith (Edie) Windsor was the 84 year-old plaintiff in the case in which the U.S. Supreme Court deemed the Defense of Marriage Act (DOMA) to be unconstitutional, therefore ensuring that same-sex marriages approved by the states be recognized at the federal level.

Historically, researchers have made assumptions about heterosexuality and vaginal-penile intercourse in their studies (Marshall 2010). This type of research and practice is heterosexist (Brick et al. 2009) because it is based on the assumption that people are heterosexual, unless specifically explicated. This construct can make heterosexism unwittingly hurtful to older LGBT adults. For example, institutionalized heterosexism has led LGBT older adults to have inadequate healthcare (National Gay and Lesbian Task Force 2011). At the macro policy level, same-sex couples are not eligible for Medicare or Social Security based on their partner's work history. Similarly, same-sex life partners have not been eligible for Social Security survivor's benefits, which results in a loss of $124 million to LGBT elders each year. Additionally, Medicaid law allows for spousal refusal, which protects a spouse's home and assets when applying for long-term care or nursing home benefits. Same-sex partners are not eligible for this exemption and may not be able to afford basic needs if their partner requires extended homecare or nursing care. Federal laws, including the Family and Medical Leave Act, do not protect LGBT older adults if they need time off from work to care for a partner (Brick et al. 2009). Hospital visitation may be denied to a same-sex partner (National Gay and Lesbian Task Force National 2011). Because of the extent of institutional heterosexism, older adults who identify as LGBT may be uncomfortable seeking healthcare from organizations that do not focus on their unique needs (Brick et al. 2009).

Physical and Mental Health

There is substantial evidence that sexual activity is correlated with better physical and mental health; various physical and mental health benefits have been associated with sexual activity (AARP 2005; Gott 2005; Lindau et al. 2006, 2007; Schick et al. 2010; Zeiss and Kasl-Godley 2001). Orgasm has been attributed to increased immune response (Barnard 2011a; Block 2008), lower rates of prostate cancer (Schwartz 2007), and relief of chronic pain (Block 2008; Schwartz 2007). In its study of adults aged 45 and older, AARP (2005) found respondents with regular sexual partners had a more positive outlook on life; they experienced less stress and had fewer symptoms of depression. Although the AARP refers to partnered sex, these positive consequences of sex are attributed to the hormone oxytocin, which allows both men and women to have orgasms, with or without a partner (Schwartz 2007). Oxytocin produces feelings of happiness. The hormone oxytocin, released when men and women have orgasms, produces feelings of happiness (Schwartz 2007). This may explain the link between continued sexual activity and lower rates of depression (Barnard 2011a). Older adults who have at least one orgasm per week have a thirty percent lower rate of mild to moderate depression. Older adults with regular sexual partners report fewer symptoms of depression and less stress (AARP 2005).

Sexual activity remains a source of pleasure and social connectedness throughout the lifecycle. It enhances a person's quality of life, personal integrity, and self-perception (Bauer et al. 2007). Because there is a close relationship between sexual identity and self worth, rejection of sexuality in older adulthood may have negative effects on self-image, mental health, and social relationships (Bauer et al. 2007). People who report poor health but are also sexually active are more likely to report problems with sexual functioning. If not addressed, these problems may lead to sexual inactivity. Health conditions correlated with sexual activity include heart disease, high blood pressure, diabetes and depression (Matthias et al. 1997; Schick et al. 2010; Westheimer and Lehu 2005; Zeiss and Kasl-Godley 2001).

Following a heart attack or other serious heart condition, specific instructions may be necessary for continued sexual activity, but some doctors fail to address these concerns. When diabetes is not properly controlled, it can result in nerve damage, thus causing difficulty with arousal in both men and women (Westheimer and Lehu 2005). Similarly, stroke survivors may continue to desire sex but may face cognitive and physical impairments that make sexual activity difficult. All types of cancer and associated treatments may affect sexual functioning and libido (Bolte 2011), particularly because fatigue and pain are often associated with cancer and its treatment. Some specific types of cancer, such as breast, cervical, and ovarian, affect women's sexual functioning and their sense of sexual desirability (AARP 2005; Westheimer and Lehu 2005). Among men, prostate cancer and its treatment can lead to decreased libido and erectile dysfunction because of surgery or radiation (Weinsberg 2011).

Approximately one third of men and women over age 60 are diagnosed with arthritis (AARP 2005). Although not life threatening, arthritis can make sex extremely painful and reduce desire as a result (Matthias et al. 1997; Westheimer and Lehu 2005). Many people incorporate a number of simple life style changes in order to remain sexually active. For example, because arthritis pain may be worse at certain times of day, one may plan sexual activity for times when pain is less severe. Other alternatives include taking a pain relieving medication or warm bath or shower before sexual activity, or changing sexual positions that may be painful.

Untreated depression negatively affects sexual arousal and desire (Westheimer and Lehu 2005) and is closely linked with erectile dysfunction (Araujo et al. 1998; Nicolosi et al. 2004; Schiavi 1999). Although many studies report a relationship between depression and erectile dysfunction, the direction of the effect is unclear because depression may cause sexual difficulties or sexual dysfunction may lead to depression. Mental health risks are higher among LGBT older adults compared to their heterosexual counterparts. The intersection of being older and a sexual minority has been identified as being detrimental to mental health (Wight et al. 2012).

There is little information available about the relationship between dementia and sexuality (Price 2011). Dementia causes impaired memory, language, and reasoning (Price 2011), which can influence sexual desire and functioning (Kuhn 2011). During the early to moderate stage of dementia, older adults may continue to desire sex and remain sexually active. However, if a partner assumes a care giving relationship, they may no longer desire sex. As the dementia progresses, sexual desire and functioning typically diminish. Some individuals with dementia may become hypersexual and demand sexual talk and activity from their partner or others (Kuhn 2011; Westheimer and Lehu 2005). Due to cognitive changes, individuals with moderate to advanced dementia may not be able to consent to sexual activity (Kuhn 2011), which leads to ethical ambiguity.

Medications, Alcohol, and Substance Use

Many commonly prescribed medications cause sexual side effects (Brick et al. 2009; Finger et al. 1997). In the U.S. in 2011, more than 4 billion prescriptions were filled (IMS Health National Prescription Audit 2012). More than half of men and women over the age of 60 take an antihypertensive to treat high blood pressure (AARP 2005). No antihypertensive medications are completely free of sexual side effects (Westheimer and Lehu 2005). Almost one in ten adults over the age of 60 report taking an antidepressant (AARP 2005). Nearly all antidepressant cause changes in sexual functioning (Finger et al. 1997). Antihypertensives and antidepressants have the most pronounced sexual side effects, however many other medications, including nonprescription medications, can affect sexual functioning (Finger et al. 1997). Over the counter pain medication or antacids, for example, may cause erectile dysfunction, lack of ejaculation, or decreased desire. More than a quarter of adults over age 60 report taking a painkiller (AARP 2005). Because of the wide array of sexual side effects from a number of medications, it is important for healthcare providers and older adults to have open discussions about medications and sexual functioning in order to ensure medication is taken as directed (Finger et al. 1997).

Alcohol or illicit substances may also affect sexual functioning among older adults. According to Han and colleagues, the prevalence of substance or alcohol misuse and substance use disorder is higher among the baby boomer cohort than previous generations (Han et al. 2009). Alcohol and substance misuse may be a factor in sexual dysfunction for men and women. Furthermore, LGBT elders have a higher risk of alcoholism and illicit substance use than heterosexual elders. This increased risk has been attributed to attempts at coping with lifelong stigma and discrimination faced by members of the LGBT community.

Sexually Transmitted Infections

The most commonly diagnosed STIs in older adults are chlamydia, gonorrhea, HPV, HIV, syphilis, genital herpes, and trichomoniasis. Prevalence studies of STIs among older adults in the U.S. are limited (Brick et al. 2009). There is, however, some indication of infection risk. One study found that approximately one out of every one hundred older adults is diagnosed with an STI (Smith and Christakis 2009). According to the CDC (2009), the number of cases of syphilis in adults aged 65 and older nearly doubled between 2004 and 2008.

In contrast to the U.S., Great Britain has focused considerable attention on STIs among older adults. They have been more aggressive in tracking STIs among older people and introducing interventions to prevent their spread. According to Family Planning Association in the UK (FPA 2010), chlamydia cases increased by 72 % in adults aged 45–64 from 2002 to 2006, and gonorrhea rates increased by 93 % in adults between 45 and 64 from 1999 to 2008. In response to these alarming statistics, the FPA released a public health awareness campaign called the "Middle-Age Spread" to promote sexual health awareness among older people.

Although HIV/AIDS infections are increasing among older adults, they are difficult to identify (Whipple and Scura 1996). Older adults do not always report symptoms during the early stages of HIV or AIDS, because they may believe these are signs of normal aging. Early symptoms of HIV include general aches and pains, coughing, headache, decreased energy, loss of appetite and weight, and short-term memory loss. Because these symptoms mimic dementia or other chronic illnesses associated with aging, HIV among older people is referred to as the *great imitator* (Whipple and Scura 1996). Health workers may delay or overlook HIV/AIDS diagnoses in older adults (Kuehn 2008) because physicians and programs that test for HIV have the false belief that older adults are not sexually active. Consequently, few medical practitioners regularly test this population for HIV/AIDS (Hooyman and Kiyak 2005).

Older people with HIV progress faster to end stage AIDS and die faster than younger people do (Cahill et al. 2000). The reason for this is not known, but potential causes may be a delay in diagnosis, normal age-related decline in the immune system, or drug interactions between HIV/AIDS medications and medications for other health ailments. Little research exists about how HIV affects older adults' immune systems differently than younger people. Although doctors prescribe HIV/AIDS medications to older people, they do so without any research foundation. The subjects in most HIV/AIDS drug trails are younger adults.

Cultural Norms and Physiological Changes

Normative values and behaviors around sexuality are expressed differently across diverse cultural dimensions. Cultures vary in their perceptions and norms around monogamy, relationships, and mating styles (Ryan and Jethå 2010; Schmitt 2005).

In Western cultures, monogamous relationships are more pervasive and recognized as the norm than in other parts of the world (Conley et al. 2012a). Little research exists on consensual nonmanogamy among older adults. As the baby boomer cohort in the United States ages, and members of this cohort are more likely to be single due to divorce or widowhood compared to previous cohorts, it is unclear how perceptions of monogamy as privileged will change.

The way people approach or participate in discussions of sexual activity or sexual expression varies greatly. Some cultures view the topic of sexuality conservatively and may be modest or reluctant to discuss sex with health or social service providers. Some countries, on the other hand, are more liberal to the extent of legalizing sexual surrogate services in which elders or younger adults with disabilities pay for assistance with sexual activity (see de la Baume 2013). Hinchliff and Gott (2004) explored the meaning of sexual relationships among long-term married couples in the United Kingdom. They found that for some couples sex within a long-term relationship is a way to express trust, love, and sharing in a physical way that complements verbal communication; this includes all expressions of intimacy and not only penetrative sex. They also argue that "sexual activity in later life may not be important to everyone, and is something we need to be mindful of when conducting research into ageing and sexuality. Indeed, it is very important not to create a new stereotype whereby all elderly people should be sexually active in order to be content" (Hinchliff and Gott 2004, p. 605).

Despite older adults' desire to have their healthcare provider breach the topic of sexuality, numerous studies confirm that healthcare providers in the U.S. do not routinely assess the sexual health of older adult patients (Dogan et al. 2008; Dunn and Cutler 2000; Gerbert et al. 1990; Mona et al. 2010; Skiest and Keiser 1997; Tangredi et al. 2008). Some factors that prevent healthcare providers from addressing sexuality may include discomfort (Agronin 2004; Burd et al. 2006; Dunn and Cutler 2000; Kligman 1991; Mona et al. 2010), fear of offending (Dunn and Cutler 2000; Mona et al. 2010), lack of education (Butler and Lewis 2002; Dogan et al. 2008; Gott 2005; Kligman 1991; Mona et al. 2010), stereotypes about aging (Bauer et al. 2007; Butler and Lewis 2002; Kligman 1991), and time (Kligman 1991; Mona et al. 2010).

Most pervasive across cultures is the heteronormative perspective that relationships primarily exist between one man and one woman. This view marginalizes not only the lesbian, gay, and bisexual communities, but also paints gender as a binary, static variable, thereby marginalizing people who are gender nonconforming, transgendered, or have a transgender history. An emerging area of research is exploring the experiences of transgendered older adults and the serious physical and mental health risks they face as a result of a lifetime of stigma and victimization (Fredriksen-Goldsen et al. 2013a; Witten 2009, 2003).

Older adults who are involved in partnerships are more likely to report being sexually satisfied than older adults who are not partnered (AARP 2005). Indeed, married older adults are nearly six times as likely to be sexually active than older adults who are unmarried (Matthias et al. 1997). Lower rates of partnered sexual activity in older adulthood may be due to the availability of a partner (Beutel et al. 2002; Kontula and Haavio-Mannila 2009; Lindau et al. 2006, 2007; Matthias et al. 1997; Schick et al. 2010; Waite et al. 2009; Zeiss and Kasl-Godley 2001).

Finally, environmental and organizational context plays a role in how older adults can express intimacy and sexuality in institutionalized settings such as nursing homes, assisted living, or residential care facilities. It is well established that sexual expression occurs in a variety of ways in long term care facilities. Despite this knowledge, facilities rarely establish policies to guide residents, direct staff, or management in responding to either appropriate or inappropriate expressions of sexual needs (Cornelison and Doll 2012; Shuttleworth et al. 2010). Furthermore, ethical challenges arise when older adults with cognitive impairment, and possibly diminished capacity for consent, become engaged in intimate or sexual relationships. Rarely do residents with dementia have the opportunity to exercise their right to consensual sex (Tarzia et al. 2012). There is a considerable need for standards of practice and training around resident rights in the area of sexuality, including but not limited to topics such as advocating for residents with dementia, discussing safe sex practices, or demonstrating cultural competence with LGBT residents.

Physiological Age-Related Changes

As men age, they produce less testosterone and experience a number of other normal changes that affect their sexual activity (Brick et al. 2009). These differences may include reduced sperm production, delayed ejaculation, less forceful orgasm, and erection changes. Younger men experience psychogenic erections, which do not require direct physical stimulation (Westheimer and Lehu 2005). However, as men age these types of erections decrease, and eventually men require direct stimulation of the penis to achieve erections. Additionally, erections may take longer to attain, are not as firm, are lost soon after orgasm, and the refractory period before another erection lengthens (Brick et al. 2009). These changes in male sexual functioning are normal.

However, even though these changes normally come with age, they are sometimes identified as dysfunctional. The clinical definition of erectile dysfunction, or ED, is the inability to attain an erection more than half of the time (Olson 2011). Nearly half of men over 65 report erectile difficulties (Waite et al. 2009), and these problems occur across countries studied (Laumann et al. 2005). However, some men may report erectile dysfunction, even when it does not meet the clinical standard. The popularization of advertising for drugs for ED may have lead to increased reports of erectile dysfunction, which grew by 26 % across age groups between 1999 and 2004 (AARP 2005).

In addition to ED, men report a number of other sexual issues (Lindau et al. 2007; Waite et al. 2009). Approximately a quarter of older men reported climaxing too quickly, anxiety about sexual performance, inability to climax, and diminished interest in sex (Lindau et al. 2007). This decreased interest in sex may relate to declining levels of testosterone because between ages 40 and 70 men experience a 30 % decline in testosterone levels (DeLamater and Sill 2005). More than one quarter of men reported that sexual problems made them avoid sex (Waite et al. 2009).

Assistance for erectile dysfunction in the form of the pill was a welcome change from injections, implants, and vacuum devices. These older treatments, though, are still used today when medications are ineffective or contraindicated due to medical concerns (Westheimer and Lehu 2005). The wide availability of ED medications has led to increased usage in older adults (Schick et al. 2010), even though it may affect the relationship.

As women age, their testosterone (Brick et al. 2009) and estrogen (Westheimer and Lehu 2005) levels decline, which lead to normal changes in female sexual functioning. Testosterone is related to libido, so declining amounts may lead to decreased desire (Brick et al. 2009). Declining amounts of estrogen lead to decreased vaginal elasticity, narrowing and shortening of the vagina, decreased size of the labia, clitoris, uterus, and ovaries (Westheimer and Lehu 2005), and thinning of the vaginal walls (Westheimer and Lehu 2005; Zeiss and Kasl-Godley 2001). Additionally, decreased blood flow to the pelvic region (Westheimer and Lehu 2005) causes post-menopausal women to have less and delayed lubrication (Price 2011; Westheimer and Lehu 2005; Zeiss and Kasl-Godley 2001). More than two-thirds of women reported difficulty with lubrication (DeLamater and Sill 2005; Lindau et al. 2007). Because of these normal changes, intercourse may become uncomfortable or painful as women age (Brick et al. 2009; Westheimer and Lehu 2005). Seventeen percent of women reported pain during intercourse (DeLamater and Sill 2005; Lindau et al. 2007). This may explain why nearly half of women report lack of interest or desire in sex (Lindau et al. 2007).

Due to continued confusion about effective treatment, physicians may be hesitant to discuss sexual dysfunction with female patients, particularly older women (Jacobson 2010). There are, however, a number of risk-free treatments available for vaginal discomfort that do not require prescriptions (Barnard 2011b; Kellogg Spadt 2011; Price 2011; Westheimer and Lehu 2005). The use of lubricants can make painful intercourse more pleasurable (Price 2011; Westheimer and Lehu 2005). The proper functioning of nerves in the vagina can be maintained by moisturizing daily, massaging the inner wall at least twice per week, and having an orgasm weekly (Barnard 2011b). Women can also eliminate number of common products, because they may cause vaginal irritation (Kellogg Spadt 2011). These products include pads, shampoo and conditioner, laundry products, douches, and shaving cream. These simple changes may reduce vaginal discomfort. Because of the continued confusion about effective treatment, physicians may be hesitant to discuss sexual dysfunction with female patients, particularly elder women (Jacobson 2010). Despite continued controversy, treatments are available for female sexual dysfunction and warrant further attention.

Masturbation

Masturbation is self-stimulation of one's body, particularly genitals, for sexual arousal with or without orgasm (Brick et al. 2009). Older adults may experience sexual pleasure through masturbation by either choice or when they do not have a partner due to death, illness, divorce, or separation (Price 2011). Even when they have partners, older adults may masturbate to supplement partnered activity (Brick et al. 2009). Couples may participate in shared masturbation or mutual masturbation, especially when sexual penetration is difficult or impossible. More than half of men and nearly a quarter of women aged 65–74 reported masturbating in the past year (Lindau et al. 2007) in a national U.S. study. Similar prevalence rates have been found across the globe in Africa, China, and Europe. For women, masturbation may help maintain sexual functioning by decreasing vaginal atrophy, promote lubrication and preserve tissue elasticity (Brick et al. 2009). As men and women age, their bodies may require different types of stimulation to attain arousal and orgasm (Price 2011).

While masturbation may be a way to explore pleasure and stimulation with or without a partner and may decrease risk of sexually transmitted infection, there remains a social taboo around scholarship and clinical practice that addresses masturbation (Shelton 2010), especially in older adults. Scholarly debates which raise questions around the benefits of masturbation are beginning to emerge, but most dialogue has been around younger adults and the use of internet pornography and masturbation.

Safer Sex Practices

Even though rates of STIs (CDC 2009; FPA 2010; Smith and Christakis 2009) and HIV/AIDS (Brick et al. 2009; CDC 2008a, b; Kuehn 2008; NYC-DOH 2010) are increasing in older adult populations, they are not regularly protecting themselves (Hooyman and Kiyak 2005; Kuehn 2008; Lindau et al. 2006; Schick et al. 2010; Stall and Catania 1994; Zablotsky 1998). One study found that seven percent of sexually active older adults were engaging in risky sexual behaviors that put them at risk for contracting an STI or HIV (Gott 2001). Of those engaging in high-risk activities, including sex with multiple partners in the past 5 years without using a condom, one-third reported related sexual health concerns and only one tenth discussed their concerns with a physician. A recent study (Schick et al. 2010) found that two-thirds of men and women had not been tested for an STI in the past year. Despite low rates of STI testing, 5 % of older adults reported that their partner had an STI. Although very little research exists on sexual risk taking behaviors in older adults (Gott 2005), studies (Hooyman and Kiyak 2005; Kuehn 2008; Lindau et al. 2006; Schick et al. 2010; Stall and Catania 1994; Zablotsky 1998) support the idea that older adults are not regularly using condoms.

Low rates of condom use among Americans over the age of 50 (Stall and Catania 1994; Schick et al. 2010) have been attributed to social as well as biological factors. After menopause, older women may assume that their male partners do not need to use a condom because they can no longer become pregnant (Hooyman and Kiyak 2005; Westheimer and Lehu 2005). Indeed, the risk of STI transmission is underestimated by older adults. Difficulty with condom use also may be due to inability to apply or use a condom due to incomplete erection or arthritic hands (Kuehn 2008), or discomfort caused by limited vaginal lubrication. For men who have difficulty attaining or maintaining an erection, a female condom may be a solution (Price 2011). For vaginal dryness, the use of a good lubricant can make sex with a condom more comfortable. Younger people have absorbed the message that condoms are important when having sex, but older adults seem resistant (Westheimer and Lehu 2005). This may reflect the failure of our society to target older adults with messages about safe sex (Price 2011).

Implications and Call to Action: Practice, Policy, and Research

The freedom of sexual expression has been established as a human right and a social justice issue for quite some time (Pangman and Seguire 2000). It is often difficult, however, for healthcare providers to discuss sexuality with their elder patients (Agronin 2004; Burd et al. 2006; Dunn and Cutler 2000; Kligman 1991; Mona et al. 2010). Concurrently, healthcare providers do not receive sufficient training to deal with sexual issues (Kligman 1991; Maes and Louis 2011; Mona et al. 2010). Compounding these challenges are the social taboos ascribed to the notion of older adults as sexual beings. This chapter introduced some key issues surrounding the current state of knowledge and practice in older adult sexuality. The research, practice, and policy implications of older adult sexuality will be discussed briefly here, with recommended actions for scholars, practitioners, educators, or policymakers to address these implications.

Research

There is a lack of research examining the role of sexuality among older adults, particularly in multicultural populations. Indeed, instruments that assess sexual health or sexual activity among older adults are outdated (i.e., developed prior to the HIV/AIDS crisis and introduction of medical treatments for sexual dysfunction) and are mostly appropriate for white heterosexual males. Literature on sexual expression among older adults today also tends to examine heterosexual and same-sex relationships separately, and there is very limited scholarship devoted to understanding sexual expression among older adults who are gender non-conforming, transgendered, or have a transgender history.

One important step researchers could take toward deepening knowledge on sexuality, intimacy, and aging would be to include questions about sexual expression, sexual orientation, and gender identity in national health surveys. This includes identifying LGBT as well as heterosexual respondents. In 2009, the Williams Institute's Sexual Minority Assessment Research Team (SMART) published *Best Practices for Asking Questions about Sexual Orientation on Surveys* (http://williamsinstitute.law.ucla.edu/wp-content/uploads/SMART-FINAL-Nov-2009.pdf) to guide researchers. Researchers have begun to explore the needs of both LGBT and heterosexual older adults together. One study of older adult sexuality attempted "to frame information in ways that are relevant to heterosexual and gay, lesbian, and bisexual individuals" (Zeiss and Kasl-Godley 2001, p. 18).

Practice

It is essential for practitioners to gain an understanding of the complex intersection of age, culture, and sexuality by discussing these issues in clinical supervision, in classrooms, and with clients as clinically appropriate (Hinrichsen 2006). Sue and Sue (2013) offer important considerations and myths for practitioners to be mindful of when counseling older adults. Clinical awareness of the resiliency that older adults from marginalized communities demonstrate in response to cumulative loss over a lifetime, and cultural humility around different norms and values related to discussing sexual activity among elders are essential for practitioners.

Typically, older adults have longstanding relationships with primary care providers (Mona et al. 2010). This may explain why adults over age fifty report a desire to have their healthcare provider raise the topic of sexuality (Tangredi et al. 2008). As elder people have increasing contact with the healthcare system (Anderson 2003; Hsiao et al. 2010; IOM 2008; NCHS 2011) and want their healthcare providers to raise the topics (Tangredi et al. 2008), healthcare providers are in a position to address issues of sexuality facing people as they age. The need for healthcare professionals to discuss sexual health with their patients is a growing area of study (Laumann et al. 2008; Lindau et al. 2006; Schick et al. 2010; Zeiss and Kasl-Godley 2001), which underscores the need for research that enhances understanding of healthcare providers' knowledge and attitudes towards addressing sexuality among their elderly patients.

Expression of intimacy and sexual expression in institutional settings is an important area of practice in which training and development is needed. In the UK, the Royal College of Nursing (2011) published a compendium of best practices called *Older People in Care Homes: Sex, Sexuality and Intimate Relationships.* The report offers guidance to staff working in long term care facilities to ensure residents' rights to sexual expression are honored and their safety is ensured, and that they are treated with dignity and respect.

The National Resource Center on LGBT Aging (NRC) offers an extensive clearinghouse of information for aging service providers to offer more culturally competent programming to older adults who are lesbian, gay, bisexual, and transgender (www.lgbtAgingCenter.org). In addition to the website, the NRC conducts trainings across the United States to assist organizations in both the aging and LGBT service fields to provide more culturally competent services that break down barriers of ageism as well as heterosexism.

Clearly related to supporting clinical practice by debunking assumptions and promoting culturally competent long term supports (that appreciate diversity of ethnicity, race, sexual orientation, gender identity, and spiritual beliefs) is the need for educating clinicians. As discussed earlier in this chapter, curricula in gerontology education are lacking appropriate content on sexual expression and attitudes around sexuality among older adults. Gerontology training programs must continue to develop curricula in this arena.

Policy

Public resources can be invaluable for building knowledge and supporting best practices for maintaining sexual health (as well as overall health) for all older adults. There has been groundwork in this area over the last decade. In 2001, Dr. David Satcher released the Surgeon General's *Call to Action to Promote Sexual Health and Responsible Sexual Behavior*, which called for more open and thoughtful conversations among health professionals, policymakers, and citizens about sexuality and risky sexual behavior. The report also underscored the benefits of sexual activity as a means to "foster intimacy and bonding as well as shared pleasure in our relationships" (Satcher 2001, p. 1). The *Call to Action* recommended the addition of curriculum content for medical personnel on sexuality and sexual history taking and training for personal bias awareness regarding sexuality among diverse populations, including the elders (Satcher 2001).

One year following the Surgeon General's *Call to Action*, the World Health Organization (WHO 2002) issued a set of recommendations in a policy paper, *Active Aging: A Policy Framework*. This action plan brought attention to increasing rates of HIV/AIDS in older adults worldwide but did not explicitly discuss sexual activity among older adults. The policy recommendations included removal of age limitations on data collection and research that focused on older adults affected by HIV/AIDS. Increased education and prevention efforts for HIV/AIDS among older adults were also recommended. Contradicting the WHO report, the Centers for Disease Control and Prevention (CDC) recommended HIV/AIDS testing exclude individuals over age 65, primarily because they accounted for only 2 % of new diagnoses (Branson et al. 2006). Though the CDC argued that the inclusion of adults aged 50–64 would increase physician awareness of the need to test their older patients and raise knowledge of HIV/AIDS risk among older adults, those over age 65 remain excluded.

In 2010, the U.S. Department of Health and Human Services (HHS 2010a) released *Healthy People 2020*, to outline U.S. health goals for the upcoming 10-years. This report serves as a national guide for health promotion and prevention of diseases. Unfortunately, it ignores the sexual health needs of older adults by excluding all adults aged 45 and older from the report's sexual health goals. The report does, however, recommend increased geriatric education for healthcare professionals (HHS 2010b). In 2009, only 2.7 % of physicians and 4.3 % of psychiatrists had specialized geriatric training. *Healthy People 2020* set a goal of a 10 percent increase in geriatric care by 2020.

It is important to note that none of the previous reports discussed raised the healthcare needs of LGBT elders. The Joint Commission on Accreditation of Healthcare Organizations (JCAHO 2011), the accrediting body for all healthcare agencies, released a field guide with recommendations for creating welcoming environments for LGBT patients. The report recommends using language that promotes cultural competence, thus making patients more comfortable discussing sexual health concerns. Although the guide does not specifically apply to older adult patients, it encourages healthcare providers to ask assessment questions in a manner that will encourage sharing of sexual orientation and gender identity.

Conclusion

This chapter introduced and discussed some of the common misconceptions around sexuality and intimacy in older adulthood that exist in many communities across the globe. Recent research and literature has been presented to offer a better understanding of this issue, particularly as it relates to the role of sexuality and intimacy in the health and well-being of older adults from diverse communities. The vignette in the beginning of this chapter is just one of many encounters in which a geriatric social worker deals with the issues of sex, intimacy and aging in her daily work. Indeed, as the baby boomer generation enters older adulthood and as researchers, policymakers, and practitioners continue to learn more and develop culturally sensitive skills, sexuality and intimacy will be viewed as a normative part of each elder's story. It is essential for the elder to guide their own narrative in this story and for the practitioner to be prepared to honor it.

References

Adam, B. D. (1995). *The rise of a gay and lesbian movement*. New York: Twayne Publishers.

Agronin, M. E. (2004). Sexuality and aging. In J. Sadavoy, L. F. Jarvik, G. T. Grossbert, & B. S. Meyers (Eds.), *Comprehensive textbook of geriatric psychiatry* (3rd ed., pp. 789–813). New York: W.W. Norton and Company.

American Association of Retired Persons (AARP). (2005). *Sexuality at midlife and beyond: Update of attitudes and behaviors*. Washington, DC: AARP.

Anderson, G. (2003). Chronic care. *Public Health & Policy, 3*(2), 110–111.
Araujo, A. B., Durante, R., Feldman, A., Goldstein, I., & McKinlay, J. B. (1998). The relationship between depressive symptoms and male erectile dysfunction: Cross-sectional results from the Massachusetts male aging study. *Psychosomatic Medicine, 60*, 458–465.
Barnard, E. (2011a). Reasons to keep your sexual self alive and functioning. In J. Price (Ed.), *Naked at our age: talking out loud about senior sex* (pp. 139–140). Berkeley: Seal Press.
Barnard, E. (2011b). Aches, pains, positions, and props: An expert responds. In J. Price (Ed.), *Naked at our age: talking out loud about senior sex* (pp. 17–18). Berkeley: Seal Press.
Basson, R., Leiblum, S., Brotto, L., Derogatis, L., Fourcroy, J., Fugl-Meyer, K., et al. (2004). Revised definitions of women's sexual dysfunction. *Journal of Sexual Medicine, 1*(1), 40–48.
Bauer, M., McAuliffe, L., & Nay, R. (2007). Sexuality, health care, and the older person: an overview of the literature. *International Journal of Older People Nursing, 2*, 63–68.
Berman, J. R., Berman, L., & Goldstein, I. (1999). Female sexual dysfunction: Incidence, pathophysiology, evaluation, and treatment options. *Urology, 54*, 385–391.
Beutel, M. E., Schumacher, J., Weidner, W., & Brahler, E. (2002). Sexual activity, sexual and partnership satisfaction in ageing men-Results from a German representative community study. *Angrologia, 34*, 22–28.
Block, J. D. (2008). *Sex over 50*. New York: Penguin Group.
Bolte, S. (2011). Sex and intimacy after cancer. In J. Price (Ed.), *Naked at our age: Talking out loud about senior sex* (pp. 200–203). Berkeley, CA: Seal Press.
Branson, B.M., Handsfield, H.H., Lampe, M.A., Janssen, R.S., Taylor, A.W., Lyss, S.B., & Clark, J.E. (2006). *Revised recommendations for hiv testing of adults, adolescents, and pregnant women in health care settings*. Retrieved from Centers for Disease Control and Prevention http://www.cdc.gov/mmwr/preview/mmwrhtml/rr5514a1.htm
Brick, P., Lunquist, J., Sandak, A., & Taverner, B. (2009). *Older, wiser, sexually smarter: 30 sex Ed lessons for adults only*. Morristown: Planned Parenthood of Greater Northern New Jersey.
Burd, I. D., Nevadunsky, N., & Bachman, G. (2006). Impact of physician gender on sexual history taking in a multispecialty practice. *Journal of Sexual Medicine, 3*, 194–200.
Butler, R. N., & Lewis, M. I. (2002). *The new love and sex after 60*. New York: Ballantine Books.
Cahill, S., South, K., & Spade, J. (2000). *Outing age: Public policy issues affecting gay, lesbian, bisexual, and transgender elders*. Washington, DC: Policy Institute National Gay and Lesbian Task Force.
Castelo-Branco, C., Huezo, M. L., & Lagarda, J. L. B. (2008). Definition and diagnosis of sexuality in the XXI century. *Maturitas, 60*, 50–58.
Centers for Disease Control and Prevention. (1996). *Pneumocystic Pneumonia—Los Angeles*. Retrieved from http://www.cdc.gov/MMWR/preview/mmwrhtml/00043494.htm
Centers for Disease Control and Prevention. (2008a). *New estimates of U.S. HIV prevalence*, Retrieved from http://www.cdc.gov/hiv/topics/surveillance/resources/factsheets/prevalence.htm
Centers for Disease Control and Prevention. (2008b). *Persons aged 50 and older*. Retrieved from: http://www.cdc.gov/hiv/topics/over50/index.htm
Centers for Disease Control and Prevention. (2009). *2008 Sexually transmitted diseases surveillance*. Retrieved from http://www.cdc.gov/std/stats08/tables/33.htm
Conley, T. D., Moors, A. C., Matsick, J. L., & Ziegler, A. (2012a). The fewer the merrier?: Assessing stigma surrounding consensually non-monogamous romantic relationships. Analyses of Social Issues and Public Policy, 1530–2415. doi: 10.1111/j.1530-2415.2012.01286.x
Cornelison, L.J. & Doll, G.M. (2012). Management of sexual expression in long-term care: Ombudsmen's perspectives. *The Gerontologist*. Advanced access published online December 10, 2012 doi:10.1093/geront/gns162
De la Baume, M. (2013). Disabled people say they, too, want a sex life, and seek help in attaining it. *New York Times*, July 4.
DeLamater, J. D., & Moorman, S. M. (2007). Sexual behavior in later life. *Journal of Aging and Health, 19*(6), 921–945.
DeLamater, J. D., & Sill, M. (2005). Sexual desire in later life. *The Journal of Sex Research, 42*(2), 138–149.

Dogan, S., Demir, B., Eker, E., & Karim, S. (2008). Knowledge and attitudes of doctors toward the sexuality of older people in Turkey. *International Psychogeriatrics, 20*(5), 1019–1027.
Dunn, M. E., & Cutler, N. (2000). Sexual issues in older adults. *AIDS Patient Care and STDs, 14*(2), 67–69.
Family Planning Association (FPA). (2010). *The Middle-age spread (STIs in the over 50s).* Retrieved From: http://www.fpa.org.uk/campaignsandadvocacy/sexualhealthweek/stisandsafersexover50/professionals
Finger, W. W., Lund, M., & Slagel, M. A. (1997). Medications that may contribute to sexual disorders: A guide to assessment and treatment in family practice. *Journal of Family Practice, 44*(1), 33–43.
Fredriksen-Goldsen, K., & Muraco, A. (2010). Aging and sexual orientation: A 25-year review of the literature. *Research on Aging, 32*, 372–413.
Fredriksen-Goldsen, K. I., Cook-Daniels, L., Kim, H.-J., Erosheva, E. A., Emlet, C. A., Hoy-Ellis, C. P., et al. (2013a). The physical and mental health of transgender older adults: An at-risk and underserved population. *The Gerontologist,*. doi:10.1093/geront/gnt021. Advance online publication.
Fredriksen-Goldsen, K. I., Emlet, C. A., Kim, H.-J., Muraco, A., Erosheva, E. A., Goldsen, J., et al. (2013b). The physical and mental health of lesbian, gay male and bisexual (LGB) older adults: The role of key health indicators and risk and protective factors. *The Gerontologist, 53*(4), 664–675. doi:10.1093/geront/gns123. Advance online publication.
Gerbert, B., Maguire, B. T., & Coates, T. J. (1990). Are patients talking to their physicians about AIDS? *American Journal of Public Health, 80*(4), 467–468.
Giami, A. (2008). Two facets of the medicalization of sexuality in the late 20th century: AIDS & Viagra. *Tidsskriff For Norsk Psykologiorening, 45*, 671–673.
Goldstein, I. (2008). Sexual dysfunction in women: What can urologists contribute? *Current Urology Reports, 9*, 475–483.
Gott, C. M. (2005). *Sexuality, sexual health, and ageing*. Berkshire: Open University Press.
Gott, C. M. (2001). Sexual activity and risk-taking in later life. *Health and Social Care in the Community, 9*, 72–78.
Han, B., Gfroerer, J. C., Colliver, J. D., & Penne, M. A. (2009). Substance use disorder among older adults in the United States in 2020. *Addiction, 104*(1), 88–96.
Hart, G., & Wellings, K. (2002). Sexual behaviour and its medicalisation: In sickness and in health. *British Medical Journal, 324*, 896–900.
Harvard Medical School. (2011). Hormone therapy: The next chapter. *Harvard Women's Health Watch, 18*(7), 1–3.
Health, I. M. S. (2012). *The use of medicines in the United States: Review of 2011*. Parsippany: IMS Institute for Healthcare Informatics.
Henderson, S. J., Bernstein, L. B., George, D. M., Doyle, J. P., Paranjape, A. S., & Corbie-Smith, G. (2004). Older women and HIV: How much do they know and where are they getting their information? *Journal of the American Geriatrics Society, 52*(9), 1549–1553.
Hirsch, D. (2011). Erectile dysfunction medication and your relationship. In J. Price (Ed.), *Naked at our age: Talking out loud about senior sex* (pp. 222–223). Berkeley: Seal Press.
Herzog, D. (2008). *Sex in crisis: The new sexual revolution and the future of American politics*. New York: Basic Books.
Hinchliff, S., & Gott, M. (2004). Intimacy, commitment, and adaptation: Sexual relationships within long-term marriages. *Journal of Social and Personal Relationships, 21*, 595–609.
Hinrichsen, G. A. (2006). Why multicultural issues matter for practitioners working with older adults. *Professional Psychology: Research and Practice, 37*(1), 29–35. doi:10.1037/0735-7028.37.1.29
Hooyman, N. R., & Kiyak, H. A. (2005). *Social gerontology: A multidisciplinary perspective* (7th ed.). Boston: Pearson Education Inc.
Hsiao, C., Cherry, D. K., Beatty, P. C., & Rechtsteiner, E. A. (2010). Ambulatory medical care survey: 2007 summary. *National Health Statistics Reports, 27*(3), 1–32.

Huang, A. J., Subak, L. L., Thom, D. H., Van Den Eeden, S. K., Ragins, A. I., Kuppermann, M., et al. (2009). Sexual function and aging in racially and ethnically diverse women. *Journal of the American Geriatric Society, 57*, 1362–1368.

Institute of Medicine (IOM). (2008). *Retooling for an aging America: Building the health care workforce*. Washington, DC: The National Academies Press.

Jacobson, S.A. (2010, November). *Doctor communication with older adults: Sexual activity and sexual health.* Poster presented at the meeting of Society for Social Work Leadership in Healthcare, Tampa, FL.

Joint Commission. (2011). *Advancing Effective communication, cultural competence, and patient- and family-centered care for the lesbian, gay, bisexual, and transgender (LGBT) community.* Retrieved from www.jointcommission.org/assets/1/18/LGBTFieldGuide.pdf

Katz, J.N. (1976). *Gay American history: Lesbians and gay men in the U.S.A.* New York: Thomas Y. Crowell Company.

Katz, J. N. (2007). *The Invention of heterosexuality*. Chicago: University of Chicago Press.

Kellogg Spadt, S. (2011). Vulvar/vaginal irritation? In J. Price (Ed.), *Naked at our age: Talking out loud about senior sex* (pp. 185–186). Berkeley: Seal Press.

Kingsberg, S. A. (2002). The impact of aging on sexual function in women and their partners. *Archives of Sexual Behavior, 31*(5), 431–437.

Kligman, E. W. (1991). Office evaluation of sexual function and complaints. *Clinics in Geriatric Medicine, 7*(1), 15–39.

Kontula, O., & Haavio-Mannila, E. (2009). The impact of aging on human sexual activity and sexual desire. *Journal of Sex Research, 46*(1), 46–56.

Kuehn, B. M. (2008). Time for "the talk"—again. *Journal of the American Medical Association, 300*, 1285–1287.

Kuhn, D. (2011). How does Alzheimer's disease affect sexuality? In J. Price (Ed.), *Naked at our age: Talking out loud about senior sex* (pp. 261–262). Berkeley: Seal Press.

Laumann, E. O., Das, A., & Waite, L. J. (2008). Sexual dysfunction among older adults: Prevalence and risk factors from a nationally representative U.S. probability sample of men and women 57–85 years of age. *Journal of Sexual Medicine, 5*, 2300–2311.

Laumann, E. O., Nicolosi, A., Glasser, D. B., Paik, A., Gingell, C., Moreira, E., et al. (2005). Sexual problems among women and men aged 40–80 y: Prevalence and correlates identified in the global study of sexual attitudes and behaviors. *International Journal of Impotence Research, 17*, 39–57.

Lawrence v. Texas, 539 U.S. 558 (2003).

Leist, A. K. (2013). Social media use of older adults: A mini review. *Gerontology, 59*, 378–384. doi:10.1159/000346818

Lindau, S. T., Leitsch, S. A., Lundberg, K. L., & Jerome, J. (2006). Older women's attitudes, behavior, and communication about sex and HIV: A community-based study. *Journal of Women's Health, 15*(6), 747–753.

Lindau, S. T., Schumm, L. P., Laumann, E. O., Levinson, W., O'Muircheartaigh, C. A., & Waite, L. J. (2007). A study of sexuality and health among older adults in the United States. *The New England Journal of Medicine, 17*, 39–57.

Loe, M. (2004). *The rise of Viagra*. New York: New York University Press.

Maes, C., & Louis, M. (2011). Nurse practitioners' sexual history-taking practices with adults 50 and older. *The Journal of Nurse Practitioners, 7*(3), 216–222.

Maes, C. A., & Louis, M. (2003). Knowledge of AIDS, perceived risk of AIDS, and at-risk sexual behaviors among older adults. *Journal of American Academic Nurse Practitioners, 15*, 509–516.

Marshall, B. L. (2010). Science, medicine and virility surveillance: 'Sexy seniors' in the pharmaceutical imagination. *Sociology of Health & Illness, 32*(2), 211–224.

Matthias, R. E., Lubben, J. E., Atchison, K. A., & Schweitzer, S. O. (1997). Sexual activity and satisfaction among very old adults: Results from a community-dwelling Medicare population survey. *The Gerontologist, 37*(1), 6–14.

McLaren, A. (1999). *Twentieth-century sexuality: A history*. Oxford: Blackwell Publishers Inc.

Mona, L. R., Goldwater, G., Syme, M. L., Cameron, R. P., Clemency, C., Miller, A. R., et al. (2010). Assessment and conceptualization of sexuality among older adults. In P. A. Lichtenberg (Ed.), *Handbook of assessment in clinical gerontology* (pp. 331–356). Boston: Elsevier.

Moynihan, R. (2003). The making of a disease: Female sexual dysfunction. *British Medical Journal, 326*, 45–47.

National Center for Health Statistics (NCHS). (2011). *Health, United States, 2010: With special feature on death and dying*. Hyattsville: U.S. Government Printing Office.

National Gay and Lesbian Task Force (2011). *Aging*. Retrieved from www.thetaskforce.org/issues/aging

National Resource Center on LGBT Aging. (2013). *Inclusive questions for older adults: A practical guide to collecting data on sexual orientation and gender identity*. New York: Author.

New York City Department of Health and Mental Hygiene (NYC-DOH). (2010). *HIV epidemiology & field services semiannual report*. Retrieved February from: http://www.nyc.gov/html/doh/downloads/pdf/dires/2010_1st_semi_rpt.pdf

Nicolosi, A., Moreira, E. D., Vila, M., & Glasser, D. B. (2004). A population study of the association between sexual function, sexual satisfaction and depressive symptoms in men. *Journal of Affective Disorders, 82*, 235–243.

Nimrod, G. (2009). Seniors' online communities: A quantitative content analysis. *Gerontologist, 50*, 382–392.

Olson, L. A. (2011). What's causing your ED? In J. Price (Ed.), *Naked at our age: Talking out loud about senior sex* (pp. 220–221). Berkeley: Seal Press.

Palacios, S. (2008). Advances in hormone replacement therapy: Making the menopause manageable. *BMC Women's Health, 8*, 22.

Pangman, V. C., & Seguire, M. (2000). Sexuality and the chronically Ill older adult: A social justice issue. *Sexuality and Disability, 18*(1), 49–59.

Parascandola, J. (2008). *Sex, sin, and science: A history of syphilis in america*. Westport: Praeger Publishers.

Price, J. (2011). *Naked at our age: Talking out loud about senior sex*. Berkeley: Seal Press.

Royal College of Nursing (2011). *Older people in care homes: sex, sexuality and intimate relationships*. London: Author. Retrieved from http://www.rcn.org.uk/__data/assets/pdf_file/0011/399323/004136.pdf

Ryan, C., & Jethå, C. (2010). *Sex at dawn: The prehistoric origins of modern sexuality*. New York: HarperCollins Publisher.

Satcher, D. (2001). *The surgeon general's call to action to promote sexual health and responsible sexual behavior*. Retrieved from http://www.surgeongeneral.gov/library/sexualhealth/call.pdf

Schiavi, R. C. (1999). *Aging and male sexuality*. Cambridge: Cambridge University Press.

Schick, V., Herbenick, D., Reece, M., Sanders, S. A., Dodge, B., Middlestadt, S. E., et al. (2010). Sexual behaviors, condom use, and sexual health of Americans over 50: Implications for sexual health promotion for older adults. *The Journal of Sexual Medicine, 7*(suppl 5), 315–329.

Schmitt, D. P. (2005). Sociosexuality from Argentina to Zimbabwe: A 48-nation study of sex, culture, and strategies of human mating. *Behavioral and Brain Sciences, 28*, 247–311.

Schwartz, P. (2007). Good lovin'. *Consumer Reports on Health, 19*(2), 6.

Shelton, J. D. (2010). Masturbation: Breaking the silence. *International Perspectives on Sexual and Reproductive Health, 36*(3), 157–158.

Shuttleworth, R., Russell, C., Weerakoon, P., & Dune, T. (2010). Sexuality in residential aged care: A survey of perceptions and policies in Australian nursing homes. *Sexuality and Disability, 28*, 187–194. doi:10.1007/s11195-010-9164-6

Skiest, D. J., & Keiser, P. (1997). Human immunodeficiency virus infection in patients older than 50 years: A survey of primary care physicians' beliefs, practices, and knowledge. *Archives of Family Medicine, 6*, 289–294.

Smith, K. P., & Christakis, N. A. (2009). Association between widowhood and risk of diagnosis with a sexually transmitted infection in older adults. *American Journal of Public Health, 99*(11), 2055–2062.

Stall, R., & Catania, J. (1994). AIDS risk behaviors among late middle-aged and elderly Americans: The national AIDS behavioral surveys. *Archives of Internal Medicine, 154*, 57–63.

Starr, B. D., & Weiner, M. B. (1981). *The Starr-Weiner report: On sex & sexuality in the mature years*. New York: Stein and Day.

Sue, D. W., & Sue, D. (2013). *Counseling the culturally diverse: Theory and practice* (6th ed.). Hoboken: John Wiley & Sons.

Tangredi, L. A., Danvers, K., Molony, S. L., & Williams, A. (2008). New CDC recommendations for HIV testing in older adults. *The Nurse Practitioner, 33*, 37–44.

Tarzia, L., Fetherstonhaugh, D., & Bauer, M. (2012). Dementia, sexuality and consent in residential aged care facilities. *Journal of Medical Ethics, 38*(10), 609–613.

The Matthew Shepard and James Byrd, Jr., Hate Crimes Prevention Act of 2009, 18 U.S.C. § 249 (2009).

U.S. Department of Health and Human Services (HHS). (2010a). *Healthy People 2020.* Retrieved from http://www.healthypeople.gov/2020/TopicsObjectives2020/pdfs/HP2020_brochure.pdf

U.S. Department of Health and Human Services (HSS). (2010b). *Health People 2020 Objectives.* Retrieved from http://www.healthypeople.gov/2020/topicsobjectives2020/pdfs/HP2020 objectives.pdf

Waite, L.J., Laumann, E.O., Das, A., & Schumm, L. (2009). Sexuality: Measures of partnerships, practices, attitudes, and problems in the national social life, health, and aging study. *Journal of Gerontology: Social Sciences, 64B*(S1), i56–i66.

Weinsberg, E. (2011). Sex after ED. In J. Price (Ed.), *Naked at our age: Talking out loud about senior sex* (pp. 227–228). Berkeley: Seal Press.

Westheimer, R.K., & Lehu, P. A. (2005). *Dr. Ruth's sex after 50: revving up the romance, passion, & excitement!* Fresno: Quill Driver Books.

Whipple, B., & Scura, K. W. (1996). The overlooked epidemic: HIV in older adults. *American Journal of Nursing, 96*, 123–126.

Wight, R. G., LeBlanc, A. J., de Vries, B., & Detels, R. (2012). Stress and mental health among midlife and older gay-identified men. *American Journal of Public Health, 102*(3), 503–510.

William, D. M., Locker, L, Jr, Briley, K., Ryan, R., & Scott, A. J. (2011). What do older adults seek in their potential romantic partners? Evidence from online personal ads. *International Journal of Aging Human Development, 72*(1), 67–82.

Witten, T. M. (2009). Graceful exits: Intersection of aging, transgender identities, and the family/community. *Journal of GLBT Family Studies, 5*(1), 35–61.

Witten, T. M. (2003). Transgender aging: An emerging population and an emerging need. *Review Sexologies, 12*(4), 15–20.

World Health Organization (WHO). (2002). *Active ageing: A policy framework.* Retrieved from http://whqlibdoc.who.int/hq/2002/WHO_NMH_NPH_02.8.pdf

Zablotsky, D. L. (1998). Overlooked, ignored, and forgotten: Older women at risk for HIV infection and AIDS. *Research on Aging, 20*(6), 760–775.

Zeiss, A. M. (1982). Expectations for the Effects of Aging on Sexuality in Parents and Average Married Couples. *The Journal of Sex Research, 18*(1), 47–57.

Zeiss, A. M., & Kasl-Godley, J. (2001). Sexuality in older adults' relationships. *Generations, 25*(2), 18–25.

Zickuhr, K., & Madden, M. (2012). Older adults and Internet use. For the first time, half of adults aged 65 and older are online. http://pewinternet.org/Reports/2012/Older-adults-and-internet-use.aspx

The Effects of Minority and Immigration Status on Elderly Health: A Community Social Capital Perspective

Qingwen Xu and Yan Du

Abstract Health disparity across ethnic and immigrant groups are well recognized. However, there is a lack of systematic understanding about health disparity among older adults particularly given the growing number and size of minority and/or immigrant elders in the United States. The chapter relies on published studies, relevant monographs, government reports and Census data; it explores possible ways in which minority and immigration status jointly or independently affects elderly health. The chapter then introduces the community social capital as a theoretical framework to understand elderly health disparity. In essence, it argues the magnitude of structural racism in the life of minority and immigrant older adults; and it discusses the role that community social capital might play to moderate the negative effects of social-economic disparity, community segregation, social discrimination, and other factors contributable to health outcomes.

While health disparities across U.S. racial, ethnic and immigrant groups are well documented and understood, we lack a systematic understanding of health disparities among older adults particularly given the growing number of minority and immigrant elders in the U.S. This chapter adopts a community social capital framework to review the existing evidence concerning health outcome disparities, and considers the complex ways in which minority and immigration status individually or jointly affect the health of the elders. This chapter also discusses the magnitude of structural racism and how it affects the lives, and in particular the health of older minority and immigrant Americans.

Q. Xu (✉) · Y. Du
Tulane University, New Orleans, USA
e-mail: qxu2@tulane.edu

H. F. O. Vakalahi et al. (eds.), *The Collective Spirit of Aging Across Cultures*,
International Perspectives on Aging 9, DOI: 10.1007/978-94-017-8594-5_7,

Health Disparity in the United States

Race relations in the United States have improved since the 1964 Civil Rights Act (Loevy 1990; Loevy et al. 1997). Racial inequality, however, remains and is expressed clearly in population health (Gee et al. 2009, 2012; Zuckerman et al. 2012). Although the collective health of the U.S. population has improved over centuries, the black/white disparity in health is quite significant. Persistently, black Americans experience poorer health status than whites (Dressler 1993; U.S. Centers for Disease Control and Prevention 2011). For example, throughout the first decade of the 21st century, the infant mortality rate was almost one and a half times higher for blacks than it was for whites (U.S. Centers for Disease Control and Prevention 2011). In 2007, black Americans comprised 13 % of the U.S. population, but they accounted for nearly 50 % of persons living with HIV/AIDS; HIV/AIDS rates (cases per 100,000) were 77 among blacks and 9.2 among whites (National Institute of Health 2010).

Across the literature, scholars have argued that the black/white health disparity is attributable to the magnitude of long-term discrimination against blacks (Williams et al. 2003) and black socioeconomic disadvantages (Harris 2010). A review by Adler and Newman (2002) summarizes that low socioeconomic status (SES) is associated with inferior health outcomes due to restricted access to healthcare and preventive health services, along with residential segregation and environmental hazards, and collective unhealthy behaviors. Evidence about these pathways is abundant (e.g., Anderson et al. 2004; Halkitis and Figueroa 2013; Pu and Chewning 2012; U.S. Centers for Disease Control and Prevention 2011; Williams and Collins 1995, 2001).

In addition, long-term experience of low SES (Baum et al. 1999), together with self-perceived discrimination against minorities (Williams and Mohammed 2009), contributes to chronic stress that is associated with negative health outcomes. The dynamic relationships between minority status and health are illustrated in Fig. 1. Studies note that the cause-effect relationships between SES and other factors are undecided; for instance, studies have found that racial residential segregation is *the* fundamental cause of the white/black health disparity as residential segregation not only has affected individual and household SES but also shaped community and neighborhood-level factors (Ward et al. 2004; Williams and Collins 1995, 2001).

In recent decades, examining health disparities in the United States has become more complex given the rapidly growing Hispanic and Asian populations. In 2010, Hispanics numbered more than 50 million or 16.4 % of the U.S. population; among them, close to 19 million Hispanics (6.1 % of the U.S. population) were foreign-born (Motel 2012). In addition, Asian Americans numbered more than 18.2 million (5.8 % of the U.S. population) in 2011 (U.S. Census Bureau 2011); among them, about 10.6 million (3.4 % of the U.S. population) were foreign born, accounting for 27.7 % of the country's 38.5 million immigrants (Batalova 2011).

Unlike the white/black disparity, white/Hispanic and white/Asian health disparities do not always favor whites; and sub-ethnic group differences are also

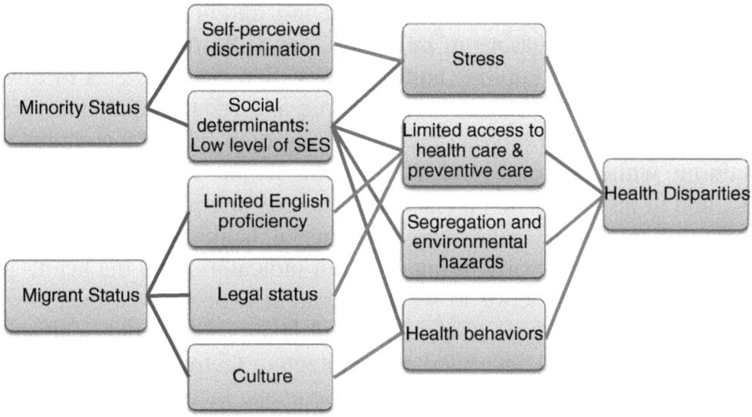

Fig. 1 Pathways between minority/immigration status and health disparity

significant. For instance, the infant mortality rate for Asian Americans in 2006 was 4.55 (deaths per 1,000 live births), or 18.5 % lower than that for non-Hispanic whites. For Hispanics, the infant mortality rate in 2006 was 5.41, 3 % lower than for whites; for sub-Hispanic groups, the infant mortality rate ranged from 8.01 for Puerto Ricans to 4.52 for Central Americans (U.S. Centers for Disease Control and Prevention 2011). Noting the heterogeneity of health outcomes across Hispanic sub-groups, substantial variations have also been observed among Asian populations (Ghosh 2003) despite a lack of comprehensive national data on health information for Asian American sub-groups. For instance, among eleven Asian American Pacific Islander sub-groups, the lifetime asthma diagnosis rate ranges from 10.9 % among Korean Americans to 23.8 % among Filipino Americans (Davis et al. 2006).

Hispanic and Asian groups face a different set of risk and protective factors. Similar to black Americans, minority status and the everyday self-perceived discrimination associated with it contributes to stress and negative health outcomes (Araújo and Borrell 2006; Gee et al. 2007; Williams et al. 2003). Unlike discussions on the black/white health disparity, health disparities of whites/Hispanics and whites/Asians have focused on factors beyond socioeconomic status. Indeed, research suggests that the relationships between socioeconomic status and health differ by race/ethnicity. One study identified significant negative effects of race on general health for blacks, Hispanics, and Asians; but when both whites and blacks reported similar education gradients for fair/poor health, the study found that education gradients for Hispanics and Asians were less steep than they were for whites (Chen et al. 2006). Another study found that the black/white disparity in obesity/overweight prevalence increased at the highest SES, while this disparity was smaller at the highest SES for the white/Hispanics and white/Asian groups (Gordon-Larsen et al. 2003). The literature does not suggest that low SES has a smaller negative effect on Hispanics and Asians than it does on black Americans, but rather that *cultural norms* for health behaviors and/or contextual factors (e.g., immigration selectivity, reasons of migration, and U.S. immigration system) may help Hispanic and Asian families buffer the negative health effects of low SES.

Given the fact that the majority of Hispanic and Asian populations are either foreign born or have an immigrant parent, immigration-related factors may be key in explaining white/Hispanic and white/Asian health disparities. Research on Hispanic, and to lesser extent, on Asian populations provides the bulk of evidence for theories of immigrant selectivity in terms of health (i.e., immigrants often surpass their U.S. counterparts in many health indicators) and the health paradox (immigrants' health is superior to what one might expect given their low socio-economic status) (Jasso et al. 2004; Singh and Hiatt 2006). Empirically, foreign-born status itself has been identified as a protective factor. Foreign-born individuals are in better health than native-born Americans, including individuals of the same race/ethnicity (Cunningham et al. 2008). For instance, foreign-born blacks with equivalent socioeconomic and demographic backgrounds had the lowest mortality risks in the U.S., followed by foreign-born Hispanics, foreign-born Asians/Pacific Islanders (APIs), U.S.-born APIs, U.S.-born Hispanics, foreign-born whites, and U.S.-born blacks (Singh and Siahpush 2002).

Nevertheless, immigrants of color often experience barriers to full participation in U.S. society based on race, ethnicity, language, and/or immigration status; thus they endure additional health risks (Derose et al. 2007; Lillie-Blanton and Hudman 2001). Research has found that lack of English proficiency largely explains the health outcome difference between Hispanic and non-Hispanic groups (Fiscella et al. 2002), as well as the difference in access to healthcare (Ponce et al. 2006) and receipt of preventive care (DuBard and Gizlice 2008). Furthermore, non-English-speaking Asians and Hispanics showed a significantly lower likelihood to adhere to physicians' orders, including follow-up medical exams (Karliner et al. 2012) and completion of prescription drugs (Gilmer et al. 2009). In addition, legal status, particularly an undocumented status in the U.S., appears to create restrictions in health insurance coverage (Goodman et al. 2005), and access to needed medical and/or social services (Rodriguez et al. 2008; Xu and Brabeck 2012). In summary, recognizing biologic differences, health disparities across minority and immigration status are largely explained by an interplay of socio-economic, cultural, and environmental factors, which we contextualize in Fig. 1.

Health Disparities Among American Elders

Health disparities remain a perplexing issue for the U.S. public health community and a difficult fact of the U.S. 65+ population. Table 1 illustrates some common health conditions affecting the elderly population. As of 2009, older Americans (65 years and above) numbered 39.6 million, of which 9 % were black/African Americans, 3 % Asian, and 7 % of Hispanic origin (U.S. Census Bureau 2011). While health disparities among older adults in the U.S. have not been fully

Table 1 Selected characteristics and health outcomes in the United States

	Total population	Black or African American	Asian[a]	Hispanic origin	Non-Hispanic white[b]
Elderly population (65 years and above)[1]	13 %	9 %	10 %	6 %	16 %
Characteristics[2]					
Education (25 years and above)					
Without high school diploma	14.7 %	18.6 %	14.7 %	39.1 %	9.6 %
College education and above	27.9 %	17.6 %	49.7 %	12.6 %	31.1 %
Family Income					
Median family income	61,082	39,587	78,529	41,423	68,390
Employment (16 years and above)					
Management professional and related occupation	35.7 %	28.0 %	48.7 %	18.6 %	39.6 %
Service occupation	17.8 %	25.3 %	16.7 %	26.3 %	14.9 %
Construction extraction and maintenance occupation	8.7 %	5.3 %	3.2 %	14.0 %	8.6 %
Poverty					
Below poverty level	14.3 %	25.8 %	11.4 %	23.5 %	10.0 %
Health outcomes[c]					
Number of death for coronary heart disease (65–74 years)[3]					
Women (per 1,000)	–	367.2	132.2	–	221.0
Men (per 1,000)	–	704.9	261.7	–	483.8
Influenza vaccination coverage (65 years and above)[4]	72 %	58.3 %	–	61.4 %	73.9 %
Colorectal cancer screening (50 years and above)[5]	–	62.9 %	62.9 %	51.2 %	66.2 %
Diabetes (67 years and above) (per 1,000)[6]	197	296	243	334	184

(continued)

Table 1 (continued)

	Total population	Black or African American	Asian[a]	Hispanic origin	Non-Hispanic white[b]
Parkinson (per 100,000 age/gender-adjusted, N. California)[7]	–	10.2	11.3	16.6	13.6
Arthritis (70 years and above)[8]	–	46.95 %	–	52.26 %	31.89 %

Note

[a] Some health statistics include Pacific Islanders

[b] Some health statistics include both Hispanic and non-Hispanic whites

[c] All statistics are from national probability samples if not specified

Sources

[1] U.S. Census Bureau, Population Division, "Monthly Resident Population Estimates by Age, Sex, Race and Hispanic Origin for the United States: April 1, 2000 to July 1, 2009," June 2010, http://www.census.gov/popest/national/asrh/2009-nat-res.html

[2] U.S. Census Bureau, 2009 American Community Survey, B15002A-I, "Sex by Educational Attainment for the Population 25 years and Over"; B24010A-I, "Sex by Occupation for the Employed Civilian Population 16 years and Over"; B19101A-I, "Family Income in the Past 12 months (In 2009 Inflation-Adjusted Dollars)"; B19113A-I, "Median Family Income in the Past 12 months (In 2009 Inflation-Adjusted Dollars)"; B17001A-I, "Poverty Status in the Past 12 months by Sex by Age"; B17010A-I, "Poverty Status in the Past 12 months of Families by Family Type by Presence of Related Children Under 18 years by Age of Related Children"; B25003, "Tenure"

[3] U.S. Centers for Disease Control and Prevention 2011, *CDC health disparities and inequalities report—United States, 2011*. Table 1, Number of deaths and age-specific death rates for coronary heart disease among adults aged ≥45 years, by age groups, sex, and race—National Vital Statistics System, United States, 2006 (p. 65)

[4] U.S. Centers for Disease Control and Prevention 2011, *CDC health disparities and inequalities report—United States, 2011*. Table 1, Influenza vaccination coverage, by race/ethnicity—Behavioral Risk Factor Surveillance System and National 2009 H1N1 Flu Survey, United States, 2009–2010 (p. 41)

[5] U.S. Centers for Disease Control and Prevention 2011, *CDC health disparities and inequalities report—United States, 2011*. Table 1, Percentage of respondents aged ≥50 years who reported receiving a fecal occult blood test within 1 year or a lower endoscopy within 10 years, by selected characteristics—Behavioral Risk Factor Surveillance System and National 2009 H1N1 Flu Survey, United States, 2008 (p. 45)

[6] McBean et al. (2004)

[7] Van Den Eeden et al. (2003)

[8] Dunlop et al. (2001)

documented and explored, scholars have asserted that ageism multiplies the negative health effects. Ageism was first coined by Butler (1969), who defined ageism as stereotyping or discriminating against individuals or groups because of their age. While both negative and positive stereotypes about older people are common in American society (Perdue and Gurtman 1990, Schaie and Willis 2011), most of these stereotypes are negative (Kite and Johnson 1988; Schaie and Willis 2011) and their influence on elders' health are much stronger than positive ones (Meisner 2012).

The implications of negative stereotypes of aging on elder health may include elder abuse (Dong 2005; Dong and Simon 2013; Dyer et al. 2000), lower quality of healthcare services (Skirbekk and Nortvedt 2012; Bowling 1999; Briggs et al. 2012), and unequal access to health services (Simon 2011) and more specifically, to preventive health services (Milaszewski et al. 2012), among other effects. In addition, studies suggest that the internalized ageism among older people may serve mediating health outcomes. Levy and colleagues (2000) found that elders primed with negative aging stereotypes demonstrated increased health concerns, such as cardiovascular response after performing stressful cognitive tasks. In contrast, elders primed with positive stereotypes were associated with better health outcomes, such as full recovery from disability (Levy et al. 2012), better functioning (Levy et al. 2002), and willingness to accept life-prolonging interventions (Levy et al. 1999–2000).

Nonetheless, the evidence regarding health disparities in later life is less clear, and inconsistent concerning whether ageism further widens the health disparity between whites and minority/immigrant groups. An early cross-sectional study based on middle-aged and aged adults found that self-assessed health of whites was better than that of other minority groups in each age stratum, with the greatest disparity among those aged 65 and older (Dowd and Bengtson 1978). Moreover, the black/white gap in disability and morbidity also increased over time (Liao et al. 1999). On the other hand, research by Ferraro and colleagues concluded that health disparities were persistent throughout the lifespan (young, middle-aged, and later life) and did not vary much with increasing age (Ferraro 1987; Ferraro and Farmer 1996; Kelley-Moore and Ferraro 2004).

The convergence of health disparities in later life could be a result of selective survival. Carreon and Noymer (2011) noted that research pointing to a widening of the gap in later life is premised upon a cumulative disadvantage perspective, that is, intersectional effects of racism and ageism, often referred to as a "double jeopardy". People are set on health trajectories as a result of disadvantages experienced across the life course and the effects aggregate over the year and readily appear in later years (Satariano et al. 1982). A recent study focusing on physical and emotional functions among older adults with cancer identified evidence of the double jeopardy effect in the sample of non-Hispanic whites, African Americans, Asian Americans and Hispanics; but the double jeopardy effect disappeared after factoring in comorbid conditions (Bellizzi et al. 2012).

Considering the framework outlined in Fig. 1, three factors require special attention relative to health disparities among older Americans across minority and immigrant groups. First, socio-economic status (SES) remains one of the most important factors influencing the health of elders across racial, ethnic minority and immigrant groups (Smith and Raynard 1998; Anderson et al. 2004). Studies indicate that older adults with low SES across all minority groups are more likely to suffer from both physical and mental illness (Amstadter et al. 2010; Angel et al. 2003; Pinquart and Sorensen 2000; Lièvre et al. 2008), and that lower SES groups are less likely to utilize health and preventive health care services (Crystal et al. 2003; Shi et al. 2012). In addition, House et al. (2005) found that the impact of SES disparities on health across minority and immigrant groups seems to be small in early adulthood, increases through middle and early old age, and then lessens again in later old age.

Secondly, despite the significant effects of culture on minority groups' health (Brach and Fraserirector 2000; Witt et al. 2002; Saha et al. 2013), certain cultural beliefs can benefit or harm elder minority and/or immigrant people, particularly for recently arrived elder immigrants who are much more likely to adhere to the health beliefs and practices of their original culture. For example, cultural beliefs surrounding memory loss and dementia among ethnic minority older adults were associated with delayed dementia diagnosis and care-seeking, compared to their non-Hispanic white counterparts (Sayegh and Knight 2013). Elder Chinese immigrants' cultural beliefs in Taoism, Buddhism and traditional Chinese medicine had influenced their healthcare utilization behaviors (Smith et al. 2013). Among Latino elders, positive perceptions of aging (cultural value emphasizing respect for people by virtue of age) and high level and quality of family support for health problems were significantly positively related to professional care and services (Beyene et al. 2002; Padilla and Villalobos 2007).

Thirdly, studies found that health of older adults is more sensitive to their immediate environment—neighborhood communities—than that of younger people. When community dwelling U.S. elders, particularly the 27 % of elders living alone (U.S. Census Bureau 2012), greatly depend on local resources and interactions for care and supports, neighborhood communities may affect their health directly. Thus, health outcomes of minority and immigrant older adults may reflect the environment in which they reside. For example, community racism can create a threatening environment and impede physical activities among older minority adults (Edwards and Cunningham 2012); whereas community affluence, a structural resource, can contribute positively to older residents' self-rated health and weaken the association between race and health (Cagney et al. 2005). Consequently, when geographic racial segregation restricts ethnic minority elders to poor neighborhoods, or when ethnic elders prefer to live in ethnic enclaves because it enables them to remain independently and they can live comfortably without speaking English (Olson 2001), in such cases, community level factors becomes essential to their health.

In summary, the available literature offers abundant evidence on the pathways between ethnic minority and immigration status and health disparities. Scholars have argued that in addition to racial-genetic differences, structural racism might be the real cause behind health disparities between whites and minority/immigrant older people; race/ethnicity could be simply a proxy for class and socioeconomic stratification (Dressler et al. 2005). Available literature also indicates the possible interactions affecting health disparity between SES and structural racism (Kawachi et al. 2005), as well as the correlations among SES, health behavior and racism (Williams and Collins 2001). To address health disparities, researchers have pointed to policies and practice interventions that focus on these root causes of disparities. The following section proposes a new way—community social capital perspective—to conceptualize and consequently to address the health disparity for ethnic minority and immigrant groups of older Americans.

Examining Elders Health Disparities via Community Social Capital as a Framework

Developed through the work of Bourdieu (1997), Coleman (1988) and Putnam (1993), social capital is a multidimensional construct that often includes cognitive (i.e., mutual trust) and structural (i.e., social networks) components—that is, what people feel and what they do in terms of social relations (Harpham et al. 2002). Typically social capital is operationalized and assessed through trust, networks and norms of reciprocity (Ferlander 2007). The public health field widely cited Putnam's (1995) definition of social capital: "features of social organization such as networks, norms, and social trust that facilitate coordination and cooperation for mutual benefit" (p. 67). Forms of social capital include *bonding*, *bridging* and *linking* social capital (Claridge 2004; Szreter and Woolcock 2004). *Bonding social capital* refers to the trusting and cooperative relations between members of a network who share similar socio-demographic descriptors such as age, ethnicity and/or education. Bridging social capital describes the heterogeneous and outward-looking connections that individuals form across social groups. Linking social capital refers to the quality of vertical ties that are formed between individuals and groups that are explicitly recognized as 'unequal' in terms of the power they wield in society such as between local governments and citizenry (Ferlander 2007).

Researchers have paid greater attention to the context and ecological understanding of social capital, through such measures as collective efficacy, trust in the community, neighborly support, community place attachment, social cohesion (Prezza et al. 2001; Perkins et al. 2002; Kawachi et al. 2004, 2008). Almedom (2005) found that, for minority and immigrant older adults, neighborhood/community level social capital is critical for their health; and community context and macro-level factors (e.g., social structures and political environments) may be more influential than individual-level social capital (Angel and Angel 2006). Evidently, institutionalized racism has perpetuated racial and ethnic stratification across the life span and has impacted individual and family health profiles; individual-level social capital may not adequately break the pattern.

The relationship between social capital and elders' health in general has been proved positive across countries and cultures (Hyyppä and Mäki 2001; Kawachi et al. 1999; Rose 2000; Schultz et al. 2008). In other words, local ties, sense of community, neighborly trust and all other psychosocial aspects of networks contribute to a better health outcome among older people. While researchers continue to debate the mechanisms between social capital and elders' health, arguably, today's baby boomers increasingly remain active in volunteering and local civic participation, thus possessing sizeable social capital (Einolf 2009). Therefore, the special characteristics of social capital nested in the minority and immigrant communities should help reduce the health gaps and offer a viable option to address the health disparity for older U.S. minority and immigrant adults living in the community.

Characteristics of Ethnic Minority and Immigrant Social Capital

Research comparing the social capital of various minority and immigrant groups is limited. Quantitative studies indicate that social capital is not equally shared across racial and ethnic groups and that significant differences appear in the social networks and their embedded resources (Lin 2000); for instance, social network diversity and size decreased from whites to Hispanics and to blacks (Marsden 1987). When race, ethnicity and class are so fundamental in defining American communities (e.g., segregated communities, immigrant enclaves and other types of minority racial enclaves), minority or immigrant elders' social capital might be localized, insular and sometimes draining, and their social capital can be described as "biased bonding" (i.e., strong bonding based on kin, extended household and people of the same race/ethnicity), "bounded bridging" (i.e., networks within a neighborhood/community boundary) (Hero 2003), and "limited linking" (i.e., restricted connections with governmental entities) (Dominguez and Watkins 2003). Lin (2000) stated that minority and immigrant groups have different access to social capital because of their advantaged and disadvantaged structural positions and associated social networks.

Thus, the health benefits of social capital for minority and immigrant elders might not be interpreted the same way as for white elders. It is noted that strong bonding capital based on kin and extended family networks enables minority and immigrant groups to pool their resources to survive and thrive; Magasi and Hammel (2004) found that this was exactly the case for disabled older African American women. Also, while "bounded bridging" capital implies that a community has limited resources and possibly experiences poverty, aspects of racial, ethnic, and immigration enclaves may well protect older adults' health, possibly because of an enhanced sense of belonging, positive social interactions and their common native language spoken in the community (Angel and Angel 2006).

Indeed, neighborly trust (trust of neighbors of the same race or ethnicity and of different races and ethnicities) was weaker in diverse communities than in homogenous communities (Putnam 2007). Mexican and Cuban older adults living in high-density Mexican–American and Cuban American neighborhoods were in better health than those who lived in lower-density neighborhoods (Eschbach et al. 2004).

In addition, minority and immigrant social capital often originates from distinct yet interconnected sub-networks based on individual characteristics such as age, education, class and reason for migration. In particular, when immigrants connect strongly to their traditional sub-network (i.e., the network based on a common place of origin), they also build and extend their new networks once resettled, through various activities such as church attendance; neighborhood interactions; friendship-making outside of their ethnic community including with classmates and workmates, and volunteering, and participating in service programs (Garcia 2005). For example, based on individual characteristics such as age, time of migration, employment history, living situation, and English capability, Liu (2003) identified four typologies for elderly Chinese immigrants in the U.S. based on their different patterns of service utilization and participation: old-age recent immigrants, old-age long-time immigrants, retired professional families, and middle-age non-professional immigrants. Recognizing the value of sub-networks, studies found that not all ties were equally beneficial; for example, network diversity (i.e., bridging capital) was unrelated to older immigrants' life satisfaction (Diwan 2008; Litwin and Leshem 2008).

Community Social Capital as a Cushion

The benefits of social capital may help cushion the negative effects of minority and/or immigration status on older adults as illustrated in Fig. 2. First, minority and immigrant older adults can benefit from their neighborhood and community ties to better cope with the stress associated with discrimination and/or chronic experience living with low socio-economic status. Ethnicity and culture play a significant role in mitigating the stress and in the coping process as ethnicity and culture can influence one's appraisal of stress events, perceptions and use of family support, and coping behaviors (Mui and Kang 2006; Aranda and Knight 1997; Utsey et al. 2002). Strong ties with formal or informal ethnic (or immigrant) community organizations not only reduce the risk of social isolation, which is potentially health-protective or -enhancing for elders, but also nurtures cultural traditions. For instance, one study indicated that older Mexican–American Catholics benefited from frequent church attendance and reported that it provided them with comfort during times of trouble (Hill et al. 2005). Under stressful life situations, Latino elders who regularly participated in programs at senior centers experienced lower levels of psychological distress than those who did not (Farone et al. 2005). In a study of older Korean Americans, Han et al. (2007) found that

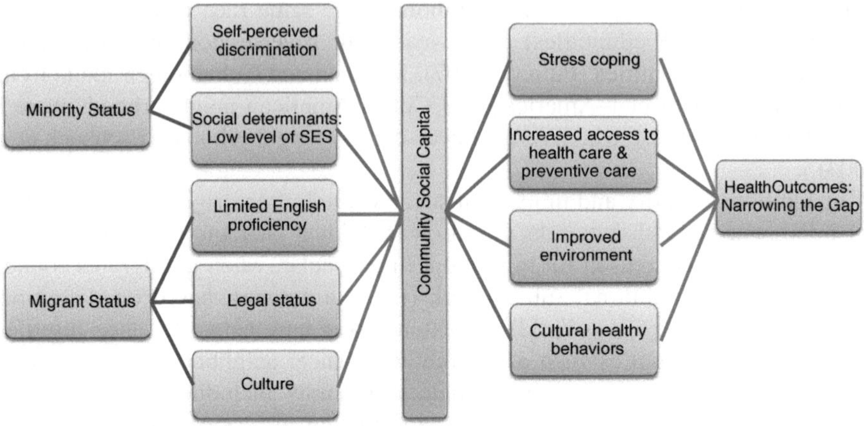

Fig. 2 Community social capital framework

lower social support levels were associated with higher depression scores when controlling for demographics and health status. Thus, when coping, "bounded bridging" capital helps older adults foster shared perceptions and strengthen ethnic and cultural identity; similarly, "strong bonding" capital offers needed financial, physical, and emotional supports and helps increase self-esteem (Cattell 2001).

Secondly, among minority and immigrant groups, community social capital functions in a particularly beneficial way that helps older adults to better access health care and preventive care services. A higher level of community engagement and ties to community groups can increase exposure to health and healthcare related information and resources, providing a critical pathway for social capital to influence health promotion (Viswanath et al. 2006). Although older minority and immigrant adults may be connected to various sub-group networks leading to varying degrees of community social capital, the literature suggests that all sub-group networks serve a similar function—to help individuals gain access to information and/or services (Garcia 2005); this is even true to communities with concentrated undocumented immigrants (Xu and Brabeck 2012). The information and resources obtained through community social capital help decrease barriers such as lack of transportation, limited health literacy and lack of health insurance to access health care and preventive care services (Gelfand 2006), especially aiding elders with limited English proficiency (e.g., Burr and Mutchler 2003; Aroian et al. 2005). For minority and immigrant older people, who often have significantly less trust in their physicians (or Western biomedicine or information from formal sources) and greater trust in informal health information sources (e.g., Musa et al. 2009), information and resources obtained through trusted ties in the community is distinctively meaningful and thus associated with increased utilization of health care and preventive care services. Besides serving as conduits for valuable information and resources, community social capital also can change collective life perceptions and traditional/cultural healthcare seeking behavior.

Studies found that elderly Chinese Americans have depended on bonding capital to meet their health needs as family and "lay others" (e.g., friends) could provide resources, especially in transportation assistance, medication requirements, and advice to avoid ineffective or even harmful self-treatment. At the same time, through interactions with community agencies elderly Chinese learned to shift their traditional expectations for assistance from family members (i.e., filial piety) to neighbors and friends, and have shown a genuine adaptability by combining Eastern and Western health care modalities (Pang et al. 2003).

Thirdly, community social capital is believed to play an important role in community life and can help strengthen the community through collective actions that ultimately benefit community dwelling elders. The literature indicates that the higher the stock of community social capital (as indicated by measures of trust and reciprocity), the higher the health level of a given community (Kawachi 1999). In a study of health outcomes of Chicago residents, neighborhood social capital was associated with lower neighborhood death rates from heart disease (Lochner et al. 2003). The dynamics of neighborhood characteristics, community social capital and health outcomes are multifaceted. On one hand, neighborhood characteristics influence network patterns and forms of social capital created. Such neighborhood factors as history, employment opportunities, local resources and opportunities for participation influence residents' ability to develop trusting relationships, and help develop community norms of cooperation and reciprocity, and patterns of mutual aid and information exchange. Communities with concentrated poverty have been associated with impoverished social networks. Residents of all races in ethnically diverse neighborhoods tend to have lower levels of trust, altruism and community cooperation, and fewer friends than residents of homogeneous neighborhoods (Putnam 2007). Community social capital differentiated by structural and cultural characteristics has impacts on psychosocial pathways involved in health effects.

On the other hand, community social capital (i.e., which people gain from involvement in clubs and other community activities, or interactions with neighbors) may affect residents' participation behavior and facilitate participation in local organizations and associations, and in more general civic arenas (Putnam 1993). A high level of community participation helps create and reinforce norms and solidarity, supports the development of ethnic and/or regional identities, and engages members in collective actions that benefit the community (Fennema and Tillie 2001). In general, community collective actions to improve its environment, such as reducing crime or improving housing conditions or overall quality of life, would ultimately strengthen residents' physical and mental well-being. Indeed, evidence has shown that when partnerships among community organizations target health education and other health promotion programs for older minority and immigrant residents, they can effectively reduce health disparities (e.g., Fernandez et al. 2008).

Lastly, community social capital helps minority and immigrant older adults preserve their cultural traditions, which in turn can help secure cultural health benefits. In theory, because social connections and interactions appear to be more available to ethnic minority elders within their communities rather than in broader society (Breton 2003), minority and immigrant networks provide a familiar environment within which to construct ethnic, cultural, and class identities. Ethnic communities require common obligations and provide support, and members' similar experiences in terms of racism help create a common understanding of their group and each other (Fennema and Tillie 2001) that aid in identity formation (Bankston and Zhou 1995). These contexts, in turn, influence how minority and immigrant elders perceive their lives, as well as the behaviors in which they engage. Ethnic and immigrant groups embody many qualities, including resourcefulness, resilience and solidarity. For minority and immigrant older adults, instead of viewing ethnic communities as places of poverty and disadvantage (Valtonen 2002), the community social capital perspective see them as a solution and thriving supportive environments for ethnic minorities elders: they offer an array of cultural knowledge, skills, abilities and contacts, possessed by and serving socially marginalized groups, rich resources that often go unrecognized and unacknowledged by society and by scholars. Various forms of capital nurtured through cultural wealth include aspirational, navigational, social, linguistic, familial and resistant capital, thus benefiting minority and immigrant elders' health.

Community Social Capital: Opportunities for the Future

The importance of community social capital in understanding and addressing health disparities among U.S. older adults has been well explored and fully acknowledged. However the application of community social capital perspective in policies and in health care practice still faces great challenges. In essence, racism has affected every aspect of U.S. ethnic and racial minority groups' lives; minority and immigrant groups face a very different opportunity structure since birth, which is constructed geographically, economically, socially and psychologically. Thus, it is not surprising that racism is associated with lower levels of social capital among ethnic and immigrant elders, as well as poorer psychological well-being and physical health (though this manifests in different ways for different ethnic groups). However, racial/ethnic and immigrant groups maintain extensive networks developed through kinship, friendship and shared communities of race, ethnicity and/or country of origin; and they rely on, and actively contribute to, the social capital banked in family and community ties for information, resources and various supports and services. To change health trajectories of African American, Asian American and Hispanic American elders, many who have immigrated to the U.S., and given our current understanding of the pathways between minority and immigration status and health outcomes, community social capital is proposed and seen as a counterweight to the negative effects of social-economic disparity, community segregation, social discrimination, and other factors that impact health outcomes.

Community social capital is also an equalizer that minimizes the health gap for elders caused by SES differences. Investments in community social capital could equalize the inequitable financial and human capital distribution. While the community social capital perspective surely brings in policy and practice implications in terms of resource relocation, arguably, the extent that community social capital can help address the health disparity across minority and immigrant elder populations hinges on changing opportunity structure. In other words, ties in neighborhoods, neighbor trust, place attachment, among other aspects of community social capital could increase minority and immigrant residents' opportunities in employment, education, and other socioeconomic and cultural events, thus fundamentally redirecting the health trajectory. Policy and programs that nurture collective spirit, foster cultural value, and expand opportunities should target on minority and immigrant people at all ages, ideally children, as they would eventually benefit minority/immigrant older adults and at the same time the minority/immigrant community development. A call to action is issued to social service providers, community activists, policy makers, and educators, for a national agenda on behalf of ethnic minorities, their children and aging parents in the U.S., couched by an old Chinese proverb: It is too late to fix the fence after a wolf stole a lamb; but better late than never.

References

Adler, N. E., & Newman, K. (2002). Socioeconomic disparities in health: Pathways and policies. *Health Affairs, 21*(2), 60–76. doi:10.1377/hlthaff.21.2.60.

Almedom, A. M. (2005). Social capital and mental health: An interdisciplinary review of primary evidence. *Social Science and Medicine, 61*(5), 943–964. doi:10.1016/j.socscimed.2004.12.025 .

Amstadter, A. B., Begle, A. M., Cisler, J. M., Hernandez, M. A., Muzzy, W., & Acierno, R. (2010). Prevalence and correlates of poor self-rated health in the United States: The national elder mistreatment study. *The American Journal of Geriatric Psychiatry, 18*(7), 615–623. doi:10.1097/JGP.0b013e3181ca7ef2.

Anderson, N. B., Bulatao, R. A. & Cohen, B. (2004). *Critical perspectives on racial and ethnic differences in health in late life*. Retrieved from http://www.nap.edu/catalog/11086.html

Angel, J. L., & Angel, R. J. (2006). Minority group status and healthful aging: Social structure still matters. *American Journal of Public Health, 96*, 1152–1159. doi:10.2105/AJPH.2006.085530.

Angel, R. J., Frisco, M., Angel, J. L., & Chiriboga, D. A. (2003). Financial strain and health among elderly Mexican-origin individuals. *Journal of Health and Social Behavior, 44*, 536–551.

Aranda, M. P., & Knight, B. G. (1997). The influence of ethnicity and culture on the caregiver stress and coping process: A sociocultural review and analysis. *The Gerontologist, 37*(3), 342–354. doi:10.1093/geront/37.3.342.

Araújo, B. Y., & Borrell, L. N. (2006). Understanding the link between discrimination, mental health outcomes, and life chances among Latinos. *Hispanic Journal of Behavioral Sciences, 28*(2), 245–266. doi:10.1177/0739986305285825.

Aroian, K. J., Wu, B., & Tran, T. V. (2005). Health care and social service use among Chinese immigrant elders. *Research in Nursing and Health, 28*(2), 95–105. doi:10.1002/nur.20069.

Bankston, C. L, I. I. I., & Zhou, M. (1995). Religious participation, ethnic identification, and adaptation of Vietnamese adolescents in an immigrant community. *The Sociological Quarterly, 36*, 523–534.
Batalova, J. (2011). *Asian immigrants in the United States*. Washington, DC: Migration Policy Institute.
Baum, A., Garofalo, J. P., & Yali, A. (1999). Socioeconomic status and chronic stress. Does stress account for SES effects on health? *Annals of the New York Academy of Sciences, 896*, 131–144. doi:10.1111/j.1749-6632.1999.tb08111.x.
Bellizzi, K. M., Aziz, N. M., Rowland, J. H., Weaver, K., Arora, N. K., Hamilton, A. S., et al. (2012). Double Jeopardy? Age, race, and HRQOL in older adults with cancer. *Journal of Cancer Epidemiology, 2012*, 478642. doi:10.1155/2012/478642.
Beyene, Y., Becker, G., & Mayen, N. (2002). Perception of aging and sense of well-being among Latino elderly. *Journal of Cross-Cultural Gerontology, 17*(2), 155–172.
Bourdieu, P. (1997). The forms of capital. In A. Halsey, H. Lauder, P. Brown, & A. StuartWells (Eds.), *Education: Culture, economy and society*. Oxford: Oxford University Press.
Bowling, A. (1999). Ageism in cardiology. *BMJ, 319*(7221), 1353–1355.
Brach, C., & Fraserirector, I. (2000). Can cultural competency reduce racial and ethnic health disparities? *A review and conceptual model, Medical Care Research and Review, 57*, 181–217.
Breton, R. (2003). *Social capital and the civic participation of immigrants and members of ethno-cultural groups*. Paper presented at the conference on The opportunities and challenges of diversity: A role of social capital? in Montreal, Canada. Retrieved in October 24, 2011 from http://www.horizons.gc.ca/doclib/OECD_RaymondBreton_E.pdf.
Briggs, R., Robinson, S., & O'Neill, D. (2012). Ageism and clinical research. *Irish Medical Journal, 105*(9), 311–312.
Burr, J. A., & Mutchler, J. E. (2003). English language skills, ethnic concentration, and household composition: Older Mexican immigrants. *Journal of Gerontology: Series B, 58*, s83–s92.
Butler, R. (1969). Age-ism: Another form of bigotry. *The Gerontologist, 9*, 243–246.
Cagney, K. A., Browning, C. R., & Wen, M. (2005). Racial disparities in self-rated health at older ages: What difference does the neighborhood make? *Journal of Gerontology: Series B, 60*(4), s181–s190. doi:10.1093/geronb/60.4.S181.
Carreon, D., & Noymer, A. (2011). Health-related quality of life in older adults: Testing the double jeopardy hypothesis. *Journal of Aging Studies, 25*(4), 371–379. doi:10.1016/j.jaging.2011.01.004.
Cattell, V. (2001). Poor people, poor places, and poor health: The mediating role of social networks and social capital. *Social Science and Medicine, 52*(10), 1501–1516. doi:10.1016/S0277-9536(00)00259-8.
Chen, E., Martin, A. D., & Matthews, K. A. (2006). Understanding health disparities: The role of race and socioeconomic status in children's health. *American Journal of Public Health, 96*(4), 702–708.
Claridge, T. (2004). Social capital and natural resource management. (Unpublished Masters Thesis). University of Queensland, Brisbane, Australia.
Coleman, J. (1988). Social capital in the creation of human capital. *American Journal of Sociology, 94*, S95–S120.
Crystal, S., Sambamoorthi, U., Walkup, J. T., & Akincigil, A. (2003). Diagnosis and treatment of depression in the elderly medicare population: Predictors, disparities, and trends. *Journal of American Geriatrics Society, 51*(12), 1718–1728.
Cunningham, S. A., Ruben, J. D., & Narayan, K. M. V. (2008). Health of foreign-born people in the United States: A review. *Health and Place, 14*(4), 623–635. doi:10.1016/j.healthplace.2007.12.002.
Davis, A. M., Kreutzer, R., Lipsett, M., King, G., & Shaikh, N. (2006). Asthma prevalence in Hispanic and Asian American ethnic subgroups: Results from the California Healthy Kids Survey. *Pediatrics, 118*(2), e363–e370. doi:10.1542/peds.2005-2687.

Derose, K. P., Escarce, J. J., & Lurie, N. (2007). Immigrants and health care: Sources of vulnerability. *Health Affairs, 26*(5), 1258–1268. doi:10.1377/hlthaff.26.5.1258.

Diwan, S. (2008). Limited English proficiency, social network characteristics, and depressive symptoms among older immigrants. *Journal of Gerontology Series B, 63*(3), S184–S191.

Dominguez, S., & Watkins, S. (2003). Creating networks for survival and mobility: Social capital among African-American and Latin-American low-income mothers. *Social Problems, 50*(1), 111–135.

Dong, X. (2005). Medical implications of elder abuse and neglect. *Clinic in Geriatric Medicine, 21*(2), 293–313. doi:10.1016/j.cger.2004.10.006.

Dong, X., & Simon, M. A. (2013). Association between elder abuse and use of ED: Findings from the Chicago health and aging project. *The American Journal of Emergency Medicine, 31*(4), 693–698. doi:10.1016/j.ajem.2012.12.028.

Dowd, J. J., & Bengtson, V. L. (1978). Aging in minority populations. An examination of the double jeopardy hypothesis. *Journal of Gerontology, 33*(3), 427–436.

Dressler, W. W. (1993). Health in the African American community: Accounting for health inequalities. Available at: http://www.jstor.org/discover/10.2307/649213?uid=3739688&uid=2134&uid=378960983&uid=2129&uid=2&uid=70&uid=3&uid=3739256&uid=60&uid=378960973&sid=21101815185553

Dressler, W. W., Oths, K. S., & Gravlee, C. C. (2005). Race and ethnicity in public health research: Models to explain health disparities. *Annual Review of Anthropology, 34*, 231–252. doi:10.1146/annurev.anthro.34.081804.120505.

DuBard, C. A., & Gizlice, Z. (2008). Language spoken and differences in health status, access to care, and receipt of preventive services among US Hispanics. *American Journal of Public Health, 98*(11), 2021–2028.

Dunlop, D. D., Manheim, L. M., Song, J., & Chang, R. W. (2001). Arthritis prevalence and activity limitations in older adults. *Arthritis and Rheumatism, 44*(1), 212–221.

Dyer, C. B., Pavlik, V. N., Murphy, K. P., & Hyman, D. J. (2000). The high prevalence of depression and dementia in elder abuse or neglect. *Journal of the American Geriatrics Society, 48*(2), 205–208.

Edwards, M., & Cunningham, G. (2012). Examining the associations of perceived community racism with self-reported physical activity levels and health among older racial minority adults. *Journal of Physical Activity and Health*, 5 November, 2012 (Epub ahead of print).

Einolf, C. J. (2009). Will the boomers volunteer during retirement? Comparing the baby boom, silent, and long civic cohorts. *Nonprofit and Voluntary Sector Quarterly, 38*(2), 181–199. doi:10.1177/0899764008315182.

Eschbach, K., Ostir, G. V., Patel, K. V., Markides, K. S., & Goodwin, J. S. (2004). Neighborhood context and mortality among older Mexican Americans: Is there a barrio advantage? *American Journal of Public Health, 94*, 1807–1812.

Farone, D. W., Fitzpatrick, T. R., & Tran, T. V. (2005). Use of senior centers as a moderator of stress-related distress among Latino elders. *Journal of Gerontological Social Work, 46*(1), 65–83.

Fennema, M., & Tillie, J. (2001). Civic community, political participation, and political trust of ethnic groups. *Connections, 24*, 26–41.

Ferlander, A. (2007). The importance of different forms of social capital for health. *Acta Sociologica, 50*(2), 115–128. doi:10.1177/0001699307077654.

Fernandez, S., Scales, K. L., Pineiro, J. M., Schoenthaler, A. M., & Ogedegbe, G. (2008). A senior center-based pilot trial of the effect of lifestyle intervention on blood pressure in minority elderly people with hypertension. *Journal of the American Geriatrics Society, 56*(10), 1860–1866. doi:10.1111/j.1532-5415.2008.01863.x.

Ferraro, K. F. (1987). Double jeopardy to health for black older adults? *Journal of Gerontology, 42*(5), 528–533.

Ferraro, K. F., & Farmer, M. M. (1996). Double jeopardy, aging as leveler, or persistent health inequality? A longitudinal analysis of white and black Americans. *Journals of Gerontology: Series B: Psychological Sciences and Social Sciences, 51*(6), S319–S328.

Fiscella, K., Franks, P., Doescher, M., & Saver, B. (2002). Disparities in health care by race, ethnicity, and language among the insured: Findings from a national sample. *Medical Care, 40*(1), 52–59.

Garcia, C. (2005). Buscando trabajo: Social networking among immigrants from Mexico to the United States. *Hispanic Journal of Behavioral Sciences, 27*(1), 3–22.

Gee, G. C., Spencer, M. S., Chen, J., & Takeuchi, D. (2007). A nationwide study of discrimination and chronic health conditions among Asian Americans. *American Journal of Public Health, 97*(7), 1275–1282. doi:10.2105/AJPH.2006.091827.

Gee, G. C., Ro, A., Shariff-Marco, S., & Chae, D. (2009). Racial discrimination and health among Asian Americans: Evidence, assessment, and directions for future research. *Epidemiologic Reviews, 31*, 130–151. doi:10.1093/epirev/mxp009.

Gee, G. C., Walsemann, K. M., & Brondolo, E. (2012). A life course perspective on how racism may be related to health inequities. *American Journal of Public Health, 102*(5), 967–974. doi:10.2105/AJPH.2012.300666.

Gelfand, D. E. (2006). *The aging network: Programs and services* (6th ed.). Berlin: Springer.

Ghosh, C. (2003). Health people 2010 and Asian Americans/Pacific Islanders: Defining a baseline of information. *American Journal of Public Health, 93*, 2093–2098.

Gilmer, T. P., Ojeda, V. D., Barrio, C., Fuentes, D., Garcia, P., Lanouette, N. M., et al. (2009). Adherence to antipsychotics among Latinos and Asians with schizophrenia and limited English proficiency. *Psychiatric Services, 60*(2), 175–182. doi:10.1176/appi.ps.60.2.175.

Goodman, D. P., Smith, J. P., & Sood, N. (2005). Legal status and health insurance among immigrants. *Health Affairs, 24*(6), 1630–1653. doi:10.1377/hlthaff.24.6.1640.

Gordon-Larsen, P., Adair, L. S., & Popkin, B. M. (2003). The relationship of ethnicity, socioeconomic factors, and overweight in U.S. adolescents. *Obesity Research, 11*(1), 121–129. doi:10.1038/oby.2003.20.

Halkitis, P. N., & Figueroa, R. P. (2013). Sociodemographic characteristics explain differences in unprotected sexual behavior among young HIV-negative gay, bisexual, and other YMSM in New York City. *AIDS Patient Care STDS, 27*(3), 181–190. doi:10.1089/apc.2012.0415.

Han, H.-R., Kim, M., Lee, H. B., Pistulka, G., & Kim, K. B. (2007). Correlates of depression in the Korean American elderly: Focusing on personal resources of social support. *Journal of Cross-Cultural Gerontology, 22*(1), 115–127.

Harpham, T., Grant, E., & Thomas, E. (2002). Measuring social capital within health surveys: Key issues. *Health Policy and Planning, 17*(1), 106–111.

Harris, A. L. (2010). The economic and educational state of black Americans in the 21st century: Should we be optimistic or concerned? *Review of Black Political Economy,*. doi:10.1007/s12114-010-9065-z.

Hero, R. E. (2003). Social capital and racial inequality in America. *Perspectives on Polities, 1*(1), 113–122.

Hill, T., Angel, J. L., Ellison, C., & Angel, R. J. (2005). Religious attendance and mortality: An eight-year follow-up of older Mexican Americans. *Journal of Gerontology: Series B, 60B*, s102–s109.

House, J. S., Lantz, P. M., & Herd, P. (2005). Continuity and change in the social stratification of aging and health over life course: Evidence from a nationally representative longitudinal study from 1986 to 2001/2002 (Americans' Changing Lives Study). *Journal of Gerontology: Series B, 60*, S15–S26. doi:10.1093/geronb/60.Special_Issue_2.S15. (Special Issue 2).

Hyyppä, M. T., & Mäki, J. (2001). Individual-level relationships between social capital and self-rated health in a bilingual community. *Preventive Medicine: An International Journal Devoted to Practice and Theory, 32*(2), 148–155. doi:10.1006/pmed.2000.0782.

Jasso, G., Massey, D. S., Rosenzweig, M. R., & Smith, J. P. (2004). Immigrant health: Selectivity and acculturation. In National Research Council of the National Academies (Ed.), *Critical perspectives on racial and ethnic differences in health in late life* (pp. 227–267). Washington: National Academies.

Karliner, L. S., Ma, L., Hofmann, M., & Kerlikowske, K. (2012). Language barriers, location of care, and delays in follow-up of abnormal mammograms. *Medical Care, 50*(2), 171–178. doi:10.1097/MLR.0b013e31822dcf2d.

Kawachi, I. (1999). Social capital and community effects on population and individual health. *Annals of the New York Academy of Sciences, 896*, 120–130.

Kawachi, I., Kennedy, B. P., & Glass, R. (1999). Social capital and self-rated health: A contextual analysis. *American Journal of Public Health, 89*(8), 1187–1193.

Kawachi, I., Kim, D., Coutts, A., & Subramanian, S. V. (2004). Commentary: Reconciling the three accounts of social capital. *International Journal of Epidemiology, 33*, 682–690. doi:10.1093/ije/dyh177.

Kawachi, I., Daniels, N., & Robinson, D. E. (2005). Health disparities by race and class: Why both matter. *Health Affairs, 24*(3), 343–352. doi:10.1377/hlthaff.24.2.343.

Kawachi, I., Subramanian, S. V., & Kim, D. (2008). *Social capital and health*. New York: Springer.

Kelley-Moore, J. A., & Ferraro, K. F. (2004). The black/white disability gap: Persistent inequality in later life? *Journal of Gerontology: Series B, 59*(1), s34–s43. doi:10.1093/geronb/59.1.S34.

Kite, M. E., & Johnson, B. T. (1988). Attitudes toward older and younger adults: A meta-analysis. *Psychol Aging, 3*(3), 233–244.

Levy, B. R., Ashman, O., Dror, I. (1999–2000). To be or not to be: The effects of aging stereotypes on the will to live. *Omega (Westport), 40*(3), 409–420.

Levy, B. R., Hausdorff, J. M., Hencke, R., & Wei, J. Y. (2000). Reducing cardiovascular stress with positive self-stereotypes of aging. *Journal of Gerontology: Series B, 55*(4), 205–213.

Levy, B. R., Slade, M. D., & Kasl, S. V. (2002). Longitudinal benefit of positive self-perceptions of aging on functional health. *Journals of Gerontology: Series B: Psychological Sciences and Social Sciences, 57*(5), 409–417. doi:10.1093/geronb/57.5.P409.

Levy, B. R., Zonderman, A. B., Slade, M. D., & Ferrucci, L. (2012). Memory shaped by age stereotypes over time. *Journals of Gerontology: Series B: Psychological Sciences and Social Sciences, 67*(4), 432–436. doi:10.1093/geronb/gbr120.

Liao, Y., McGee, D. L., Cao, G., & Cooper, R. S. (1999). Black-white differences in disability and morbidity in the last years of life. *American Journal of Epidemiology, 149*, 1097–1103.

Liѐvre, A., Alley, D., & Crimmins, E. (2008). Educational differentials in life expectancy with cognitive impairment among the elderly in the United States. *Journal of Aging and Health, 20*(4), 456–477. doi:10.1177/0898264308315857.

Lillie-Blanton, M., & Hudman, J. (2001). Untangling the web: Race/ethnicity, immigration, and the nation's health. *American Journal of Public Health, 91*(11), 1736–1738. doi:10.2105/AJPH.91.11.1736.

Lin, N. (2000). Inequality in social capital. *Contemporary Sociology, 26*(6), 785–795.

Litwin, H., & Leshem, E. (2008). Late-life migration, work status, and survival: The case of older immigrants from the Former Soviet Union in Israel. *International Migration Review, 42*(4), 903–925.

Liu, Y.-L. (2003). Aging service need and use among Chinese-American seniors: Intragroup variations. *Journal of Cross-Cultural Gerontology, 19*, 273–301.

Lochner, K. A., Kawachi, I., Brennan, R. T., & Buka, S. L. (2003). Social capital and neighborhood mortality rates in Chicago. *Social Science and Medicine, 56*, 1797–1805.

Loevy, R. D. (1990). *To end all segregation: The politics of the passage of the Civil Rights Act of 1964*. Lanham: University Press of America.

Loevy, R. D., Humphrey, H. H., Rauh, J. L, Jr, & Stewart, J. G. (1997). *The Civil Rights Act of 1964: The passage of the law that ended racial segregation*. Albany: State University of New York Press.

Magasi, S., & Hammel, J. (2004). Social support and social network mobilization in older African American women who have experienced strokes. *Disability Studies Quarterly, 24*(4).

Marsden, P. (1987). Core discussion networks of Americans. *American Sociological Review, 52*, 122–131.

McBean, A. M., Li, S., Gilbertson, D. T., & Collins, A. J. (2004). Differences in diabetes prevalence, incidence, and mortality among the elderly of four racial/ethnic groups: Whites, blacks, Hispanics, and Asians. *Diabetes Care, 27*(10), 2317–2324. doi:10.2337/diacare.27.10.2317.

Meisner, B. A. (2012). A meta-analysis of positive and negative age stereotype priming effects on behavior among older adults. *Gerontology: Series B: Psychological Sciences and Social Sciences, 67*(1), 13–17. doi:10.1093/geronb/gbr062. (Epub 2011 July 11).

Milaszewski, D., Greto, E., Klochkov, T., & Fuller-Thomson, E. (2012). A systematic review of education for the prevention of HIV/AIDS among older adults. *Journal of Evidenced-Based Social Work, 9*(3), 213–230. doi:10.1080/15433714.2010.494979.

Motel, S. (2012). *Statistical portrait of Hispanics in the United States, 2010*. Washington, DC: Pew Research Hispanic Center.

Mui, A. C., & Kang, S.-Y. (2006). Acculturation stress and depression among Asian immigrant elders. *Social Work, 51*(3), 243–255. doi:10.1093/sw/51.3.243.

Musa, D., Schulz, R., Harris, R., Silverman, M., & Thomas, S. B. (2009). Trust in the health care system and the use of preventive health services by older black and white adults. *American Journal of Public Health, 99*(7), 1293–1299. doi:10.2105/AJPH.2007.123927.

National Institute of Health. (2010). *Fact sheet*: *Health disparities*. Washington, DC: National Institute of Health Retrieved from http://report.nih.gov/nihfactsheets/Pdfs/HealthDisparities%28NIMHD%29.pdf

Olson, L. K. (Ed.). (2001). *Age through ethnic lenses: Caring for the elderly in a multicultural society*. Lanham: Rowman & Littlefield Publishers Inc.

Padilla, Y. C., & Villalobos, G. (2007). Cultural responses to health among Mexican American women and their families. *Family and Community Health, 30*(Supplement 1), s24–s33.

Pang, E. C., Jordan-Marsh, M., Silverstein, M., & Cody, M. (2003). Health-seeking behaviors of elderly Chinese Americans: Shifts in expectation. *The Gerontologist, 43*(6), 864–874.

Perdue, C. W., & Gurtman, M. B. (1990). Evidence for the automaticity of ageism. *Journal of Experimental Social Psychology, 26*, 199–216. doi:10.1016/0022-1031(90)90035-K.

Perkins, D., Hughey, J., & Speer, P. (2002). Community psychology perspectives on social capital theory and community development practice. *Journal of Community Development Society, 33*(1), 33–52.

Pinquart, M., & Sorensen, S. (2000). Influences of socioeconomic status, social network, and competence on subjective well-being in later life: A meta-analysis. *Psychology and Aging, 15*, 187–224. doi:10.1037/0882-7974.15.2.187.

Ponce, N. A., Hays, R. D., & Cunningham, W. E. (2006). Linguistic disparities in health care access and health status among older adults. *Journal of General Internal Medicine, 21*(7), 786–791. doi:10.1111/j.1525-1497.2006.00491.x.

Prezza, M., Amici, M., Roberti, T., & Tedeschi, G. (2001). Sense of community referred to the whole town: Its relations with neighboring, loneliness, life satisfaction, and area of residence. *Journal of Community Psychology, 29*(1), 29–52.

Pu, J., & Chewning, B. (2012). Racial difference in diabetes preventive care. *Research in Social and Administrative Pharmacy,*. doi:10.1016/j.sapharm.2012.11.005.

Putnam, R. (1995). Bowling alone: America's declining social capital. *Journal of Democracy, 6*(1), 65–78. doi:10.1353/jod.1995.0002.

Putnam, R. D. (1993). *Making democracy work: Civic traditions in modern Italy*. Princeton: Princeton University Press.

Putnam, R. D. (2007). E Pluribus Unum: Diversity and community in the twenty-first century. *Scandinavian Political Studies, 30*(2), 137–174. doi:10.1111/j.1467-9477.2007.00176.x.

Rodriguez, M. A., Bustamante, A. V., & Ang, A. (2008). Perceived quality of care, receipt of preventive care, and usual source of health care among undocumented and other Latinos. *Journal of General Internal Medicine, 24*(3 Supplement), 508–513.

Rose, R. (2000). How much does social capital add to individual health? A survey study of Russians. *Social Science and Medicine, 51*(9), 1421–1435.

Saha, S., Korthuis, P. T., Cohn, J. A., Sharp, V. L., Moore, R. D., & Beach, M. C. (2013). Primary care provider cultural competence and racial disparities in HIV care and outcomes. *Journal of General Internal Medicine,*. doi:10.1007/s11606-012-2298-8.

Satariano, W. A., Albert, S., & Belle, S. H. (1982). Race, age and cancer incidence: A test of double jeopardy. *Journal of Gerontology, 37*(6), 642–647. doi:10.1093/geronj/37.6.642.

Sayegh, P., & Knight, B. G. (2013). Cross-cultural differences in dementia: The sociocultural health belief model. *International Psychogeriatrics, 25*(4), 517–530. doi:10.1017/S104161021200213X.

Schaie, K. W., & Willis, S. L. (Eds.). (2011). *Handbook of the psychology of aging* (7th ed.). San Diego: Elsevier.

Schultz, J., O'Brien, A. M., & Tadesse, B. (2008). Social capital and self-rated health: Results from the US 2006 social capital survey of one community. *Social Science and Medicine, 67*(4), 606–617.

Shi, D., Yin, M. T., Shi, Q., & Hoover, D. R. (2012). Associations of education level and bone density tests among cognitively intact elderly white women in managed medicare. *Current Gerontology and Geriatrics Research, 2012*, 179150. doi:10.1155/2012/179150.

Simon, A. (2011). Geriatric patients in intensive care medicine: Ethical aspects. *Medizinische Klinik Intensivmedizin und Notfallmedizin, 106*(1), 24–28. doi:10.1007/s00063-011-0022-9. (Epub 2011 October 7).

Singh, G. K., & Hiatt, R. A. (2006). Trends and disparities in socioeconomic and behavioural characteristics, life expectancy, and cause-specific mortality of native-born and foreign-born populations in the United States, 1979–2003. *International Journal of Epidemiology, 35*(4), 903–919. doi:10.1093/ije/dyl089.

Singh, G. K., & Siahpush, M. (2002). Ethnic-immigrant differentials in health behaviors, morbidity, and cause-specific mortality in the United States: An analysis of two national data bases. *Human Biology, 74*(1), 83–109. doi:10.1353/hub.2002.0011.

Skirbekk, H., & Nortvedt, P. (2012). Inadequate treatment for elderly patients: Professional norms and tight budgets could cause "Ageism" in hospitals. *Health Care Analysis*. doi: 10.1007/s10728-012-0207-2 (Epub ahead of print).

Smith, J. P. & Raynard, K. (1998). The Relationship between the socioeconomic status and health of the elderly. Santa Monica, CA: RAND Corporation. Available at http://www.rand.org/pubs/research_briefs/RB5020/index1.html

Smith, A., Macentee, M. I., Beattie, B. L., Brondani, M., Bryant, R., Graf, P., et al. (2013). The influence of culture on the oral health-related beliefs and behaviours of elderly Chinese immigrants: A meta-synthesis of the literature. *Journal of Cross Culture Gerontology, 28*(1), 27–47. doi:10.1007/s10823-013-9187-4.

Szreter, S., & Woolcock, A. (2004). Health by association? Social capital, social theory and the political economy of public health. *International Journal of Epidemiology, 33*, 650–667. doi:10.1093/ije/dyh013.

U.S. Census Burean (2009). American Community Survey, B09017, "Relationship by household type (Including Living Alone) for the population 65 years and over"; B18101, "Sex by age by disability status"; B18102, "Sex by age by hearing difficulty"; B18103, "Sex by age by vision difficulty"; B18104, "Sex by age by cognitive difficulty"; B18105, "Sex by age by ambulatory difficulty"; B18106, "Sex by age by self-care difficulty"; B18107, "Sex by age by independent living difficulty." Accessed Dec 2010.

U.S. Census Bureau. (2011). *2011 population estimates*. Retrieved from "http://www.census.gov/popest/data/national/asrh/2011/index.html

U.S. Census Bureau. (2012). *Persons 65 years and over—Living arrangements and disability status: 2009*. Retrieved from http://www.census.gov/compendia/statab/2012/tables/12s0035.pdf

U.S. Centers for Disease Control and Prevention. (2011). *CDC health disparities and inequalities report—United States, 2011*. Washington, DC: US Department of Health and Human Services Centers for Disease Control and Prevention Retrieved from http://www.cdc.gov/mmwr/pdf/other/su6001.pdf.

Utsey, S. O., Payne, Y. A., Jackson, E. S., & Jones, A. M. (2002). Race-related stress, quality of life indicators, and life satisfaction among elderly African Americans. *Cultural Diversity and Ethnic Minority Psychology, 8*(3), 224–233. doi:10.1037//1099-9809.8.3.224.

Valtonen, K. (2002). The ethnic neighborhood: A locus of empowerment for elderly immigrants. *International Social Work, 45*, 315–323. doi:10.1177/0020872802045003358.

Van Den Eeden, S. K., Tanner, C. M., Bernstein, A. L., Fross, R. D., Leimpeter, A., Bloch, D. A., et al. (2003). Incidence of Parkinson's disease: Variation by age, gender, and race/ethnicity. *American Journal of Epidemiology, 157*(11), 1015–1022. doi:10.1093/aje/kwg068.

Viswanath, K., Steele, W. R., & Finnegan, J. R, Jr. (2006). Social capital and health: Civic engagement, community size, and recall of health messages. *American Journal of Public Health, 96*, 1456–1461. doi:10.2105/AJPH.2003.029793.

Ward, E., Jemal, A., Cokkinides, V., Singh, G., Cardinez, C., Ghafoor, A., et al. (2004). Cancer disparities by race/ethnicity and socioeconomic status. *CA: A Cancer Journal for Clinicians, 54*(2), 78–93.

Williams, D. R., & Collins, C. (1995). U.S. socioeconomic and racial differences in health: Patterns and explanations. *Annual Review of Sociology, 21*, 349–386.

Williams, D. R., & Collins, C. (2001). Racial residential segregation: A fundamental cause of racial disparities in health. *Public Health Reports, 116*, 404–416.

Williams, D. R., & Mohammed, S. A. (2009). Discrimination and racial disparities in health: Evidence and needed research. *Journal of Behavioral Medicine, 32*(1), 20–47.

Williams, D. R., Neighbors, H. W., & Jackson, J. S. (2003). Racial/ethnic discrimination and health: Findings from community studies. *American Journal of Public Health, 93*(2), 200–208.

Witt, D., Brawer, R., & Plumb, J. (2002). Cultural factors in preventive care: African-Americans. *Primary Care, 29*(3), 487–493.

Wright, S. (2005). *The Civil Rights Act of 1964: Landmark anti-discrimination legislation*. New York: The Rosen Publishing Group. ISBN 1-4042-0455-5.

Xu, Q., & Brabeck, K. (2012). Service utilization for Latino children in mixed-status families. *Social Work Research, 36*(3), 209–221. doi:10.1093/swr/svs015.

Zuckerman, R. B., Tinsley, L. J., Hawk, H., & Cohen, B. (2012). Perceived reactions to race and health status in the Massachusetts behavioral risk factor surveillance system survey. *Ethnicity and Disease, 22*(4), 492–496.

A Profile of Rural African American Lesbian Elders: Meeting Their Needs

Debra A. Harley, Kim L. Stansbury, Marva Nelson and Christina T. Espinosa

Abstract Elderly African American lesbians in rural society represent diversity across numerous demographic characteristic and social positions that require specific responses to their needs. The population must address issues related to age, race/ethnicity, gender, and sexual orientation. In this chapter a trans-disciplinary examination of issues relevant to this population is examined. In order to do so, a profile of elderly African American lesbians is presented. Additional information is presented on the concepts of feminism and womanism, the construct of positive marginality, characteristics of rural society and how African American women and lesbians exist in such a context, cultural homophobia in the black community, and racism in the white gay community. Psychosocial adjustment considers the intersection of sexual orientation, age, and race statuses. Finally, healthcare, education, and public policy issues are examined with respect to elder African American lesbians.

The exact number of African American or Black (used interchangeably in this chapter) lesbian elders is not known because the exact number of the total lesbian, gay, bisexual, transgender, and questioning (LGBTQ) (sexual minorities) population of all ages in the United States is unknown. A Gallup poll survey reports that 3.4 % of Americans identify as lesbian, gay, bisexual, or transgender (LGBT), and 4.6 % of African Americans identify as such with 3.6 % of African American women identifying as lesbian (Carary 2012). There are 16,638,000 African American women in the United States (U.S. Census Bureau 2011). It is estimated that between 1 and 2 % of African American women identify as lesbian

D. A. Harley (✉) · C. T. Espinosa
University of Kentucky, Lexington, KY, USA
e-mail: dharl00@email.uky.edu

K. L. Stansbury
Eastern Washington University, Cheney, WA, USA

M. Nelson
Parkland College, Champaign, IL, USA

H. F. O. Vakalahi et al. (eds.), *The Collective Spirit of Aging Across Cultures*, International Perspectives on Aging 9, DOI: 10.1007/978-94-017-8594-5_8,

(Gates 2011). Applying this prevalence to African Americans, there are at least 166,380 African American lesbians (Dibble et al. 2012). However, it is difficult to glean how many of the 166,380 are lesbian elders because of the rate of non-disclosure. The difficulty in determining the number of African American lesbian elders is due to several reasons, including limited research on sexual minorities of color, a lack of national surveys that ask about sexual orientation or gender identity, reluctance to coming out, and variation in responses to surveys due to methodological issues (Bostwick 2007; DeBlaere et al. 2010).

The purpose of this chapter is to present a trans-disciplinary examination of issues relevant to African American lesbian elders in rural society. Information is presented in several key areas. A profile of African American elders provides a contextual framework of this population in contemporary society. The theoretical concepts of feminism and womanism, and the construct of positive marginality are presented to explain identity development of African American lesbians. Characteristics of rural society and African American women and lesbians demonstrate how they exist in those contexts. Cultural homophobia in the black community and racism in the gay community describes discrimination from several societal perspectives. A discussion of psychosocial adjustment looks at the intersection of sexual orientation, age, and race statuses. Healthcare, education, and public policy issues are examined with respect to elder African American lesbians. In addition, a sample syllabus of a trans-disciplinary course is included. Finally, select community resources are provided.

Overview

Elder (age 65 and over) African American lesbians in rural society represent diversity across age, race, gender, sexual orientation, socioeconomic status, ability levels, education, family composition, geography, and experiences. Each of their identities dictates certain *positionalities* (social position) (Maher and Tetreault 1993) in society and some of their identities further relegate them to positions of marginality. For example, from a cultural perspective elder African Americans are revered and entitled to respect within the African American community, and a position of age carries with it a high level of cultural capital. On the other hand, African Americans continue to face discrimination in general, and within the African American community sexual minorities are shunned and homosexuality is viewed as a European phenomenon and not indigenous to African culture (Johnson 2001; Savage and Harley 2005). But, in fact, lesbians often face ridicule and gay bashing in the Black community. Based on their multiple subordinate-group identities, African American lesbians have "intersectional invisibility" as victims of sexism, heterosexism, and racism within both the dominant culture and Black community (Purdie-Vaugns and Eibach 2008, p. 377). Ramsey et al. (2010) summarize the experiences of African American lesbians as that of simultaneous resilience and disadvantage in the face of extreme obstacles.

Clearly, African American lesbians are confronted by challenges from multiple avenues. That is, they must address challenges as a double minority based on the dual effect of race and sexual orientation. With the addition of gender, African American lesbians are required to respond to triple jeopardy status. The layers of jeopardy increase when other variables such as age and disability enter the matrix (Battle and Bennett 2000; Harley et al. 2013).

For all adults, later life is known as a period of both growth and decline, with studies overwhelmingly focused on the latter (de Vries et al. 2011). We know very little about the health care needs of older sexual minorities, however as African American lesbians age, they must address health issues specific to aging, disability, health, and sexuality (Comerford et al. 2004; Dibble et al. 2012). The health and well-being of elder African American lesbians is one that has received limited attention. Although research has begun to focus on elder lesbians, elder African American lesbians continue to be excluded in much of mainstream research, which focus primarily on White lesbians or include African American lesbians more as a comparison group (Dibble et al. 2012). Elder African American lesbians in rural communities are further marginalized and excluded. In fact, over two decades ago Deevey (1990) identified older lesbians as the invisible minority and, today many researchers, scholars, and academicians continue to omit them from curricula content to the extent as if older [African American] lesbians do not exist (Hall and Fine 2005). LGBT elders have not been the subject of extensive research and, just as aging is not frequently considered in LGBT research, the field of gerontology has not frequently considered LGBT aging issues (Fredriksen-Goldsen and Muraco 2010; de Vries et al. 2011; Scherrer 2009). The continued limited focus on older LGBT populations is surprising particularly as LGBT elders are gaining visibility with the aging of LGBT baby boomers, the first generation of LGBT people to have lived openly gay in large numbers (Baker and Krehely 2012). Thus, one may ask the question, what happens to them as they age?

A Profile of the Black Experience in the United States

African Americans represent a diverse group of people. A variety of names have been used to identify them, including "Colored," "Negro," "Black," and a series of degrading labels (e.g., nigger, darkie). During the 1980s, the term African American became important in the Black community to reflect the cultural distinctiveness of Blacks in the United States, and to equalize this cultural distinctiveness to that of other ethnic groups (e.g., Italian Americans). The early existence of Blacks in the United States has traditionally been a population whose base was in the South. After the Civil War and as the hopes for Reconstruction faded, sharecropping and Jim Crow became common features for ex-slaves in the South (Gelfand 2003). With the passage of the 13th Amendment to the U.S. Constitution, slavery became illegal, and the 14th Amendment declared all persons born in the United States to be citizens. The status of African Americans did not improve greatly with the legal end to slavery because they continued to be targets of racialized terrorism and sanctioned discrimination. In efforts to escape the tyranny of the South and in search of better jobs, many Blacks relocated to northern cities (Billingsley 1992; Hill 2001; Martin and Martin 1978).

For African Americans, the Black identity as a lived experience is manifested as suggested by Boykin's (1986) Triple Quandary Theory (TQT) one in which they must learn to negotiate three worlds or types of experiences: (a) the world of marginalization, insult, discrimination, and racism, (b) the world of the mainstream, and (c) the world that constitutes the Black community itself. Boykin's TQT was further elaborated on by Strauss and Cross (2005) through five transactional modalities that operationalize identity as a lived experience in everyday life: (a) *buffering* (transacting racism and discrimination), (b) *code-switching* (moving in and out of the mainstream), (c) *bridging* (transacting friendships across social categories), (d) *bonding and attachment* (transacting relationships within the Black community), and (e) *individuality* (transacting one's sense of individuality). In this model, the five forms of transactions convey not five different identities within the same person, but a coherent identity that constructs a repertoire of expressions. In essence, the person is the same person from one situation to the next; however, the demand characteristics of the situation determine what form of identity expression or mix of expressions the person will apply (Balcazar et al. 2010).

The Black family, school, and church provided the base from which the civil rights movement of the 1950s and 1960s evolved (Calhoun-Brown 2000; Moore 1991). Those movements are important elements in the lives of the current cohort of Black elders because many of them are children of families who gained access to job opportunities in manufacturing, education, and skilled labor. However, due to job insecurity, lower income jobs, and discrimination, many of the current cohorts of Black elders enter their later years with inadequate income, poor health status, or lower levels of education (Jones 2009; Lekan 2009; U.S. Social Security Administration 2011). African American elders have higher rates of arthritis, hypertension, diabetes, and cancer (with the exception of breast cancer) and, lower survival rates for cancer and hypertension. Many African Americans' higher cancer mortality rates appear to be related to occupational hazards and poor living conditions. African American women ages 20–94 were asked to identify their most serious health problems. These women listed the Center for Disease Control's (CDC) top four causes of mortality as their top four most serious health threats (Sadler et al. 2005). In addition, African Americans have higher rates of HIV and AIDS, many of whom were infected at a younger age and are now living longer (Baker and Krehely 2012). For many older persons, the major problems in life are conditions that impair their functioning. In comparison to White elders, African Americans have severe limitations in daily tasks requiring assistance with housework, personal care, and preparing meals.

The circumstances for elder African American lesbians are difficult to discern and separate from the general Black population because of their covert existence in the Black community. Nevertheless, one can glean that Black lesbians, especially lesbian leaders, feel an ownership of Black historical movements because of their race and membership in the Black community. African American LGBT people, just as other African Americans continue to struggle for acceptance, equality, and fairness in the United States.

Feminism, Womanism, and Positive Marginality

There is a consensus in the literature that LGBT populations in the United States are oppressed; however, the intersection of sexual orientation, gender, race, and age compounds the marginalization of elder African American lesbians. "Older lesbians are not a uniform discrete community but rather, cohort groups who may share some common cultural experiences. Making meaning of their experiences and modes of self-definition and expression necessitates an understanding of their development within social, historical, cultural, and geographic contexts" (Comerford et al. 2004, p. 420). Several theories have been proposed to explain identity development for women (e.g., Downing and Roush 1985) and lesbian and gay persons (e.g., Cass 1979; McCarn and Fassinger 1996). These theories, however, have been criticized for having a singular focus on identity that does not consider the fact that many people are members of more than one subordinate group. The feminist identity development model is also seen as falling short of addressing the intersecting identities of racial/ethnic minority women (Moradi 2005). Feminist theory aims to understand the nature of gender inequality through an examination of women's social roles, experience, interest, and feminist politics. The basic concepts of feminist theory include (a) viewing all people regardless of their gender or culture within the context of society, (b) promoting egalitarian relationships in all settings, and (c) realizing that social and personal identities are intertwined (Jodry and Trotman 2008).

Two other concepts that lend themselves to providing insight into identity development of elder African American lesbians are womanist identity development model (Helms 1990) and positive marginality (Mayo 1982). Womanist identity development model draws from extant literature on gender identity, black feminist writers (e.g., bell hooks, Patricia Hill-Collins, Audre Lorde, Alice Walker), and conceptualizations of racial identity development, and is characterized by women moving from an externally based to an internally based definition of oneself as a woman (Moradi 2005). Womanist identity development consists of four stages or sets of attitudes (i.e., pre-encounter, encounter, immersion-emersion, internalization) toward adaptation of internal standards of womanhood. Womanist is more than the combination of blackness and feminist; it signifies a unity in the experience of black women where their race and gender are fused into a singular identity, thereby creating her unique space (Fryer 2008). One noteworthy area of womanist theory is its challenge to the Black church community's homophobia. Womanist theologians confirm a commitment to the survival and wholeness of the Black people, male and female, and heterosexual and homosexual. Womanist

Mandy Carter is one of the leading African American lesbian activists in the United States. She has a 42-year movement history of social, racial, and "*lesbigaytrans*" justice organizing since 1968. Ms. Carter helped found two ground breaking organizations: Southerners on New Ground (SONG) and the National Black Justice Coalition (NBJC). SONG integrates work against homophobia into freedom struggles in the South. She served as it Durham-based Executive Director from 2003-2005. NBJC was founded in 2003 and is the only current national civil rights organization of concerned black lesbian, gay, bisexual, and transgender individuals and allies dedicated to fostering equality by ending racism and homophobia. NBJC actively pursues ways to counter anti-gay organizing within African American communities.

Ms. Carter was named the American Civil Liberties Union-North Carolina's 2011 Porter Graham Award winner. In addition, she was nominated for the Nobel peace Prize as part of the "100 Women for the Nobel Peace Prize 2005". During the 2008 presidential elections, Ms. Carter was one of the five National Co-Chairs of Obama.

Ms. Carter is an "out", southern, African American.

Learn more about her at http://www.wakewaa.wordpress.com/mandy-carter/
Or http://www.bilerico.com/2008/02/black_history_lgbt_profiles_day_2_mandy.php

Fig. 1 Spotlight on an Advocate for African American Lesbians

theologian Cannon (1995) remarked, womanist scholars are compelled by their very womanist identity to interrogate homophobic attitudes and heterosexist systems and structures, as they exist within the Black church community, in an effort to debunk and dismantle them.

The concept of positive marginality suggests "people situated at social margins do not necessarily internalize their exclusion but instead embrace difference as a strength and sometimes as a source of critique and action" (Hall and Fine 2005, p. 177). People living on the margins of society utilize their status to understand the strategies of those in the mainstream and to use them to their own advantage (Lorde 1984). Positive marginality is viewed as a means of activism and self-advocacy to dissolve the barriers created by the margin. In fact, positive marginality is recognized as one of the cornerstone of the Black experience because it provides psychological and political tools that are used to teach survival skills and coping styles for multiple generations (Hall and Fine). Advocacy is an inherent part of positive marginality. A noteworthy advocate for the rights of African American lesbians is Mandy Carter (see Fig. 1 for a spotlight of this advocate).

Hall and Fine (2005) applied the concept of positive marginality through the use of narrative analysis to chronicle the lives of two older Black lesbians ages 73 and 85 years old. The results revealed how the two women created affirming lives on the margins and how they redirected their experiences with racism, sexism, and homophobia to make opportunities, create safe space, and live long satisfying lives. Shapiro et al. (2010) used grounded theory to gain insight into lesbian sexual identity development. They found through women's narratives a dynamic, non-linear, and contextualized representation of sexual identity development that may more accurately describe lesbian identity in terms of intersecting identities, national and local contexts, and personal and professional relationships. Saunders (2010) found similar issues of marginality among Cuba's LGBT community, with Black lesbians being the most socially marginalized population. Several intersecting factors were identified as contributing to such marginalization: (a) culturally based discourses concerning race, (b) gender, (c) sexuality, (d) limitations of post-revolutionary citizenship discourse to address these forms of social inequality, and (e) the economic downturn beginning in 2008 called the "Special Period".

Regardless of which individual or collective approach to identity development and social position one takes, it is clear that Black lesbians are marginalized in society. Older African American lesbians have had to develop multiple identities to live in a country that is premised on racism, sexism, ageism, homophobia and heterosexism. Each dimension of identity (e.g., race/ethnicity, age, class, sexual orientation) may alter Black women's sense of womanhood, self, and identity (Moradi 2005). African American lesbians are confronted by both an interracial and intra-racial struggle for acceptance (Moore 2010).

Characteristics of Rural Society and African American Women and Lesbians

In the United States, a rural area is defined by dimensions such as size and population density of the community, any geographic area outside cities and towns, and social values, norms, and sanctions of its residents (Comerford et al. 2004). Ten percent of the U.S. geography is urban and the remaining 90 % is rural. Nearly three out of ten Americans live in a rural area or a very small city (U.S. Census Bureau 2010). Rural America today has moved predominantly to a post-agrarian society, large intergenerational families are a thing of the past, many rural residents no longer work at local jobs, and many commute longer distances to work. In addition to those characteristics, the elders in rural areas are more likely than their urban counterparts to have lower incomes, live in poverty, have less formal education, be in poorer health, own their own homes, which are more likely to be substandard dwellings, have less access to health and social services and transportation, and their health and long-term care needs are less likely to be met (Fowles and Greenberg 2011).

For many African American elders, those issues are compounded by the intersection of factors such as racism and discrimination, generational biases, and historical referents (e.g., Jim Crow). A substantial number of elder African

Americans live in southern states, which have disproportionate and persistent poverty rates. The states with the largest percentage of Blacks in the state population are Mississippi (38 %), Louisiana (33 %), Georgia (32 %), South Carolina (29 %), and Alabama (27 %) (U.S. Census Bureau 2010). According to Brown and Swanson (2006), counties that are classified as "persistent poverty counties" are located across the heart of the old plantation South, referred to as the Black Belt, and contain approximately 80 % of the rural African American population and 45 % of all African Americans in the U.S.

The literature suggests that several details are known about women, elder African American women, and lesbians in rural communities. First, lesbians are shunned from rural society and experience heightened stigma because of the traditional values, fundamentalist religious beliefs, and strong conservative ideas found within these communities. A sense of cultural heterosexism is common among rural residents and within African American communities. Second, the anonymity often found in large cities is non-existent in the rural areas. The closeness and interdependence in rural areas and small towns often blur boundaries of privacy and the behavior of residents is easily known and information shared through local venues throughout the community. Thus, sexual minorities tend to remain "in the closet" (i.e., not reveal their identity) because of fears of harassment or violence. In a comparison of supports for LGBT aging in rural versus urban areas, Lee and Quam (2013) found rural individuals reported lower levels of "outness", guardedness with people including siblings and close friends, and lower levels of household income. Third, lesbians face potential isolation due to the lack of a visible gay community and less opportunities for social connection (Bostwick 2007; Hastings and Hoover-Thompson 2011).

Fourth, rural elder African American women are among the poorest of the poor and those with disabilities experience the most severe poverty (Brault 2012; Fowles and Greenberg 2011). Black women are one group for whom Social Security remains especially important (IWPR 2011). According to IWPR, of the 52.5 million Social Security beneficiaries of all ages in the United States 5.7 million are Black, of which 2.8 million are Black women. An explanation of the role of Social Security in the lives of Black women is presented in Table 1. For many African American women who are the oldest of the old (i.e., age 70 and over), poverty may affect them disproportionately because much of their experience spans from The Great Depression of the 1930, the 1940s, 1950s and 1960s when employment for Black women was primarily relegated to unskilled and semiskilled positions, often excluded from the social security tax system.

A fifth detail is that African American women often experience early onset of disease and disability and increased mortality because of health disparities. African American women experience stress and health disadvantages related to an interaction and multiplicative effects of race, gender, class, and age (Lekan 2009). The most frequently occurring health issues among African Americans include hypertension, diabetes, heart disease, stroke, cancer, asthma, and unintentional injuries (Centers for Disease Control 2013). Racial and ethnic disparities among elder African American women with disabilities include increased hospital days,

Table 1 Social security and black women

Black women aged 62 and above receive on average $961 per month in Social Security benefits as retired workers and $911 per month as disabled workers (U.S. Social Security Administration 2011)
Twenty-six percent of Black women who are Social Security beneficiaries receive disability benefits (U.S. Social Security Administration 2011)
Social Security is the most common source of income for older Black women (IWPR 2011)
Eighty-three percent of older Black women aged 65–74 receive Social Security income (IWPR 2011)
Eighty-eight percent of older Black women aged 75 and older receive Social Security income (IWPR 2011)
Social Security benefits are modest in size, but constitute the largest share of income received by older Black women (IWPR 2011)
Social Security is vital to Black women's economic stability. Three out of four Black women (74 %) aged 74 and older rely on Social Security for at least half of their income and more than half (51 %) relay on Social Security for all of their income (IWPR 2011)
The poverty rate of older Black women aged 75 and older with Social Security is 26 % and without Social Security is 61 % (IWPR 2011)

poor health days, hospitalizations, being functionally dependent, and having increased primary and repeat amputations when compared to their White counterparts (Jones 2009). Research suggests that older lesbians have fears of discrimination by health care providers (Baker and Krehely 2012; Butler and Hope 1999; Cochran and Mays 1988; Travis 2011; van Dam et al. 2001), especially in rural areas that are unlikely to provide knowledgeable, compassionate care to LGBT consumers (Butler and Hope 1999; Comerford et al. 2004; Lindhorst 1997; National Healthcare Disparities Report 2011).

In interviews with late middle-aged and old lesbians in a rural area, Butler and Hope (1999) found that their fears included panic when they think about ever having to go to a nursing home because of heterosexual assumptions of the whole institution; being punished for being a lesbian; being old and fearful of having to deal with extremely hostile prejudice of the radical political right; and denying themselves the healthcare they needed because of lack of affordability. In fact, older lesbians may feel that revealing their identity is not important if it means receiving inferior health care services (Hunter 2005). See Table 2 for a list of fears of discrimination. In a study of lesbians deciding who to see for health and mental health care, Saulnier (2002) found that lesbians encountered a continuum of provider reactions that shaped their decisions: (a) homophobia, (b) heterosexism, (c) tolerance, (d) lesbian sensitivity, and (e) lesbian affirmation.

Psychosocial and economic stressors are exacerbated in rural areas due to lower levels of education, limited opportunities, lack of availability of services, lack of transportation, and isolation in general and more so for elder African American women (Fowles and Greenberg 2011). Seventh, many people in rural society experience mental illnesses and do not seek care. For elder African American women, seeking counseling may be seen as a violation of the cultural norm of privacy and resolution of problems within the family, through the church, or

Table 2 Fears of discrimination from healthcare providers

Hostility
Rejection
Invisibility
Deny care
Reduced care
Stigmatization
Anti-gay violence/Safety
Inadequate/substandard health care
Refrain from touching a patient who is lesbian
Careless management of private information and identity disclosure
Inappropriate verbal and/or nonverbal responses from providers and office staff
Refusal of service providers and health care systems to recognize extended families within the gay community

Adapted from Saulnier (2002)

kinship and community networks (Harley 2005). Eighth, many lesbians have limited access to health care, are uninsured, underinsured, and lack financial resources for essential and preventive health care (Baker and Krehely 2012; Hunter 2005). Finally, lesbians who live in rural areas face the same issues (e.g., poverty, lower quality of housing, scarce health and mental health resources) that heterosexual residents face resulting from political and economic inequities between rural and urban areas that perpetuate inequities in the system (Comerford et al. 2004). In addition, elder African American lesbians must confront racism (Bostwick 2007). These multiple levels of oppression and the experience of being a minority within a minority may contribute to an increased vulnerability to health issues and mental illness, particularly depression and anxiety (Dibble et al. 2012; Greene 1994).

Elder lesbians, African American lesbians, and any combination thereof in rural areas may become more isolated from their families of origin than others. Because elder lesbians are more likely than their non-lesbian counterparts to be single, childless, and estranged from family members, they must often rely on friends and "informal families of choice" (Baker and Krehely 2012, p. 19). As they age, elder lesbians may lose their visibility within the lesbian community (having little- to-no voice in the political agenda of the LGBT community), which can exacerbate the endemic loneliness of women in rural communities. They may be excluded or marginalized by the younger LGBT community from community discussions and issues pertinent to them and absent from the mainstream LGBT political agenda. The experience of ageism in the LGBT community may compound psychological, victimization, and despair issues. Many experience the loss of a long-term partner and may have to hide the grieving process from others as to hide the relationship (Hunter 2005). Thus, as elder lesbians in rural communities become more isolated, they eventually transform into the "truly invisible" (Comerford et al. 2004, p. 420).

Cultural Homophobia and Societal Racism

The African American community is known as a collective society that provides support and refuge to its people. The cultural characteristics of the community consist of strong kinship bonds, valuing education, strong religious orientation, high achievement orientation, strong work ethic, self-reliance, and adaptability of family roles (Brown Wright and Fernander 2005; Hill 2001; Martin and Martin 1978). Although African American heterosexual women and lesbians experience racism and sexism in general, "lesbianism is largely considered incompatible with role expectations in the Black community" (Hunter 2005; Loiacano 1989, p. 21). In addition, several cultural values of the Black community are in direct conflict with values held by the larger LGBT community. First, the Black community values privacy, which is in contrast to the dominant LGBT community's value of "coming out". Second, strong family and kinship ties stress that marriage and family always come first, and the family may present a united front against the lesbian member or disown her, resulting in a loss of the sense of unity that helps the lesbian member form cultural and/or race identity (Savage and Harley 2005).

Third, as a historically oppressed group, African Americans have placed great importance on reproductive sexuality to ensure continue existence of the group in face of racist, genocidal practices by the dominant White group (Greene and Boyd-Franklin 1996). Thus, Black lesbians are seen as a threat to the social structure of the family (Boykin 1996). Finally, the Black church, with its heterogeneous character, is more monolithic in its attitude toward homosexuality. According to Douglas (2006) "the Black church community, even with all of their diversity, the Black church people are regarded as strikingly similar in their attitudes toward non-heterosexual sexualities. They are viewed as not simply homophobic but more homophobic than other populations of society" (p. 12). Older African American lesbians and gay men perceive social discrimination and alienation within their families, the larger Black community, organized religion, and feel isolated because of sexual orientation (Woody 2011). While these views are commonly held in the rural south, greater opportunities for social integration are present in the northeastern regions of the United States, which tend not to carry the same level of overt bias.

Although non-acceptance of sexual minorities remains in the Black community, in recent years ambivalence, tolerance, and acceptance have emerged in African American families. Acceptance is often in the form of family instructions such as "be silent and invisible," which may allow the family to accept the person without having to deal with sexual orientation and the issues associated with it (Hunter 2005). Many older African American LGBT individuals never revealed or discussed their sexuality with family member, even though the families knew they were gay. Black families function within their own code of "don't ask, don't tell". This code suggests that sexual minorities are accepted within African American families as long as they do not label themselves or acknowledge publicly that they engage in same-sex relationships (Miller 2011). Despite the sometime disapproving attitudes and religious condemnation, the majority of Black LGBT people remain in predominantly Black communities and social contexts and negotiate daily with family and community. Even when they do not find needed psychological and social support, they remain because they trust in racial solidarity and racial group membership (Moore 2010). For African American LGBT individuals, their same-race communities help them maintain a positive identity in the face of racism and oppression (Robinson-Wood 2009).

Elder African American lesbians often have experienced discrimination or prejudice on the basis of race, gender, sexual orientation, age, and geography. For these women, the intersection of *positionalities* is influenced by the dynamics of white and black patriarchy, and because of the degradation of black female heterosexuality in America, older African American lesbian sexuality is more devalued and more marginal in White and Black America (West 1993). In addition, the types and degree of discrimination may also be based on generational status. That is, elder African American lesbians in rural areas more than likely have been exposed to Jim Crow practices, whereas their younger counterparts have not. However, this is not to say that younger lesbians do not experience discrimination based on race. In a study of the correlates of well-being among African American lesbians ages 22–79, Dibble et al. (2012) found that substantial numbers of women reported that they had received poor health care related to their race and ethnicity, and not their sexual orientation because they were not out to their health care provider.

Although the rate of homophobia and heterosexism is high in the Black community, most African American LGBT indicate that they still find more support and refuge in the Black community, especially against the tyranny of racism in the White gay community (Boykin 1996; Greene 1994; Moore 2010; Savage and Harley 2005). For African American LGBT people "a perceived link that connects its members regardless of other differences that might also exists" (Moore 2010, p. 17). In fact, some researchers suggest that African American lesbians, having learned to handle their ethnic minority status, have developed a great deal of resilience and personal strength and may be better equipped to also handle their status as a sexual minority (Cooper-Lewter 2007; Dibble et al. 2012; Hall and Fine 2005).

Experiences of racism by ethnic minority LGBT people in the White LGBT community are well documented (Asanti 1999; Boykin 1996; Brown 2008; Loiacano 1989; Plummer 2007; Stansbury et al. 2010). According to Parham et al. (1999), "irrespective of how one comes to understand the concept of racism, there is little doubt that its origins, promotion, and continuation are anchored in the context of how Whites relate to African-descent people and other people of color on individual, institutional, and cultural levels" (p. 134). It has and remains necessary to dehumanize African ancestral people and to cast them as inferior beings to enforce white, heterosexual superiority (Gibson 2009). Gibson explored the behavioral and psychological strategies used by lesbians of African descent to negotiate relationships within their families of origin while simultaneously developing and maintaining an affirmative lesbian identity. The results showed that lesbians of African descent negotiated multiple identities of race, sexual orientation, disability, and gender through application of several identity management strategies (e.g., cultivate LGBT community and support systems, educate others about lesbian identity, maintain visibility, engage in LGBT activism), including ways to manage conflicting loyalties between the LGBT community and Black community without any loss of significant relationships and cultural ties.

In a similar study, Moore (2010) examined strategies Black LGBT people used in Black environments to proclaim a gay identity that is simultaneous with a Black identity. Moore found three distinct features: (a) Black gay protest takes on a particular form when individuals are also trying to maintain solidarity with the racial group despite the distancing that occurs as a result of their sexual minority status, (b) Black sexual minorities who see their self-interests linked to those of other Blacks use cultural references to connect their struggles to historical efforts for Black equality and draw from nationalist symbols and language to frame their political work, and (c) they believe that increasing their visibility in Black spaces will promote a greater understanding of gay sexuality as an identity status that can exist alongside, rather than in competition with race.

Psychosocial Adjustment to Intersection of Identities

The literature on the psychosocial adjustment of older LGBT persons is mixed. On the one hand, the literature suggests that lesbian, gay, bisexual, and transgender people in later life typically are well adjusted and mentally healthy (de Vries et al. 2011). When asked to identify strengths and challenges of being a sexual minority, older LGBT persons said that being LGBT has helped them prepare for aging in terms of personal/interpersonal strengths and in overcoming adversity, being more resilient, having greater self-reliance, and being more careful in legal and financial matters. They cited drawbacks of being older LGBT to include fewer opportunities to find a new relationship, fear of being doubly discriminated against as they age, feeling vulnerable with healthcare providers, and having fewer opportunities for social activities (MetLife Mature Market Institute 2010). On the other hand, the literature indicates that due to factors such as low rates of health insurance coverage, high rates of stress due to systematic harassment and discrimination, and lack of cultural competency in the healthcare system, LGBT people are at higher risk of cancer, mental illnesses, and other diseases, and are more likely to smoke, drink alcohol, use drugs, and engage in other risky behaviors.

LGBT racial/ethnic minorities often face the highest level of health disparities (Center for American Progress 2009). Evidence suggests higher levels of depression and psychological distress among midlife and older lesbians, which is attributed to the accumulated effects of a lifetime of stigmatization (Baker and Krehely 2012; de Vries 2008). Among older minority LGBT persons, minority stress (chronic stress related to stigmatization and actual experience of discrimination and violence) has been found to increase loneliness (Kuyper and Fokkema 2010). Thus, one can glean that older African American lesbians may experience multiple levels of minority stress based on race, gender, and sexual orientation, in addition to biases found in rural settings and in the Black community.

African American older lesbians in rural areas are complex. They live their lives from several introspections: (a) at the intersection of multiple identities experienced over the life span, (b) in a culture steeped in racism, sexism, ageism, heterosexism and homophobia, (c) at the margins and alienated within their own racial and ethnic culture, (d) in condemnation and skepticism, and (e) in dependence upon the institutions within society that discriminate against them (Hall and Fine 2005; Woody 2011). On the intersection of identities, Myer (2010) argues that lesbian, gay and bisexual (LGB) people of color do not have an inherent conflict between racial/ethnic and gay identities because they do not form a different culture; they are "very much surrounded by, contribute to, shape, and are affected by mainstream American culture" (p. 444). However, Myer addresses the issues of stress and resilience within the larger perspective of minority stress theory, and indicates that the study of stress and resilience among LGB people of color is relevant to core questions about social stress as a cause of mental disorders. Myer concludes that minority stressors require group-level resources because only the group can repair societal prejudicial norms, promote affirmative social support, counter internalized homophobia, and improve well-being.

As LGBT persons age they may need assistance in overcoming their own as well as others' negative attitudes about aging (Hunter 2005). Ageism against older persons is a unique form of oppression in a youth-oriented society. Both the LGBT community and society at large need to work on eliminating ageist stereotypes, language, and behavioral codes. In the final analysis, service providers should be aware that the effects of a lifetime of stigma, discrimination, and violence put LGBT older adults at greater risk for physical and mental illnesses. For African American lesbians those risks are complicated further by societal stressors related to racism.

Healthcare, Education, and Public Policy

Older adults in the general population and among LGBT groups tend to be the most frequent users of healthcare services in the United States. In the case of access and utilization, limited attention has been given to healthcare among elder LGBT persons. However, LGBT people of all ages are much more likely than heterosexual adults to delay or not seek medical care (de Vries et al. 2011). Lesbians experience many life circumstances differently as compared to their heterosexual counterparts, including seeking, accessing, and using human services (Hash and Netting 2009; Maccio and Doueck 2002). For older LGBT adults health concomitants of aging may be exacerbated by factors associated with gender and sexual orientation (de Vries et al. 2011). In 2001, the U.S. Surgeon General called for Americans to respect a "diversity of sexual values", to understand that sexual minorities are human beings, and to recognize that harassment has consequences on the mental health of gays and lesbians. The translation of the call to health and human service providers of older LGBT populations is not yet fully realized. In fact, the federal government does not collect data about the health of LGBT older adults or about the interactions between sexual orientation, gender identity and expression, and aging. The most widely referenced federal health survey, the *National Health Interview Survey*, excludes LGBT people (Sage and MAP 2010).

In a study of interactions between community-based aging service providers and LGBT older adults, Hughes et al. (2011) found very few services specific to the needs of older LGBT adults and very little outreach to this community. Knochel et al. (2012) surveyed leaders of Area Agencies on Aging, in which half of the existing agencies in the United Stated participated, to understand their services, training, and beliefs about serving LGBT older adults. The results showed that few agencies provided LGBT outreach services, one-third had trained staff around LGBT and four-fifths were willing to offer training. Moreover, those agencies that provided, or were willing to provide training, were more likely to be urban-based.

Sexual minorities encounter two unique obstacles in navigating healthcare, social, and human service: homophobia and heterosexism. Navigating those services can be further complicated by the degree to which lesbians self-disclose to others (Maccio and Doueck 2002). Providers along the healthcare continuum of caregivers ranging from doctors to pharmacists to hospitals and nursing home staff, may be hostile toward LGBT elders, untrained to work with them, unaware that older LGBT adults even exist, may lack knowledge about health disparities affecting LGBT people, and may lack skills of appropriate behavior dealing with closeted LGBT individuals (Public Advocate for the City of New York 2008). In addition, many LGBT older adults are not in positions to advocate on their own behalves, and they and their families, and facility staff are unaware of federal protections under the *Nursing Home Reform Act of 1987*.

The report on *Improving the Lives of LGBT Older Adults* (Sage and MAP 2010) identified the key challenges facing LGBT elders as social stigma and prejudice, reliance on informal families of choice, and unequal treatment under laws, programs, and services. Those three challenges impede LGBT elders' successful aging through reduced financial security, health/healthcare, and community support. LGBT elders as a group are poorer and less financially secure than American elders as a whole. The lifetime of discrimination faced by LGBT elders and the resulting effects on financial security is compounded by major laws and safety net programs (e.g., Social Security, Medicaid, tax-qualified retirement plans, retiree health insurance benefits, veterans' benefits, inheritance laws) that fail to protect and support them equally with their heterosexual peers (Baker and Krehely 2012). Typically, social support and community challenges result in social isolation at higher rates among LGBT older adults than in the general population of elders. LGBT elders are more likely to live alone, and more likely to feel unwelcome or be welcome in healthcare and community settings. The harmful effects of this type of social isolation include higher depression, poverty, re-hospitalization, delayed care-seeking, poor nutrition, and premature mortality.

Table 3 Affordable Care Act impact on elders

Reduce prescription drug costs in Medicare Part D
Provide a free annual wellness visit for all Medicare beneficiaries
Provide free Medicare coverage of vital preventive services
Encourage better care coordination
Expand coverage for seniors under age 65
Protect patient rights and lower costs in the private health insurance market
Provide new options for long-term care
Implement the Elder Justice Act

Adapted from Baker and Krehely (2012)

Healthcare reform is anticipated to improve access to healthcare and health outcomes for LGBT populations. "The *Affordable Care Act* is the most significant and far-reaching reform of America's health system since the creation of Medicare and Medicaid in the 1960s" (Baker and Krehely 2012, p. 21). The *Affordable Care Act* (ACA) introduces new protections and options for consumers in the private health insurance market, expands access to more comprehensive benefits and services, focuses on improving the nation's health, emphasizes lowering health care cost, and removes barriers for preexisting health conditions. For LGBT groups, the ACA is also key to efforts such as (a) expanding cultural competency in the healthcare workforce to include LGBT issues, (b) improving data collection to better identify and address health disparities, and (c) recognizing the increasing diversity of America's families. Especially for seniors, including LGBT elders, the ACA includes numerous other provisions (see Table 3) (Baker and Krehely 2012). The ACA is viewed as a historic opportunity to change the way people conceptualize how the health care industry provides care and promotes wellness, and is seen as the strongest foundation for closing LGBT disparities (Baker and Krehely 2012; National Coalition for LGBT Health 2012).

The delivery of healthcare services to elder African American lesbians should be an approach of understanding the "person-in-environment" or "person-in-situation" (Wheeler 2003). It is important that health care providers understand elder African American lesbians cannot be understood apart from multifaceted contexts (e.g., familiar, social, political, spiritual, economic, and physical) of their environment. With a person-in-environment framework, health care providers will be uniquely equipped to assess both the medical and psychological aspects in an elder and intervene in a culturally competent manner. This is especially critical given the complex psychosocial issues underlying unmet healthcare needs, life style-related medical conditions and treatment non-adherence challenges that pose significant barriers in elder African American lesbians receiving optimal healthcare. To facilitate an understanding of the needs of older African American lesbians, especially in rural areas, an integrated, transdisciplinary approach to training is needed (see Fig. 2 for a sample syllabus) to reduce the discrepancy between pedagogy of multicultural training and the execution of producing culturally competent professionals across the disciplines.

Trans-disciplinary Approach to Understanding African American Lesbians in Rural Settings

Course Description:
This course is designed to assist service providers in understanding and developing an inclusive atmosphere for racial/ethnic minority elderly LGBT populations.

Objectives: During and upon completion of this course, students should be able to:

1. Identify key concepts and theories related to aging and LGBT populations.
2. Understand the intersecting identities of race, ethnicity, and culture.
3. Address misconceptions about LGBT groups.
4. Understand the perspectives and values of a trans-disciplinary approach to working effectively across disciplines in elder service delivery to older LGBT populations.
5. Formulate a continuum of service delivery for elderly LGBT populations, including cultural factors and resource identification.

Session 1: *Course Overview*
During this session, the instructor will introduce terminology, constructs, and theories.

Session 2: *African American Elderly and African American History in the United States*
Explore backgrounds of black elderly life experience, sociopolitical experiences, …

Session 3: *Cultural and Ethnic Values and Beliefs*
The focus is on respect of cultural diversity, addressing the cultural values and beliefs of older African American adults and composition of families.

Session 4: *Identity Development*
Review race, gender, lesbian, and intersection of identities. Multiple oppressions such as heteronormativity, homophobia and heterosexism will be examined in the context of identity development.

Session 5: *Interviewing Skills, Engagement, and Assessment*
The focus of this session is to introduce students to culturally appropriate interviewing and counseling skills, assessment, and evidence-based practices.

Session 6: *Healthcare Status*
During this session the focus is to explore current health and mental health challenges, barriers to optimal health care, and identify gaps in the literature and areas for further research.

Session 7: *Laws and Public Policy*
This session will focus on history of gay rights in the United States and the impact of current federal and state laws, and their impact on the well-being of LGBT elders.
Rural African American Lesbian Elders

Session 8: *Service Continuum*
This session will examine the process for effective healthcare service delivery to elderly LGBT populations. Attention will be given to the unique needs of rural populations and racial/ethnic minority groups.

Fig. 2 Sample syllabus. Each session will have assigned readings to further enhance understanding of each of these areas

Conclusion

Older African American lesbians are a group with intersecting identities and who face discrimination from society at large and within the Black community. The intersection of race, age, culture, geography, and sexual orientation highlights the importance of examining these constructs as they overlap and interact. African American lesbians are confronted with both a practical and perceived dichotomy of allegiances based on race and sexual orientation. Neither the Black community nor the LGBT community is a uniform one. The needs of elder African American lesbians in rural areas have to be considered within the context of their lived experiences.

Education and training will need to take a proactive stance to include issues on LGBT populations and develop effective training models that address multicultural issues at a programmatic level. Beyond a broad social-political context, there is a point at which education, training, and service delivery of healthcare needs of elder African American lesbians should become "infused, self-sustaining, and self-perpetuating" (Horton-Ikard and Munoz 2010, p. 168). Eventually, the desire is that educators, counselors, health care providers, and policy makers will acknowledge that the African American, elder, rural, and LGBT communities are relevant. In some ways these identities are split and, in others they are intersecting.

References

Asanti, T. (1999). Racism in the 21st century: How will the lesbian and gay community respond. *Lesbian News, 24*(9), 24.

Baker, K., & Krehely, J. (2012). How heath care reform will help LGBT elders. *Public Policy & Aging Report, 21*(3), 19–23.

Balcazar, F. E., Suarez-Balcazar, Y., Taylor-Ritzler, T., & Keys, C. B. (2010). *Race, culture and disability: Rehabilitation science and practice*. Boston: Jones and Bartlett Publishers.

Battle, J., & Bennett, M. (2000). Research on lesbian and gay populations within the African American community: What have we learned? *African American Research Perspectives, 6*, 35–46.

Billingsley, A. (1992). *Climbing Jacob's ladder: The enduring legacy of African-American families*. New York: Simon & Schuster.

Bostwick, W. B. (2007). *Disparities in mental health treatment among GLBT populations*. Arlington, VA: National Alliance on Mental Illness.

Boykin, A. W. (1986). The triple quandary and the schooling of Afro-American children. In U. Neisser (Ed.), *The school achievement of minority children* (pp. 57–92). Hillsdale, NJ: Lawrence Erlbaum.

Boykin, K. (1996). *One more river to cross: Black and gay in America*. New York: Anchor Books.

Brault, M. W. (2012). Americans with disabilities: 2010. Household economic studies. Retrieved January 12, 2013, from www.census.gov/prod/2012pubs/p70-131pdf

Brown, C. E. (2008). Racism in the gay community and homophobia in the Black community: Negotiating the gay Black male experience. Dissertation, Virginia Polytechnic Institute and State University, Blacksburg, VA.

Brown, D. L., & Swanson, L. E. (2006, January). The changing face of rural America. *Issue Brief: Challenges for Rural America in the Twenty-first Century*. Rural Sociological Society.
Brown Wright, L., & Fernander, A. (2005). The African American family. In D. A. Harley & J. M. Dillard (Eds.), *Contemporary mental health issues among African Americans* (pp. 19–34). Alexandria, VA: American Counseling Association.
Butler, S. S., & Hope, B. (1999). Health and well-being for late middle-aged and old lesbians in a rural area. *Journal of Gay & Lesbian Social Services, 9*(4), 27–46.
Calhoun-Brown, A. (2000, June). Upon this rock: The Black church, nonviolence, and the Civil Rights movement, pp. 169–174. *Political Science & Politics*.
Cannon, K. G. (1995). *Katie's cannon: Womanism and the soul of the Black community*. New York: Continuum Publishing.
Carary, D. (2012, September). *Gallup study: 3.5 percent of U.S. adults are LGBT*. WTOP. Associated Press. Retrieved January 13, 2013 from www.wtop.com/?nid=893&sid=3083789
Cass, V. C. (1979). Homosexuality identity formation: A theoretical model. *Journal of Homosexuality, 4*(3), 219–235.
Center for American Progress. (2009, December). *How to close the LGBT health disparities gap: Disparities by race and ethnicity*. Washington, DC: Author.
Centers for Disease Control. (2013). Black or African American populations. Retrieved February 30, 2013 from http://www.cdc.gov/minority/health;populations/REMP/black.html
Cochran, S. D., & Mays, V. M. (1988). Disclosure of sexual preference to physician by Black lesbian and bisexual women. *The Western Journal of Medicine, 149*, 616–619.
Comerford, S. A., Henson-Stroud, M. M., Sionainn, C., & Wheeler, E. (2004). Crone songs: Voices of lesbian elders on aging in a rural environment. *Affilia, 19*(4), 418–436.
Cooper-Lewter, N. (2007). Conceptualizing soul as a mental health resource in the Black community. In S. M. L. Logan, R. W. Denby, & P. A. Gibbson (Eds.), *Mental health care in the African-American community* (pp. 213–231). New York: Haworth press.
de Vries, B. (2008). Lesbian, gay, bisexual, and transgender persons in later life. In D. Carr (Ed.), *Encyclopedia of the life course and human development* (pp. 161–165). Farmington Hills, MI: Gale Publishing.
de Vries, B., et al. (2011). *The health of lesbian, gay, bisexual, and transgender people: Building a foundation for better understanding*. Washington, DC: The National Academies Press.
DeBlaere, C., Brewster, M. E., Sarkees, A., & Moradi, B. (2010). Conducting research with LGBT people of color: Methodological challenges and strategies. *The Counseling Psychologist, 38*(3), 331–362.
Deevey, S. (1990). Older lesbian women: Invisible minority. *Journal of Gerontological Nursing, 16*, 35–39.
Dibble, S. L., Eliason, M. J., & Crawford, B. (2012). Correlated of wellbeing among African American lesbians. *Journal of Homosexuality, 59*, 820–838.
Douglas, K. B. (2006). Black church homophobia: What to do about it. Available at www.yale.edu/divinity/publications/reflections/spring06/12_17.pdff
Downing, N. E., & Roush, K. L. (1985). From passive acceptance to active commitment: A model of feminist identity development for women. *The Counseling Psychologist, 13*, 695–709.
Fowles, D. G., & Greenberg, S. (2011). A profile of older Americans: 2011. Washington, DC: Administration on Aging, U.S. Department of Health and Human Services
Fredriksen-Goldsen, K. I., & Muraco, A. (2010). Aging and sexual orientation: A 25-year review of the literature. *Research on Aging, 32*(3), 372–413.
Fryer, D. R. (2008). African-American queer studies. In L. R. Gordon & J. A. Gordon (Eds.), *A companion to African-American Studies* (pp. 305–329). Carlton, Victoria, Australia: Blackwell Publishing.
Gates, G. J. (2011, April). *How many people are lesbian, gay, bisexual, or transgender?* Retrieved January 12, 2013 from http://www.lgbttobacco.org/files/How-many-people-are-LGBT-Final.pdf

Gelfand, D. E. (2003). *Aging and ethnicity: Knowledge and services*. New York: Springer Publishing.

Gibson, D. (2009). *Negotiating multiple identities: The experiences of African-American lesbians*. Dissertation, Seaton Hall University, New York

Greene, B. (1994). Ethnic-minority lesbians and gay men: Mental health and treatment issues. *Journal of Consulting and Clinical Psychology, 62*, 243–251.

Greene, B., & Boyd-Franklin, N. (1996). African American lesbian couples: Ethnocultural considerations in psychotherapy. *Women & Therapy: A Feminist Quarterly, 19*(3), 49–60.

Hall, R. L., & Fine, M. (2005). The stories we tell: The lives and friendship of two older black lesbians. *Psychology of Women Quarterly, 29*, 177–187.

Harley, D. A. (2005). The Black church: A strength-based approach in mental health. In D. A. Harley & Dillard, J. M. (Eds.), *Contemporary mental health issues among African Americans* (pp. 191–203). Alexandria, VA: American Counseling Association.

Harley, D. A., Stansbury, K., & Nelson, M. (2013). Sisters gone missing: The lack of focus on African American lesbians in mental health counseling and research. In H. Jackson-Lowman (Ed.), *Afrikan American women: Living at the crossroads of race, gender, class, and culture* (pp. 287–304). San Diego, CA: Cognella Academic Publishing.

Hash, K. M., & Netting, F. E. (2009). It takes a community: Older lesbians meeting social and care needs. *Journal of Gay and Lesbian Social Services, 21*(4), 326–342.

Hastings, S. L., & Hoover-Thompson, A. (2011). Effective support for lesbians in rural communities: The role of psychotherapy. *Journal of Lesbian Studies, 15*(2), 197–204.

Helms, J. E. (1990). *Black and white racial identity: Theory, research, and practice*. New York: Greenwood Press.

Hill, R. (2001). *The strengths of African American families: Twenty-five years later*. Lanham, MD: Rowman and Littlefield.

Horton-Ikard, R., & Munoz, M. L. (2010). Addressing multicultural issues in communication sciences and disorders. *Contemporary Issues in Communication Science and Disorders, 37*, 167–173.

Hughes, A. K., Harold, R. D., & Boyer, J. M. (2011). Awareness of LGBT aging issues among aging services network providers. *Journal of Gerontological Social Work, 54*(7), 659–677.

Hunter, S. (2005). *Midlife and older LGBT adults: Knowledge and affirmative practice for the social services*. New York: Haworth Press.

Institute for Women's Policy Research. (2011, May). *Fact sheet: Social security and Black women* (IWPR #D496). Washington, DC: George Washington University.

Jodry, J., & Trotman, F. (2008). A call to the profession: Incorporating feminist competencies into professional counseling. Retrieved September 27, 2012 from http://counselingoutfitters.com/vistas/vistas08/Jodry.htm

Johnson, C. A. (2001). Hearing voices: Unearthing evidence of homosexuality in precolonial Africa. In D. Costantine-Simms (Ed.), *The greatest taboo: Homosexuality in Black communities* (pp. 132–148). Los Angles, CA: Alyson Books.

Jones, D. A. (2009). Disabilities in older African American women: Understanding the current state of the literature. *Journal of National Black Nurses Association, 20*(2), 55–64.

Knochel, K. A., Croghan, C. F., Moone, R. P., & Quam, J. K. (2012). Training, geography, and provisions of aging services to lesbian, gay, bisexual, and transgender older adults. *Journal of Gerontological Social Work, 55*(5), 426–443.

Kuyper, L., & Fokkema, T. (2010). Loneliness among older lesbian, gay, and bisexual adults: The role of minority stress. *Archives of Sex Behavior, 39*(5), 1171–1180.

Lee, M. G., & Quam, J. K. (2013). Comparing supports for LGBT aging in rural versus urban areas. *Journal of Gerontological Social Work, 56*(2), 112–126.

Lekan, D. (2009). Sojourner syndrome and health disparities in African American women. *Annuals of Advance Nurse Science, 32*(4), 307–321.

Lindhorst, T. (1997). Lesbians and gay men in the country: Practice implications for rural social workers. *Journal of Gay and Lesbian Social Services, 7*(3), 1–11.

Loiacano, D. K. (1989). Gay identity issues among Black Americans: Racism, homophobia, and the need for validation. *Journal of Counseling & Development, 68*, 21–25.
Lorde, A. (1984). *Sister outsider*. Trumansburg, NY: The Crossing Pressing.
Maccio, E. M., & Doueck, H. J. (2002). Meeting the needs of the gay and lesbian community: Outcomes in the human services. *Journal of Gay & Lesbian Social Services: Issues in Practice, Policy & Research, 14*(4), 55–73.
Maher, F. A., & Tetreault, M. K. (1993). Frame of positionality: Constructing meaningful dialogues about gender and race. *Anthropological Quarterly, 66*(3), 118–126.
Martin, E. P., & Martin, J. M. (1978). *The Black extended family*. Chicago, IL: University of Chicago Press.
Mayo, C. (1982). Training for positive marginality. In C. L. Bickman (Ed.), *Applied social psychology annual, 3* (pp. 57–73). Beverly Hills, CA: Sage.
McCarn, J. E., & Fassinger, R. E. (1996). Revisioning sexual minority identity formation: A new model of lesbian identity and its implications. *The Counseling Psychologist, 24*, 508–534.
MetLife Mature Market Institute. (2010, March). *Still out, still aging: The MetLife study of lesbian, gay, bisexual, and transgender baby boomers*. New York, NY: Metropolitan Life Insurance Company.
Miller, S. J. (2011). African-American lesbian identity management and identity development in the context of family and community. *Journal of Homosexuality, 58*(4), 547–563.
Moore, T. (1991). The African American church: A source of empowerment, mutual help, and social change. *Prevention in Human Services, 10*, 147–167.
Moore, M. R. (2010). Articulating a politics of (multiple) identities: LGBT sexuality and inclusion in Black community life. *Du Bois Review, 7*(2), 1–20.
Moradi, B. (2005). Advancing womanist identity development: Where we are and where we need to go. *The Counseling Psychologist, 33*(2), 225–253.
Myer, I. H. (2010). Identity, stress, and resilience in lesbians, gay men, and bisexuals of color. *The Counseling Psychologist, 38*(3), 442–454.
National Coalition for LGBT Health. (2012). *The Affordable Care Act*. Retrieved March 15, 2013 from http://lgbthealth.webolutionary.com/affordable-care-act
National Healthcare Disparities Report. (2011). Rockville, MD: Agency for Healthcare Research and Quality. Retrieved May 7, 2013 from http://www.ahrq.gov/research/findings/nhqrdr11/index.html
Parham, T. A., White, J. L., & Ajamu, A. (1999). *The psychology of Blacks: An African-centered perspective*. Upper Saddle River, NJ: Prentice Hall.
Plummer, M. D. (2007). *Sexual racism in gay communities: Negotiating the ethnosexual marketplace*. Thesis, University of Washington. Available at http://hdl.handle.net/1773/9181
Public Advocate for the City of New York. (2008). *Improving lesbian, gay, bisexual, and transgender access to health care at New York City health and hospital corporation facilities*. New York: Author.
Purdie-Vaugns, V., & Eibach, R. (2008). Intersectional invisibility: The distinctive advantages and disadvantages of multiple subordinate-group identities. *Sex Roles, 59*(5–6), 377.
Ramsey, F., Hill, M. J., & Kellam, C. (2010). *Black lesbians matter: An examination of the unique experiences, perspectives, and priorities of the Black lesbian community*. Sacramento, CA: Zuna Institute.
Robinson-Wood, T. L. (2009). *The convergence of race, ethnicity, and gender: Multiple identities in counseling*. Upper Saddle River, NJ: Merrill.
Sadler, G. R., Escobar, R. P., Ko, C. M., White, M., Lee, S., Neal, T., et al. (2005). African-American women's perceptions of their most serious health problems. *Journal of the National Medical Association, 97*(1), 31–40.
Sage & MAP. (2010, March). *Improving the Lives of LGBT Older Adults*. Available at www.lgbtmap.org and www.sageusa.org
Saulnier, C. F. (2002). Deciding who to see: Lesbians discuss their preferences in health and mental health care providers. *Social Work, 47*(4), 355–365.

Saunders, T. L. (2010). Black lesbians and racial identity in contemporary Cuba. *Black Women, Gender, and Families, 4*(1), 9–36.
Savage, T. A., & Harley, D. A. (2005). African American lesbian, gay, and bisexual persons. In D. A. Harley & J. M. Dillard (Eds.), *Contemporary mental health issues among African Americans* (pp. 91–105). Alexandria, VA: American Counseling Association.
Scherrer, K. S. (2009). Images of sexuality and aging in gerontological literature. *Sexuality Research and Social Policy: A Journal of the NSRC, 6*(4), 5–12.
Shapiro, D. N., Rios, D., & Stewart, A. J. (2010). Conceptualizing lesbian sexual identity development: Narrative accounts of socializing structures and individual decisions and actions. *Feminism & Psychology, 20*(4), 491–510.
Stansbury, K., Harley, D. A., Allen, S., Nelson, N. J., & Christensen, K. (2010). African American lesbians: A selective review of the literature. *African American Research Perspective, 11*(1), 99–108.
Strauss, L. C., & Cross, W. E, Jr. (2005). Transacting Black identity: A two-week daily-diary study. In G. Downey, J. S. Eccles, & C. M. Chatman (Eds.), *Navigating the future: Social identity, coping, and life tasks* (pp. 67–95). New York: Russell Sage.
Travis, L. A. (2011, May). *Elder care: A resource for interprofessional providers. What you should know about LGBT older adults*. Phoenix, AZ: Arizona Geriatric Education Center, University of Arizona.
U.S. Census Bureau. (2010). Retrieved January 13, 2013 from www.census.gov/geo.reterence/urban-rural-2010.html
U.S. Census Bureau. (2011). *Current survey, annual social and economic supplement, 2010*. Retrieved January 12, 2013 from http://www.census.gov/population/www/socdemo/race/pp1-bc10.html
U.S. Social Security Administration. (2011). *Annual Statistical Supplement to the Social Security Bulletin, 2010*. Retrieved January 15, 2013 from http://www.ssa.gov/policy/docs/statcomps/supplement/
U.S. Surgeon General. (2001, July 9). The Surgeon General's call to action to promote sexual health and responsible sexual behavior. Retrieved January 30, 2013 from http://www.surgeongeneral.gov
van Dam, M. A. A., Koh, A. S., & Dibble, S. L. (2001). Lesbian disclosure to health care providers and delay of care. *Journal of the Gay and lesbian Medical Association, 5*(1), 11–19.
West, C. (1993). *Race matters*. New York: Vintage Books.
Wheeler, D. P. (2003). Methodological issues in conducting community-based health and social services research among urban Black and African American LGBT populations. In W. Meezan & J. I. Martin (Eds.), *Research methods with gay, lesbian, bisexual, and transgender populations* (pp. 65–78). New York: Haworth Press.
Woody, I. (2011). *Lift every voice: a qualitative exploration of ageism and heterosexism as experienced by older African American lesbian women and gay men when addressing social services needs*. Dissertation, Capella University, Minneapolis, MN.

Websites

American Society on Aging: www.asaging.org
Funders for Lesbian-Gay Issues: www.lgbtfunders.org
Gay & Lesbian Medical Association (GLMA): http://www.glma.org
Gay Lesbian Bisexual and Transgender Health Access Project: www.glbthealth.org
LGBT Aging Issues Network (LAIN): www.asaging.org/lain
Movement Advancement Project (MAP): www.lgbtmap.org
National Black Justice Coalition: www.nbjc.org
National Coalition for LGBT Health: http://lgbthealth.webolutionary.com
National Resource Center on LGBT Aging: www.lgbtagingcenter.org

National Senior Citizen's Law Center (NSCLA): www.nsclc.org
Services & Advocacy for GLBT Elders (SAGE and SAGECAP) Caring and Planning: www.sageusa.org
Survivor Project: LGBT Elder Abuse and Neglect Issues: http://www.survivorproject.org/elderabuse.html
Zuna Institute: www.zunainstitute.org

Resources

Funders for LGBT Issues. (2005). Out for change: Racial and economic justice LGBT issues. Available at http://www.lgbtracialequity.org

Grant, J. M. (2010). *Outing age 2010: Public policy issues affecting lesbian, gay, bisexual, and transgender elders*. Washington, DC: The National Gay and Lesbian Task Force Policy Institute. Available at www.thetaskforce.org/downloads/reports/reports/outingage_final.pdf

National Resource Center on LGBT Aging. (2009). *2009 directory on LGBT People of color organization and projects in the U.S.* Available at http://www.lgbtagingcenter.org/resources/resources.cfm?r=58

National Resource Center on LGBT Aging. (2013). *Inclusive questions for older adults: A practical guide to collecting data on sexual orientation and gender identity*. New York: Author.

Next Step in Care. (2011). A Guide to LGBT Caregiving. United Hospital Fund. Available at www.nextstepincare.org

Services & Advocacy for Gay, Lesbian, Bisexual & Transgender Elders (2011). *A guide to LGBT caregiving*. Available at www.nextstepincare.org

Travis, L. A. (2011). *Elder care: A resource for interprofessional providers. What you should know about LGBT older adults*. Phoenix, AZ: Arizona Geriatric Education Center.

Intergenerational Dynamics Related to Aging and Eldercare in Asian American Families: Promoting Access to Services

Suzie S. Weng and Jacqueline Robinson

Abstract As the number of older Asian Americans continues to grow so do the number of issues they face related to aging. As older adults age, many may face issues related to aging such as daily care, physical health, mental health, and access to services. These issues are complicated by differences between Eastern and Western cultures within Asian American families and the different acculturation levels between the generations within the family. This chapter uses the intergenerational ambivalence paradigm (Luescher and Pillemer 1998) to guide the discussion of intergenerational dynamics related to aging and eldercare in Asian American families. The chapter begins with a brief description and comparison between Eastern and Western cultures and an exploration of the differences in cultural values that affect familial caregiving. Caregiving among Asian American families is then put into the contexts of structural and individual ambivalences. Caregiving options from an Asian American perspective are presented next. The chapter concludes with interventions and suggestions for community service providers.

As the number of older Asian Americans continues to grow so do the number of issues they face related to aging. As older adults age, many may face issues related to aging such as daily care, physical health, mental health, and access to services. In American families, these issues are complicated by differences between Eastern and Western cultures, and the different acculturation levels between the generations within the family. Culture shapes how individuals view their environment and make sense of the world, and thus influences what are deemed as acceptable caregiving options for aging parents. Donovan and colleagues (2011) state that "cultural factors are highly relevant to the caregiving experience, given that

S. S. Weng (✉)
University of North Florida, Jacksonville, USA
e-mail: s.weng@unf.edu

J. Robinson
Virginia Commonwealth University Partnership for People with Disabilities, Richmond, USA
e-mail: robinsonjb3@vcu.edu

H. F. O. Vakalahi et al. (eds.), *The Collective Spirit of Aging Across Cultures*, International Perspectives on Aging 9, DOI: 10.1007/978-94-017-8594-5_9,

they shape not only household structures and individual and group identity but norms, beliefs and traditions around illness/health, caregiving, dying, death and bereavement" (p. 339).

The cultural dissonance that can arise from the differences between Western and Eastern cultures can create friction in the family between the generations and may force Asian American families to no longer assume that parents will be cared for by their adult children. The situation is further complicated when economic issues contribute to caregiver burden, and these issues in particular may be difficult to resolve. The result is that the family is forced to explore different caregiving options for their aging parents, including options that may be stigmatized by Eastern cultures. As family members try to balance the need to care for older adults while simultaneously maneuvering the various caregiving options available in light of their culture, it may result in adverse mental health outcomes for all family members due to their different philosophies about elder care. The situation within Asian American families is further exacerbated by the lack of culturally appropriate resources and professionals to assist this population.

When compared to other race groups, the Asian American population had the highest percentage of growth between 2000 and 2010. Asian Americans currently make up about 5.6 % of the total U.S. population and are projected to grow representing 9.2 % by 2050 (U.S. Census Bureau 2010). In 2010, about 3.5 % of the U.S. older population was Asian American which is projected to make up about 7.8 % of the older population by 2050. Older Asian Americans are more likely to be foreign born (i.e., 70 % compared with 9 % for non-older Asian Americans) and more likely to speak a language other than English at home (i.e., 80 % compared with 12 % for non-Asian Americans) (Capezuti et al. 2007, p. 61). This means that the majority of older Asian Americans is first generation and may not be English proficient. Among the Asian American population, it is important to recognize that it encompasses 43 distinct subgroups who use 100 different languages (U.S. Department of Health and Human Services [USHHS] 2001). Further, each subgroup has its own socio-cultural background and history that may affect how they view caregiving and help-seeking (Braun et al. 1995).

When discussing caregiving and eldercare (used interchangeably in this chapter), it is important to define what it involves. In the literature, caregiving activities are often grouped into two categories: activities of daily living (ADL) and instrumental activities of daily living (IADL). ADL are "tasks that involve the care of one's own body to maintain health and well-being" (Capezuti et al. 2007, p. 8). These can involve bathing, eating, dressing, and getting out of the bed (Bonnel et al. 2011). IADL involve tasks related to the individuals' social environment. These activities include transportation, shopping, preparing food, and managing the household (Bonnel et al. 2011).

The intergenerational ambivalence paradigm (Luescher and Pillemer 1998) is used to guide the discussion of intergenerational dynamics related to aging and eldercare in Asian American families. The theory is useful to describe and explain the complexities of the parent–child relationships as it relates to caregiving. The theory delineates two types of ambivalence: (1) structural ambivalence based on an individual's place in a social structure and (2) individual ambivalence related to feelings experienced by individuals when dealing with structural ambivalence. Luescher and Pillemer (1998) define ambivalence as "…contradictions in relationships between parents and adult offspring that cannot be reconciled" (p. 416).

The chapter begins with a brief description and comparison between Eastern and Western cultures and how the differences in cultural values affect familial caregiving. Caregiving among Asian American families is then put into the contexts of structural and individual ambivalences. Caregiving options from an Asian American perspective will be addressed, concluding with a focus on interventions and recommendations for community service providers. The authors are keenly aware that the discussion on Asian Americans imposes an artificial homogeneity on an extremely heterogeneous population; therefore, these generalizations should be taken as a starting point that needs to be modified in accordance to the specific population and individual in order to avoid stereotyping.

Eastern and Western Influences

First generation older Asian Americans are more likely to have spent the majority of their lives living in their native countries. The possibility that they may lack English proficiency may also mean that they are more influenced by their Eastern cultures. At the same time, their children may have been raised in the U.S. and are more likely to be influenced by Western cultures. To better understand cultural values of Asian Americans, and how both Asian and American cultures may have conflicting values and influences, a comparison of Eastern and Western cultural values is provided.

Traditionally, Eastern cultures have promoted close family ties in which the values reflect familial collectivism, where the individual views him/herself as part of the whole family, rather than a separate entity. These values are influenced by Confucian ideals where "traditional beliefs center around the unity and survival of the family" with an emphasis on "interpersonal relationships and interdependence" (Braun and Browne 1998). This group orientation is historically reinforced by the practice of extended families living together or near each other (Braun et al. 1996). Additionally, individual behaviors are heavily influenced by the social pressure to conform to the family expectations (Triandis 1995). The collectivistic identity socializes family members to be concerned about the appearance of the family and to uphold the family name (Weng and Nguyen 2011). Therefore, personal mistakes are viewed as a negative reflection on the entire family which causes shame and loss of face (Braun and Browne 1998). Individuals are socialized to not disclose private family matters to outsiders. Problems such as mental illness and poverty are considered shameful and a loss of face for the family (Sue 1994). Traditional ways of coping with problems included endurance in living with the problem, ignoring the problem, and/or looking the other way as if the problem does not exist (Huang 1991).

In general, Western cultural values are centered on individualism in which members are seen as individuals who are not necessarily responsible for the overall well-being of the family unit. From a young age, children in Western cultures are raised to be self-sufficient and achieve personal goals (Lee 1997). Being independent from parents is "…encouraged and expected once children reach adulthood. They are expected to create their own lives and support themselves" (Weng and Nguyen 2011, p. 330). Western cultures emphasize self-worth and competitiveness (Kalavar and Van Willigen 2005). This emphasis fragments the individual from the family unit as a separate entity, not always working in harmony.

Differences in Caregiving

From an Eastern cultural perspective, older people are highly respected in the family and old age is seen as a blessing (Braun et al. 1996). Scholars believe the Confucian belief of filial piety, devotion to and respect for parents, may be a major influence on Asians not seeking help from outside the family (Braun and Browne 1998), which has translated into a cultural norm for many Asian societies. In fact, Singapore and China laws require adults to care for their parents (Lum 2012). This notion translates to caring for parents into old age and reciprocating the care from their parents (Asai and Kameoka 2005).

The value of individualism has a direct effect on caretaking for older adults in Western cultures. The cultural values of independence continue as the individual enters into the latter phases of their lives and older adults continue to strive to be independent for as long as they can (Lai 2007). It is often expected that once an individual retires, they will make plans to spend their late life in retirement homes that continue to value independence. Additionally, in Western cultures, older adults strive to live alone and remain active as long as they can, which creates less dependence on their children (Weng and Nguyen 2011). Asian American adult children may feel increased burden because of the ambiguity between Eastern values that expect them to care for their parents, and Western values emphasizing that it is not their responsibility.

Factors that Contribute to Caregiving Choices

Filial obligations may influence children's decision to care for their aging parents but factors in today's global society may make the obligation difficult to uphold. Factors that challenge the assumption that eldercare can be provided by the extended family is categorized in this section as structural and individual ambivalence according to Luescher and Pillemer (1998). Structural ambivalence explains the ambiguity of one's place in a social structure. Within the context of Asian American families and caregiving, these structures can include family dynamics, housing, and workforce participation issues. Likewise, individual ambivalence explains the ambiguity of one's feelings or sentiments when dealing with structural ambivalence which may include isolation, health status of older adults, mental health, and caregiver burden. These structural and individual ambivalences indicate the need to avoid assumptions about and expectations of familial caregiving due to filial obligations. In other words, simply having values about family-based care does not guarantee that the family will actually provide the care because of the various factors related to structural and individual ambivalences (Finch and Mason 1993).

Structural Ambivalence

Relationships within families are not completely static (Klein and White 1996), but consisting of harmony, conflict, and ambiguity. The marked differences between Eastern and Western cultural values can create ambiguity and conflict for Asian Americans who are trying to acculturate themselves in a culture that has very different values. Cordero and Kurz (2006) define acculturation as a social process by which immigrants keep their native culture while at the same time, adopt the values, beliefs, practices, attitudes, and norms of the new culture. Younger generations are found to acculturate faster than older generations (Osako and Liu 1986). Different levels of acculturation within the family can be a source of stress. With caregiving for example, the younger generation may adopt a more Western view of aging, with parents caring for themselves, while the aging parents may hold onto their Eastern view of expecting their children to uphold their filial obligations. Differences among generations in terms of language proficiency may also be a barrier to communication within the family and may also cause intergenerational ambiguity or conflict (Hwang and Wood 2009).

Housing has long-term implications for family dynamics (Clark et al. 2000). Burr and Mutchler (2012) found many differences in housing characteristics among older Asian American subgroups. Housing characteristics are an indication of well-being that has been consistently linked to better health, longevity, and life satisfaction (Rohe et al. 2001). Housing can also provide the physical structure in which social support may be provided. Residential density, the number of persons per room in the home, has recently increased in Asian American families due to immigration. However, there is evidence of decreased residential density with more time in the U.S. (Myers and Liu 2005). Residential density limits privacy and may contribute to physical and mental health problems (Evans et al. 2002). For immigrant families, it may take some time to acquire the necessary resources to attain living spaces adequate for all members and/or to purchase homes (Clark et al. 2000). Burr and Mutchler (2012) found older Asian Americans to report lower homeownership rates and more residential crowding compared to older European Americans. Homeownership may also provide status within the community and wealth accumulation that can be a potential source of income.

Household and older adults' personal finances allow for economic mobility and increased options for eldercare. When immigrants arrive in the host country later in life and do not engage in paid employment, they do not have the opportunity to accumulate savings and assets, nor are they entitled to social security or other retirement pensions. If older adults are employed, they may be underemployed due to their language skills or the undervaluation of their previous education and experience. These disadvantages result in greater dependence on the family or other entities for support (Patel 1993). Caregivers, for example, often assist with the expenses of care recipients, the most common being food, transportation, and medications (USHHS 2003).

Many adults in Asian American families may need to work, including women who may be less part of the labor market in some Asian countries. Having to care for aging parents may mean adult children taking time off work or going from full-time to part-time employment. These actions may have consequences for adult children's careers and incomes. Occupational migration within the U.S. may also mean adult children may no longer live near their parents thus removing the option of caring for them. If the decision is made to have the parents live with the adult children, it would require uprooting the parent from an environment in which they are familiar as well as their existing support system. If family members are scattered throughout the world, this may result in fewer family members being available to share caregiving responsibilities.

Individual Ambivalence

Older adults are particularly vulnerable to isolation due to structural ambivalence related to their role in the family, ageism, and community infrastructure. Individual ambivalence may also be a factor for isolation in reference to older adults' economic stability, mental health, and physical health. Many Asian Americans, especially those who settle in ethnic enclaves, areas where people of similar nationality live near each other, may be further isolated from mainstream society and its available services. Older adults may be left alone at home as other members of the family go to work or school, which can decrease feelings of self-worth and sense of belonging to community and society (Rumbaut 1995), as well as cause mental health problems and exacerbate physical health issues (Gardner et al. 2005).

The physical health of older adults may create circumstances in which it may be more difficult for adult children to care for their parents in the home. The adult children may not be adequately equipped or possess the knowledge to effectively perform caregiving tasks. As the aging Asian American population in the U.S. continues to grow, so does the percentage of individuals who report having a disability (which includes both physical and mental disabilities). Of Asian Americans ages 65 and older, 31.9 % reported having a disability. For individuals who are 75 and older, there is a marked increase in reports of disability, with 49.6 % of the population having one (Cornell University 2010).

When one considers socio-economic factors such as knowing the need for or cost of accommodations like hearing devices or walkers, this can further complicate issues because not having these accommodations can negatively affect one's quality of life. Older adults' social health may be impacted due to a decrease in health if they are not able to participate in social activities, thus leading to social exclusion and isolation. Emotional and mental health can be impacted and potentially lead to depression because of the stigmas surrounding disability. It is important to note that the impact of many health problems can be reduced by increasing environmental responsiveness that includes the availability of appropriate accommodations such as absence of stairs to prevent falls.

Mental illness is stigmatized in some Asian cultures and viewed as retribution for sins, imbalance of yin and yang, and possession by evil spirits (Elliot et al. 1996). Therefore help-seeking for stigmatized issues may be delayed and formal services may be used as a last resort. Older adults who develop physical problems may experience mental distress which has implications for older adults and their ability to make caregiving decisions.

Among the adult children, Hwang and Wood's (2009) study suggest cultural expectations of caring for aging parents may be an added source of stress among Asian Americans. Pitsenberger (2006) reports a third of caregivers experience emotional stress with the degree of stress linked to whether the individual chose the role. According to Lai (2007), a significant source of stress and guilt may be what the care entails and the inability to meet filial expectations (Lai 2007). Feeling trapped in a caregiving situation, experiencing shame, fearing criticism, or losing face were found to be significantly related to depression among caregivers (Martin et al. 2006). Among Korean Americans, research has shown that caregiving leads to poor physical health (Kim and Knight 2008), depression, and anxiety (Chun et al. 2007); whereas another study found that Chinese family caregivers experienced poor physical and psychological health (Ngan and Cheng 1992).

There are social and economic factors that have been found to be potential predictors of burden. Pinquart and Sorensen (2003) found the number of caregiving tasks and level of physical impairment of the older adult is associated with caregiver burden. In studies with Korean Americans, gender and spousal caregivers (Casado and Sacco 2012) and care recipients' functional dependency (Pinquart and Sorensen 2003) were found to be related to caregiving burden. Chan and Chui (2011) also established that being a female caregiver and having stronger family values were correlated with a higher level of caregiver burden. Likewise, Casado and Sacco (2012) concluded that having a large family support network alleviates caregiving burden because of greater availability of support.

Caregiving Options

The cultural dissonance between the generations as well as structural and individual ambivalence factors discussed thus far may force older adults and their children to no longer assume that parents will be cared for by their children, but to explore different caregiving options. The different types of caregiving options are organized into two categories: (1) family care in the home which includes family members providing care and reliance on community supports to provide care for aging parents at home, and (2) community-based care in which older adults are living outside of the home in residential type facilities and care is provided by non-family members. It is important to note that economic factors affect the type of caregiving option that is chosen, as some caregiving options are more expensive than others (Lum 2012).

Care in the Home

Due to the influence of Eastern culture and filial obligation of caring for one's family, Asian American adult children may choose to provide caregiving for their aging parents. This section highlights some of the caregiving options for older adults that are provided in the home, allowing older adults to spend the rest of their life in a familiar environment.

The first caregiving option is one in which the older adult lives in their home and their family cares for them. At times, when both aging parents are alive and at least one parent is in good health, one spouse may provide assistance for the other with the help of their adult children. Other times if one of the spouses passed away, the older adult may move into one of their adult children's homes. Lum (2012) found this caregiving option, living in the adult children's home, to be common among aging adults in Asia. Similarly, Asian Americans may choose to care for their aging parents due to filial expectations. This caregiving option is most at risk of inducing caregiver burden (Pinquart and Sorensen 2003).

A similar caregiving option is caregiving by neighbors or friends. Among Asian Americans, this is more likely to occur in ethnic enclaves. This is sometimes used in combination with family caregiving or as the sole caregiving if adult children have moved away from the aging parent. The importance of neighborhood support should not be underestimated because it can reduce the risk of social exclusion by increasing feelings of connection to others and environmental safety (Yuan and Ngai 2012).

One of the options for outside help while still allowing aging adults to remain in their homes is respite or short term care. A caregiver comes to the home and provides the older adult with services ranging from personal assistance to medical care. Respite care is used to relieve the familial caregivers from caregiving tasks that they normally perform for the elder. Respite care has been found to "promote the life quality of both caregiver and care receiver" and has been shown to "prevent the early institutionalization of older people" (Huang et al. 2009, p. 192). Despite evidence proving their effectiveness, this type of care is not frequently used by family caregivers (Huang et al. 2009). In cities with a large Asian American population, respite caregivers who are of similar ethnicity as the older adults are available which allow the older adult to be more comfortable in interacting with someone who understands his/her culture.

Another option that provides respite care while still allowing older Asian Americans to remain in their homes are senior care centers. They provide support for older adults outside of the home for part of the day thus providing relief for the family caretakers. The services range from speech and physical therapy to recreational activities allowing adults the opportunity to get out of the home and socialize with their peers (Capezuti et al. 2007). In cities where there is a large Asian American population, senior centers specifically for older Asian Americans are available. These centers are preferred by Asian Americans because they eliminate the language barrier and the food, activities, and other services are more culturally appropriate. Because the senior centers offer a variety of services, the use of some stigmatizing services such as counseling could be minimized and hidden because the older adult is already at the center for non-stigmatizing activities.

Home health agencies can also provide care by professionals like nurses and physical therapists. These services are frequently provided through Medicare with a focus on helping older adults remain in their homes with limited assistance (Capezuti et al. 2007). In general, physical health services are less stigmatizing for Asian Americans, particularly at the recommendation of physicians. Therefore, physical health services are likely welcomed in the care of aging adults in Asian American homes.

Lastly, when the aging parent requires extensive care and the family can no longer care for the older adult on their own, some Asian American families may choose to hire domestic caregivers to live in their home and provide caregiving tasks for the older adult. Lum (2012) found this caregiving option to be common among families in Asia who have the economic means. For Asian American families who have the financial ability and who want the aging parent to remain in the home, this may be an option.

Community-Based Care

The structural and individual ambivalences in Asian American families may influence decisions to seek care outside the home. Choosing not to be the caregiver or discontinuing the role of caregiver for some Asian Americans may be seen as going against family values and expected roles. Therefore, this decision will not be made lightly. This section highlights some of the more frequently used community-based care options that are available to aging adults that Asian American families may consider.

Some older adults who need minimal care can live relatively well on their own in an independent living community. These are housing communities where individuals who are 55 (the minimum age can vary) and older can live but they do not provide specific caregiving to the resident. These communities provide a variety of services for their members such as recreational activities, transportation, and homes that are designed for aging adults. This caregiving option may be good for Asian American families who are not residing in the same city and believe that uprooting the aging parent would result in a loss of their support network. The ideal independent living community would be close to the older adult's previous residence where they are already familiar with the environment. In addition, the independent living community would have friends who currently live there, or people who are similar in cultural background to the older adult.

Another similar option for older adults is adult foster care homes, which are residences where older adults live together and unrelated caregivers provide assistance for them. The limit on the number of adults who can live in one home varies by state regulations that range from three to five individuals (Capezuti et al. 2007). Adults who live in this type of setting will typically share common areas and have their own room.

For aging adults who need more involved caregiving services, there are a variety of options available. Assisted living facilities combine various aspects of the different caregiving options and allow the individual to be more independent and live in a community where help is close by. The type of care the individual receives depends on their individual needs, as well as the type of assisted living facility one resides in. Assisted living provides personal caregiving services, recreational and health services, and around the clock supervision. Individuals who live in assisted living cannot live alone, but do not need around the clock medical supervision. Assisted living communities provide more services and are more involved in the lives of older adults than independent living communities.

Similar to assisted living facilities, nursing homes provide care for older adults, yet these older adults are in need of around the clock monitoring. This caregiving option is for individuals who need more intensive and specialized care than they would receive in assisted living facility. This option should be considered by Asian American families if they are no longer able to care for their aging parents in the home. Because mental health is stigmatized in many Asian cultures, Asian American older adults with advanced stages of dementia may enter the nursing homes. Similar to utilization of mainstream services related to other areas, the families are likely to do whatever they could, using the nursing home as a last resort.

For the foster care home, assisted living, and nursing home options, they may be best for Asian American older adults if there are other Asian Americans in the facility with whom they feel comfortable. Another consideration that may be important for Asian American families is that the food, language, and activities provided in the facility are culturally appropriate. The caregivers and staff in the facilities need to be able to communicate with older Asian Americans. Activities older adults regularly participate in prior to entering the home must be incorporated. For Western activities, they should not require language or cultural fluency so that Asian American older adults can also participate. Finally, for older adults who practice a religion, the facility must allow for and accommodate for their specific religious practices.

When it comes time for Asian American families to make caregiving decisions as the older adult is nearing the end of life, they have a few options. Palliative care "strives to relieve suffering and support the best possible quality of life for patients with advanced chronic or life-threatening illnesses" (Bonnel et al. 2011, p. 3). Palliative care involves health care services that are designed to make the individual as comfortable as possible near the end of his/her life. Another option is hospice care, designed to provide services for terminally ill individuals. It is like palliative care in that it is focused on providing care for the individual when he/she is close to death. However hospice care normally occurs when the individual has less than 6 months of life left to live (Bonnel et al. 2011). The services that individuals can receive are designed to address the "social, emotional, and spiritual needs of terminally ill individuals and their families" (Capezuti et al. 2007, p. 28). For Asian American families, end of life care is an option if physicians strongly recommend it. In general, Asian Americans tend to respect authority and if physicians make a good case for this option, it will be a strong factor in the decision making process.

Interventions and Recommendations

Systems of care in the U.S. are set up with presumptions that the consumer is European American, literate, and fluent in English, resulting in exclusion of other groups "...from the array of welfare rights while providing a legitimation for this exclusion" (Ahmad and Walker 1997, p. 143). A study examined participation of Asian Americans in mainstream programs for aging adults and found that Asian Americans felt they were "uninformed about, ignored by, and at times unwelcome" (Bonnel et al. 2011, p. 61).

Underutilization of formal services is well-documented for the Asian American population. Adverse effects of caregiving may be greater among Asian American than European American caregivers due to the multiple barriers to utilizing formal services for Asian Americans (Han et al. 2008). The use of formal services may be culturally-specific, whereby those with stronger traditions towards family care may avoid external help (Lai 2007). This may be a result of stigma and shame for being unable to cope with familial responsibilities, belief that services will not be culturally appropriate, or due to other barriers such as language (Donovan and colleagues 2011). The low service use is compounded by health and human services entities not being prepared to serve individuals who do not speak fluent English, a category which many older Asian Americans fall under. The onus should not be on the client to provide translators because those translators are most likely family members, including children who may be too young to learn about the situation. Instead, service providers should work with the local Asian American community to develop a list of volunteer translators who can be on call (Weng 2013).

Helping professionals should make services more culturally appropriate for Asian Americans. Culturally appropriate services mean practitioners are knowledgeable about the cultural values and perceptions of the population and therefore adjust their understanding in accordance to what they learn and observe with the individuals they directly work with. Practitioners should also be empathic toward the clients' circumstances and thought processes. Practitioners must be sensitive in their behaviors and communication skills to facilitate help-seeking. Finally, appropriate language should be used regarding the problem, appropriate criteria for determining problem resolution, and indigenous helping resources that are within the group's definition of the program as well as ideas about appropriate interventions should be incorporated (Mokuau and Shimizu 1991).

Helping professionals should also partner with the Asian American community to provide more culturally appropriate services. One option is to recruit cultural brokers, someone who is bicultural in terms of the dominant culture and the culture of the client's native country in service planning and delivery (Egli 1987). The cultural broker's role would be to explain to service providers and clients how things work and where the other is coming from. In order to be an effective facilitator, the cultural broker must be trained to remain objective throughout the process and not take sides.

Cultural brokers can also help to build awareness and decrease intergenerational ambivalence among the Asian American community and educate them about the differences between Eastern and Western cultures. Weng and Nguyen (2011) recommend educating older adults about U.S. systems and values to help them better understand their adult children's perspective. At the same time, the adult children should learn where their elder parents are coming from.

Education should include awareness of the different caregiving options and work to decrease the stigma of utilizing those options for the benefit of all family members involved. In a study in Taiwan, Huang and colleagues (2009) found 65 % of caregivers did not know about respite care. Because parts of the Asian American elder population may not be literate in or speak English, educational delivery methods should be appropriate. Videos can be recorded by community members in the appropriate languages and shown at community gatherings. Group discussions and community forums can be held not only to educate but also to address concerns and questions. Literature should be translated to the languages of the target populations. Asian American community members should be involved from the beginning of the process in order to provide guidance.

In contrast to Western ideals, Asian cultures favor a family centered approach to care. With involvement of the family, Lai (2007) suggests members may filter information as a way of protecting individuals from bad news or losing hope. This conflicts with many service professionals' code of ethics in providing full disclosure, individual autonomy, and self-determination. Cultural brokers should be brought in for consultation if service professionals suspect family members are keeping information from others.

While considering the aging adults' best interests, it is also important to be sensitive to the needs of the entire family and caregiving system to reduce caregiving burden. Empirical evidence has shown that burden can be lessened with caregiver resources that include perceived health, relationship to care receiver, coping strategies, and social support (Kramer 1997). Because a large family support network has been found to alleviate caregiving burden (Casado and Sacco 2012), interventions can include approaches designed to build family support and decrease family conflict.

Finally, in the area of policy, Ube and Sue (1991) recommended three approaches to increasing service use for Asian Americans. The first is ethnic specific services in which parallel services are developed specifically for the target group. This method is the most expensive but for some groups, also most effective (Braun and Browne 1998). The second approach is to add bilingual workers to programs that are facilitated by university admission policies and financial aid programs that support ethnic minorities to pursue advanced degrees in health and social welfare. The third approach is for mainstream society to include cross-cultural awareness and sensitivity training as part of the education curriculum that is then carried on in health and social service agencies.

In sum, this chapter explored the dilemmas within Asian American families regarding elder caregiving options influenced by cultural, structural, and individual factors in hopes of increasing the cultural humility of helping professionals working with this population. Helping professionals are urged to engage the Asian American community in this endeavor of developing and delivering culturally relevant service options that would benefit older Asian Americans in the 21st century.

References

Ahmad, W. I. U., & Walker, R. (1997). Asian older people: Housing, health and access to services. *Ageing and Society, 17*(2), 141–165.

Asai, M. O., & Kameoka, V. A. (2005). The influence of sekentei on family caregiving and underutilization of social services among Japanese caregivers. *Social Work, 50*(2), 111–118.

Bonnel, W., Calvery-Carman, J., Davis, A., Dean, M., DeGroot, L., Ferris, M., et al. (2011). *Lippincott's nursing guide to expert elder care* (1st ed.). Philadelphia: Lippincott Williams & Wilkins.

Braun, K. L., & Browne, C. V. (1998). Perceptions of dementia, caregiving, and help seeking among Asian and Pacific Islander American. *Health and Social Work, 23*(4), 262–274.

Braun, K. L., Takamura, J. C., Forman, S., Sasaki, P., & Meininger, L. (1995). Developing and testing outreach materials on Alzheimer's disease for Asian and Pacific Islander Americans. *The Gerontologist, 35*, 122–126.

Braun, K. L., Takamura, J. C., & Mougeot, T. (1996). Perceptions of dementia, caregiving, and help-seeking among recent Vietnamese immigrants. *Journal of Cross-Cultural Gerontology, 11*, 213–228.

Burr, J. A., & Mutchler, J. E. (2012). Housing characteristics of older Asian Americans. *Journal of Cross Cultural Gerontology, 27*, 217–237.

Capezuti, E., Siegler, E., & Mezey, M. (2007). *Encyclopedia of elder care: The comprehensive resource on geriatric and social care* (2nd ed.). New York: Springer Publishing.

Casado, B., & Sacco, P. (2012). Correlates of caregiver burden among family caregivers of older Korean Americans. *The Journals of Gerontology, Series B: Psychological Sciences and Social Sciences, 67*(3), 331–336.

Chan, C., & Chui, E. (2011). Association between cultural factors and the caregiving burden for Chinese spousal caregivers of frail elderly in Hong Kong. *Aging & Mental Health, 15*(4), 500–509.

Chun, M., Knight, B. G., & Youn, G. (2007). Differences in stress and coping models of emotional distress among Korean, Korean-American and white-American caregivers. *Aging & Mental Health, 11*, 20–29.

Clark, W. A. V., Deurloo, M., & Dieleman, F. (2000). Housing consumption and residential crowding in U.S. housing markets. *Journal of Urban Affairs, 22*, 49–63.

Cornell University. (2010). *Disability statistics*. Retrieved from http://www.disabilitystatistics.org/reports/acs.cfm?statistic=1

Cordero, A., & Kurz, B. (2006). Acculturation and the mental health of Latina women in the women, infant, and children program. *Affilia, 21*(1), 46–58.

Donovan, R., Williams, A., Stajduhar, K., Brazil, K., & Marshall, B. (2011). The influence of culture on home-based family caregiving at end-of-life: A case study of Dutch reformed family care givers in Ontario, Canada. *Social Science & Medicine, 72*(3), 338–346.

Egli, E. (1987). *The role of bilingual workers without professional mental health training in mental health services for refugees*. Minneapolis: Department of Psychology, University of Minnesota.

Elliot, K. S., Di Minno, M., Lam, D., & Tu, A. M. (1996). Working with Chinese families in the context of dementia. In G. Yeo & D. Gallagher-Thompson (Eds.), *Ethnicity and the dementias* (pp. 89–108). Bristol: Taylor & Francis.

Evans, G., Kantrowitz, E., & Eshelman, P. (2002). Housing quality and psychological well-being among the elderly. *Journal of Gerontology: Psychological Sciences, 57B*, 381–383.

Finch, J., & Mason, J. (1993). *Negotiating family responsibilities*. London: Routledge.

Gardner, W., Pickett, C., Jefferis, V., & Knowles, M. (2005). On the outside looking in: Loneliness and social monitoring. *Personality and Social Psychology Bulletin, 31*(11), 1549–1560.

Han, H., Choi, Y. J., Kim, M. T., Lee, J. E., & Kim, K. B. (2008). Experiences and challenges of informal caregiving for Korean immigrants. *Journal of Advanced Nursing, 63*, 517–526.

Huang, K. (1991). Chinese Americans. In N. Mokuau (Ed.), *Handbook of social services for Asian and Pacific Islanders* (pp. 79–96). New York: Greenwood Press.

Huang, H., Shyu, Y., Chang, M., Weng, L., & Lee, I. (2009). Willingness to use respite care among family caregivers in Northern Taiwan. *Journal of Clinical Nursing, 18*, 191–198.

Hwang, W. C., & Wood, J. J. (2009). Acculturative family distancing: Links with self-reported symptomatology among Asian Americans and Latinos. *Child Psychiatry Human Development, 40*, 123–138.

Kalavar, J., & Van Willigen, J. (2005). Older Asian Indians resettled in America: Narratives about households, culture and generation. *Journal of Cross Cultural Gerontology, 20*(3), 213–230.

Kim, J., & Knight, B. G. (2008). Effects of caregiver status, coping styles, and social support on the physical health of Korean American caregivers. *The Gerontologist, 48*(3), 287–299.

Klein, D. M., & White, J. M. (1996). *Family theories*. Thousand Oaks: Sage.

Kramer, B. (1997). Differential predictors of strain and gain among husbands caring for wives with dementia. *Gerontologist, 37*, 239–249.

Lai, D. W. L. (2007). Cultural predictors of caregiving burden of Chinese-Canadian family caregivers. *Canadian Journal on Aging/La Revue Canadienne DuVieillissement, 26*(1), 133–148.

Lee, E. (1997). Overview: The assessment and treatment of Asian American families. In E. Lee (Ed.), *Working with Asian Americans: A guide for clinicians* (pp. 3–36). New York: Guilford Press.

Luescher, K., & Pillemer, K. (1998). Intergenerational ambivalence: A new approach to study of parent-child relations in later life. *Journal of Marriage & the Family, 60*(2), 413–425.

Lum, T. (2012). Long-term care in Asia. *Journal of Gerontological Social Work, 55*(7), 563–569.

Martin, Y., Gilbert, P., McEwan, K., & Irons, C. (2006). The relation of entrapment, shame and guilt to depressions, in carers of people with dementia. *Aging and Mental Health, 10*, 101–106.

Mokuau, N., & Shimizu, D. (1991). Conceptual framework for social services for Asian and Pacific Islander Americans. In N. Mokuau (Ed.), *Handbook of social services for Asian and Pacific Islanders* (pp. 21–36). New York: Greenwood Press.

Myers, D., & Liu, C. (2005). The emerging dominance of immigrants in the US housing market 1970–2000. *Urban Policy and Research, 23*, 347–365.
Ngan, R., & Cheng, I. C. K. (1992). The caring dilemma: Stress and needs of caregivers for the Chinese frail elderly. *Hong Kong Journal of Gerontology, 6*, 34–41.
Osako, M. S., & Liu, W. T. (1986). Intergenerational relations and the aged among Japanese Americans. *Research on Aging, 8*, 128–155.
Patel, N. (1993). Health margins: Black elders' care models, policies and prospects. In W. Ahmad (Ed.), *'Race' and Health in Contemporary Britain*. Buckingham: Open University Press.
Pinquart, M., & Sorensen, S. (2003). Associations of stressors and uplifts of caregiving with caregiving burden and depressive mood: A meta-analysis. *Journal of Gerontology: Psychological Sciences, 52B*, P112–P118.
Pitsenberger, D. J. (2006). Juggling work and elder caregiving: Work-life balance for aging American workers. *Continuing Education, 54*(4), 181–185.
Rohe, W., Van Zandt, S., & McCarthy, G. (2001). *The social benefits and costs of homeownership: A critical assessment of the research*. Boston: Joint Center for Housing Studies, Harvard University.
Rumbaut, R. G. (1995). Vietnamese, Laotian, and Cambodian Americans. In P. G. Min (Ed.), *Op. cit* (pp. 232–270). Thousand Oaks: Sage.
Sue, S. (1994). Mental health. In N. W. S. Zane, D. T. Takeuchi, & K. N. J. Young (Eds.), *Confronting critical health issues of Asian and Pacific Islander Americans* (pp. 266–288). Thousand Oaks: Sage Publications.
Triandis, H. C. (1995). *Individualism and collectivism*. Boulder: Westview Press.
Ube, L., & Sue, S. (1991). Nature and scope of services for Asian and Pacific Islander Americans. In N. Mokuau (Ed.), *Handbook of social services for Asian and Pacific Islanders* (pp. 3–19). New York: Greenwood Press.
U.S. Census Bureau. (2010). *American FactFinder*. Retrieved January 6, 2012, from http://factfinder2.census.gov/faces/tableservices/jsf/pages/productview.xhtml?src=bkmk
U.S. Department of Health and Human Services. (2001). *Mental health: Culture, race, and ethnicity—A supplement to mental health: A report of the surgeon general*. Retrieved July 2, 2009, from http://download.ncadi.samhsa.gov/ken/pdf/SMA-01-3613/sma-01-3613A.pdf
U.S. Department of Health and Human Services, Administration on Aging. (2003). *What we do makes a difference (Annual Report)*. Washington, DC: U.S. Government Printing Office.
Weng, S. S. (2013). *Dimensions of informal support network development in an Asian American community in the new south: A grounded theory*. Saarbrücken, Germany: Scholar's Press.
Weng, S. S., & Nguyen, P. V. (2011). Factors affecting elder caregiving in multigenerational Asian American families. *Families in Society: The Journal of Contemporary Human Services, 92*(3), 329–335.
Yuan, R., & Ngai, S. (2012). Social exclusion and neighborhood support: A case study of empty-nest elderly in urban Shanghai. *Journal of Gerontological Social Work, 55*(7), 587–608.

Seeing the Social in Technology for Older Adults: Making the Implicit Explicit Through a Multidisciplinary Lens

Clara Berridge

Abstract Technology is profoundly human, shaping and shaped by our social practices. It causes change, engages diverse stakeholders, raises ethical concerns, and embodies values through its design and use. In one form or another, technology is ubiquitous in the lives of elders. Technologies to support connectedness, safety and well-being are fascinating topics of study not only for what they can do and promise to do for older adults, but for what they and their associated practices could tell us about social perceptions and expectations of old age. The aim of this chapter is to feature the diversity of lenses applied to the study of how technology practices intersect with age, as well as implications for future research and policy making processes. Intersections of age and ability, gender, socioeconomic status, and the understudied intersections of technology, age and culture are areas that can be viewed through a single disciplinary lens, but truly illuminated through a multidisciplinary approach.

Older adults are a widely heterogeneous population with a variety of opinions, perceptions, preferences, experiences with and levels of access to technology. Technology is also a heterogeneous category with diverse purposes, embedded values, risks and benefits. On the topic of aging and technology, people often think of one particular type of technology, for example, the Internet, computer, or even the broad category of patient-centered technology. However, forms of technology are involved in each topic in this book: aging and health, spirituality/religiosity, intimacy, care, and abuse. Technology, in one form or another, is ubiquitous in the lives of elders.

Interest in the study of technology and old age and aging is growing in a range of fields and disciplines. Centers dedicated to this work in the U.S. include the Center for Research and Education on Aging and Technology Enhancement, the LeadingAge Center for Aging Services Technology, and the Center for

C. Berridge (✉)
University of California-Berkeley, Berkeley, CA, USA
e-mail: clarab@berkeley.edu

H. F. O. Vakalahi et al. (eds.), *The Collective Spirit of Aging Across Cultures*, International Perspectives on Aging 9, DOI: 10.1007/978-94-017-8594-5_10,

Technology and Aging that promotes deployment of patient-centered technologies. A North American chapter of the International Society for Gerontechnology was launched in 2013. The international society "encourages and promotes technological innovations in products and services that address older peoples' ambitions and needs on the basis of scientific knowledge about ageing processes including cultural and individual differences" (ISG website 2013). Gerontechnology is an interdisciplinary field of research that seeks to apply insights from aging processes to technological innovation in order to meet the needs and aspirations of older adults (Herman Bouma Fund for Gerontechnology 2013).

The aim of this chapter is to feature the diversity of lenses applied to the study of how technology practices intersect with age. It does not cover all applications given the limited space in the chapter; however, it focuses on technologies that are both prevalent in the literature and likely to receive considerable treatment in the foreseeable future. This selection typifies the ways in which well-being, connectedness, and safety can be enhanced by technologies, while others may raise concerns about surveillance, control and responsibility for care. The overarching theme is that technology is profoundly human and should not be mistaken for a value-neutral object that is somehow applied to, yet autonomous from human practices like care, communication, or aging (Lehoux 2006; Neven 2011). In the case of technology products for older adults, technologies are designed by people for other people to use in specific ways for specific purposes. In this way, values and expectations are embedded in the products themselves. Technologies both shape and are shaped by our social practices. They are fascinating topics of study not only for what they can do and promise to do for older adults, but for what they and their associated practices could tell us about social perceptions and expectations of old age.

To begin, two distinct ways of framing older adults' relationships to technology are presented along with trends in gadget ownership and internet use. This is a useful entrée into technology and aging because hardware like smart phones and computers are increasingly serving as the foundation to other technology-based services that previously relied on purpose-built devices (CTA 2013b). Examples are provided of existing programs that promote education and Internet access for older adults. The section that follows narrows the focus on technologies to support aging in place and independent living. Telehealth and passive monitoring are presented along with single and multidisciplinary approaches to their evaluation.

Generational Incompetence and the Technogenarian

One concept that is often referenced, though not always explicitly, about reasons why the elder population is considered to be behind the technological curve is generational incompetence. This idea is based on a nearsighted perception of technology development and adoption over the lifespan, as well as misperceptions about older adults' abilities and willingness to learn how to use new technologies if they perceive that incorporating them into their lives would be of benefit. The incompetence storyline is hard to disentangle from ageist notions about competencies in general; it often overlooks the importance of agented choices, the weighing of perceived costs of new technology adoption, such as time and investment to learn, and can lead to the dismissal of older adults as "paranoid" or "stuck in their ways" if they reject a particular technology. The generational incompetence framing is counterproductive if the goal is to achieve greater value from technologies and uptake among older adults.

An alternative framing is one that has been explored through ethnography by a sociologist who writes about "technogenarians." Loe's (2010) research describes the ways in which women in their 90s (nonagenarians) employ technologies ranging from phone trees and timers to electronic scooters, cell phones and computers to maintain their health, connectedness and well-being. Loe shows that they are technogenarians by examining technology use inside and outside the home, concluding that "the line between health and non-health technologies, as well as low and high tech, are blurred when elders are put at the centre of analysis" (Loe 2010, p. 320). Cast as resourceful and technologically literate, these older women chose and creatively used technologies in their daily lives to meet their individual needs.

The Digital Divide

The digital divide is typically described using survey data about internet access and use (Jaeger 2012). It should be noted that inclusion in the digital world is not a static condition; technology can change rapidly, requiring new skills and learning to remain digitally included (Jaeger 2012). In the context of what has been called the Information Society (and also contested, see Webster 2002), the Internet and information communication technologies (ICT) increasingly enable remote communication and connectedness to information, people, communities, commerce and recreation (CTA 2013b). Marginality in these practices of information exchange minimizes opportunities for older adults to participate in shaping them. Applying a constructivist approach, Jaeger (2004) explains that when older people use technologies, possibly in ways that may differ from others' use in the form of information searches, interests or needs, they are positioned to become "a part of the construction of the technology" because these new interpretations can motivate new products and uses of ICT (p. 11). People who have lived most of their lives with a disability and have learned to use assistive technology such as voice recognition software and magnification products and are now experiencing old age often have valuable ideas about how their shifting needs could be better met through innovation and modification. For example, some members of the Deaf Community use assistive listening devices like auto-captioning, video conferencing, and light or vibration to signal the ringing of a telephone (Ladner 2010; Baker and Bass 2003). These and the next generation of these devices will be critical tools for people aging with a disability as well as those experiencing age-related sensory loss.

In 2012, 69 % of Pew Internet and American Life Project survey participants ages 65 and older compared with 89 % of adults 18–64 reported owning a cell phone (Zickuhr and Madden 2012), an increase in the older population of more than 10 % since 2010. Cell phone ownership among people over 75 has grown to 56 % (Zickuhr and Smith 2012). Internet use is also growing with internet or email users representing more than half (53 %) of people 65 and older compared with 82 % of those 18–64. One-third (34 %) of these older adult internet users are on Facebook and other social networking sites (Zickuhr and Madden 2012). On-line social networking may be gaining appeal among older adults because it can help them reconnect with the past, participate in communities that provide support with chronic conditions, and bridge generational gaps (Madden 2010). Access and ability to use the internet can open channels of communication for older adults, whether the purpose is to maintain existing social ties or build new ones, gather information about health or social services, or for entertainment or educational attainment.

Age is only one of several demographic factors correlated with access to these online resources. The Pew survey results show that adults with disabilities are much less likely to use the Internet than those without disabilities (54 vs. 81 %), as are those with household incomes less than $30,000 (62 %) compared to those with household incomes above $75,000 (Zickuhr and Smith 2012). Those without a high school education are less likely to use the internet than high school graduates (43 vs. 71 %). In the Pew study, Internet use was slightly less prevalent among African American (71 %) and Latino (68 %) respondents compared with White (80 %) respondents (Zickuhr and Smith 2012).

The presence of high-speed broadband connection greatly influences internet adoption, making it a significant issue that nearly twice as many younger adults than older adults have broadband at home (60 vs. 30 %) (Zickuhr and Smith 2012). Disparities in home broadband rates by educational attainment and income are greater than those in Internet use rates. Differences in access by race disappear when other demographic factors are controlled for (66 % of whites, 49 % of African Americans, and 51 % of Latinos have it) (Zickuhr and Smith 2012), but these data have not yet been explored for the older population.

Internet use and broadband access is lower among those 75 and older. As of April 2012, 34 % used the internet and only 21 % had home broadband (Zickuhr and Madden 2012). The primary reason given in 2010 by nonusers of the Internet or email in this age group was that they're "just not interested" (38 %). In response to being asked if they felt they knew enough about computers and technology to start the using internet and email on their own, the majority (68 %) said they were not confident enough and would need assistance getting online (Zickuhr and Madden 2012).

If older adults did not gain experience with new technologies at work or in school, it may be that these differences in Internet use and mobile phone ownership are a result of cohort effects. If, on the other hand, technologies are designed in such a way that people who are experiencing changes in sensory or motor functioning will have difficulty using them, age differences in technology uptake may persist. Prior experience with technology promotes technology use (Charness et al. 2001). Furthermore, designers and developers of electronic services have not considered older adults to be potential users of technology and have generally failed to obtain and integrate knowledge about age-related changes or interests into their products (Czaja and Lee 2007; Jaeger 2004).

It should come as no surprise that barriers to usability have resulted from this lack of attention on older potential users. Barriers identified through research with older adults include but are not limited to screen design, input device design, complex commands and operating procedures, as well as inadequate training and instructional support (Czaja and Lee 2007). The field of human factors seeks to improve design and usability by applying an understanding of perceptual, cognitive, and motor functioning limitations (Fisk et al. 2009). Research laboratories like the Human Factors and Aging Laboratory of the Georgia Institute of Technology produce knowledge about everything from factors that influence use of ehealth websites and ways older adults respond to home maintenance challenges, to video game use to improve cognition. These research projects are invested in pushing methodological rigor beyond fixed-choice survey questions about technology use in order to gain a deeper understanding of variability: the how, how often, why and why not of technology use. These questions imagine the digital divide as a continuum rather than a binary model (Jaeger 2012), making them all the more likely to inform better design and outreach programs (Olson et al. 2011).

Attempts to close the digital divide include programs to enhance access and provide targeted education for older adults, such as offering classes with peers who want to learn how to use technologies at a speed that does not assume lifelong or workplace computer experience. Older Adults Technology Services (OATS) has been partnering with non-profit organizations since 2004 to provide technology training to older adults. They have provided more than 12,000 classes to older adults in New York City using a 10-week course module at no cost to participants. An evaluation of the program found that participants increased contact with friends and family, went online to research health information, and regularly used the computer 6 months after exiting the program (Gardner et al. 2012). In collaboration with the City of New York's Department of Information Technology and Telecommunications and U.S. Department of Commerce, OATS has developed 23 public computer centers throughout New York City for older adults to use for free with 300 state-of-the-art PCs (OATS website 2013). It has also opened an innovative technology-themed center for older adults called the Senior Planet Exploration Center, equipped with computers, digital cameras and other devices.

The Virtual Senior Center (VSC) is designed for older adults who may be unable to attend activities and classes in person. It was developed in 2010 by an independent living organization, Selfhelp Community Services, in partnership with Microsoft and the New York City Department for the Aging with funding from UJA-Federation of New York. Its focus is remote access to interactive general interest and chronic disease management classes for low-income older adults who face obstacles to leaving their homes. Each participant has a 22-inch touch-screen monitor equipped with an easy-to-use interface and webcam so they can join live classes and discussions about everything from chronic disease and pain management to singing groups. For more on the VSC, see the spotlight below on Dr. Raik.

The Virtual Senior Center is one program developed to reduce social isolation and enhance the experience of living at home for low-income elders. It is also one of many interventions that depend on Internet access, a reality that has led its developers and people who study technogenarians to call on policy makers to consider issues of cost and availability of broadband in initiatives for older adults (FCC 2013; Loe 2010; Dring 2013 "personal communication").

In the Spotlight

Dr. Barrie Raik is a member of the Weill Cornell University's Division of Geriatrics and Gerontology who spent a six-month sabbatical working with the Virtual Senior Center at Selfhelp Community Services. A computer programmer prior to studying medicine, Dr. Raik has been a life-long techie. While serving a culturally diverse patient population as a member of the Division of General Medicine at Columbia University Medical Center, she took on a side project as medicine consultant to a team of nurses, social workers, and psychiatrists who made house calls to adults who could not leave their homes in the Washington Heights neighborhood of New York City. Unaware that she was stepping on a path to geriatrics, Dr. Raik was drawn in by the complexity, multidisciplinarity, diverse needs, and ethics involved in work with older adults. In 1999 she became a geriatrician and joined Cornell as the first Director of the Geriatric Fellowship Program. She is currently teaching a self-management module virtually to older adults who join the interactive class through the Virtual Senior Center. This community involvement is something she has enjoyed so much that she plans to support the expansion of the VSC to other states, develop additional VSC curricula, and engage the geriatric fellows at her university in the program in a mutually beneficial learning experience.

Technology-Based Services for Living at Home

Two interrelated concepts represent for policy makers and practitioners the gold standard of aging today: 'aging in place' and independent living (Katz and Marshall 2004; Vasunilashorn et al. 2012; Wahl et al. 2012). Built into each is the idea that the ideal place for older adults to live and receive care if necessary is in a private home as opposed to a residential care setting. This belief is supported by concerns over costs as well as research findings that indicate that the vast majority of older adults wish to remain living in familiar environments, even if they acquire an illness, and prefer living independently (AARP 2005; Gibson 2000; Rubinstein et al. 1992).

Long-term care innovation has become a focal issue as many countries are experiencing an aging population and significantly higher rates of chronic illness and disability (Brodsky et al. 2002). The underlying economic logic of aging in place and independent living is tied to the forecasted "care crisis" or "care gap." In the context of discourses on the neo-liberalization of healthcare and the demographic shift to an aging population, technology-based home care services are positioned as a possible solution to the problem of decreasing resources and increasing demand for care for chronic conditions (Oudshoorn 2011). Examples of crisis framing in technology reports and research literature abound as lead-ins to the conclusion that innovative methods like technology-based services are required to enable people to age in place or remain living independently (Kodner 2003).

The preservation, enhancement or enabling of independence may be the most commonly cited potential benefit of monitoring technologies for use in the home (Demiris 2009), yet there is no consensus on the meaning of independence in the broader literature and it is often left undefined in the literature on remote monitoring (Demiris 2009). Despite its widespread use, the concept of independence is complex and highly relative. Self-reliance, while an important goal in the U.S., is not a universal value. For example, in a study in Hong Kong, successful aging was not associated with independent living, but instead with one's needs being met by family (Pecchioni et al. 2004). Cohen's (1998) work in India on senility describes the social importance of elders' dependence on family and friends. While the geographical boundaries of gerontological discourse and knowledge are shifting in a globalized world, there is no universal meaning of or value placed on independence or independent living.

Even within the U.S., the meaning of independence varies by context. The philosophy underlying the Independent Living Movement, for example, resisted the dominant medical model that aims to improve the patient through medical or therapeutic intervention (Burchardt 2004; Krogh 2004). One of the primary concerns of the movement was the unmet need for assistance required by people to live in their communities and engage in society in desired, meaningful ways. Support in the form of assistance with activities and instrumental activities of daily living, medication, communication and assistive technologies was framed as a requirement for individuals with disabilities to contribute to society as full citizens (Krogh 2004; Scala and Nerney 2000), and independence was embedded in a rights-based approach. Advocates of the Independent Living Movement asserted that independence means having control over how and where assistance is provided rather than having no need for it (Morris 2001). Independence does not mean having control over one's body or bodily needs. In fact, that idea is exposed by a disability rights framework as the "myth of control" (Grenier 2003).

No comparative movement has emerged among elders and their families (Kane 2009; Ruggiano 2012). In a review of the literature on independence in old age, Secker et al. (2003) find that the literature on service provision relies heavily on the idea of the ability to function unassisted. In another review, Demiris (2009) found that independence was most commonly defined in relation to functional ability; that is, the ability to perform activities of daily living (ADLs). Thus, independence is more closely associated with the goal of avoiding dependence rather than a value of choice and control over the provision of support (Grenier 2003). In contrast to the paradigm offered by the disability rights movement, this paradigm involves the treatment of individual deficiencies that inhibit independence (Demiris 2009). Grenier (2003) notes that the measurement of functional impairment, or "functional independence" in the home care industry, presents a problem for elders with disabilities within a context of service restrictions brought about through policy reform. Service restriction is antithetical to the provision of support, but personal assistance services in the home are not treated as a right. This has important implications for how the word independence is used (Grenier 2003), and it is frequently used in political, commercial, and policy contexts without a definition. Without a consistent, agreed-upon definition, "independence" is a mutable concept, and an abstract goal for new technology-based services.

Over the past two decades, technology has increasingly been featured in the literature on 'aging in place' (Vasunilashorn et al. 2012). Torres-Gil and Hofland (2012) caution that extreme wealth disparities limit the choices Latino and African American elders in the United States have for where they age. Insufficient home equity and financial resources raise their risk of being "stuck in place" in old age (Torres-Gil and Hofland 2012). Aging in place is a fairly straight-forward concept compared with independence, but one that also warrants consideration of context. Aging in place and independent living are the dominant normative themes in gerontology research and public policy applied for the promotion of in-home technologies. Concepts so widely cited as the primary goal in technology products and services for older adults should be unpacked within diverse contexts of needs, priorities and lived experiences so that their multiple meanings and implications for other services can be discerned (Demiris 2009).

Telehealth and Telehealth-Based Care Management

A variety of technologies have been developed for the purpose of helping to meet challenges older adults may face when aging in place, such as social isolation, access to emergency services in the case of a fall, or easing the care work of monitoring the health and safety of an elder who is living on their own with chronic conditions or mobility constraints. A significant minority of older adults (estimates range from 17 to 43 %) report that they are lonely (Theeke 2010; Perissinotto et al. 2012). These older adults are two times as likely as those who are not lonely to have functional limitations, which could compromise their ability to remain living at home (Perissinotto et al. 2012). Loneliness and social isolation are issues targeted by multipurpose connective technologies like the Virtual Senior

Center. A related challenge is transportation to healthcare service providers, particularly among rural-dwelling older adults where bus and taxi services are unavailable. In addition to transportation difficulties, rural-dwelling older adults have cited lack of quality health care, limited health care supply, social isolation and financial constraints as barriers they face to obtaining health care (Goins et al. 2005). Appropriate health or social services may not be available in rural communities and visits out of town can be lengthy excursions for which recruiting a volunteer driver may be impossible (Goins et al. 2005). Older adults who live in urban areas may also have difficulty getting to medical care appointments, depending on the accessibility of their environment and their physical limitations.

For all of these reasons combined with concerns about cost containment, use of telehealth and telehealth-based care coordination is likely to expand. Telehealth-based care coordination involves remote exchange of symptom and vital sign data through a telehealth appliance. Patients are presented with interactive content on a daily basis so that their information can be collected and transmitted for monitoring and identifying barriers to self-management and medication adherence. Care managers receive alerts when data deviate from the norm, and then follow-up by contacting the appropriate service provider. One of its advantages over telephonic case management is that it makes daily monitoring economical; care managers respond to only those patients who need assistance (CTA 2013a). There is evidence from large randomized control trials that these programs can reduce hospitalizations and days spent in inpatient care, as well as costs. The Veterans Health Administration's Care Coordination/Home Telehealth program is the most commonly cited example in the U.S. It has reduced hospitalizations by 19 % and inpatient care days 25 % while maintaining high patient satisfaction (Darkins et al. 2008). The world's largest trial to date is the Whole Systems Demonstrator Program in the United Kingdom, which reported a 45 % reduction in mortality rates (Department of Health 2011). Organizations like Centura whose service population includes rural and mountain areas of Colorado have also seen improved outcomes with their Health at Home Integrated Telehealth Program on a variety of measures (Broderick and Steinmetz 2013). This kind of evidence for improved outcomes and savings is critical if these devices and services are to become widely reimbursed by third party payers.

Passive Monitoring

Passive remote monitoring is another expanding branch of technology to support independent living. It is the primary subject of writing on technological innovation to support independent living and aging in place, with the majority of these publications focusing on motion detection (Brownsell et al. 2011). Most people are aware of PERS (personal emergency response systems), infamous for the "I've fallen and I can't get up" commercials. A significant innovation is the shift from these actively triggered social alarms to passive monitoring systems that are continuous, requiring no action by the person being monitored. The monitoring occurs through sensors placed throughout an individual's living environment. Sensor-based passive monitoring systems target a combination of indicators, including but not limited to motion patterns that can detect medication adherence and sleep disturbance, falls, as well as environmental temperature, stove on/off status and other safety indicators (CTA 2013b). Sensor system technologies that enable passive monitoring of location and movement behaviors have been proposed as a way to revolutionize home healthcare by enabling earlier detection and prevention of health events (Demiris 2010; Skubic et al. 2009).

Quantifiable outcomes and cost effectiveness have been the dominant criteria in evaluating the emerging practices of technology use to support independent living (Lehoux 2006) as seen in the case of telehealth-based care coordination where studies focus on number of hospital admissions and cost containment. Qualitative research guided by "why" and "how" questions, including user needs and perspectives, are also emerging in the literature on passive monitoring systems in response to the problem that the development of and research on remote monitoring systems has largely been divorced from social and ethical concerns (Bowes et al. 2012; Courtney et al. 2008; Demiris and Hensel 2008; EFORTT 2011).

A variety of disciplinary lenses, from sociology and feminist science and technology studies to nursing and engineering, are applied to questions surrounding passive monitoring of older adults, including its impact on identity, sense of control, privacy, relationships, worry and the work of providing care and assistance. A useful reminder for the limited scope of the overview offered in this chapter is to highlight the often-overlooked fact that technology is embedded in the social; they are mutually involved in design and practice and thus mutually constitutive (Lehoux 2006, p. xiii). Not only is technology introduced in the context of existing services, but practices of technology use are social practices. Its effectiveness is influenced by emotions, values, knowledge, cultural practices, and organizational processes (Lehoux 2006).

Lehoux (2006) provides an example of how health technology is involved in human practices through the technology-enabled collection, analysis, and transmission of data that informs the expertise of healthcare professionals, thus helping to shape their professional practice with patients. Likewise, data derived from passive monitoring systems help shape care practices, and the meanings assigned to these data are shaped by social norms. Passive monitoring sensor systems involve algorithms that detect deviations from an individual's normal movements. The systems' developers identify and embed these behavioral norms in the system. For example, if one is in the bathroom for longer than an hour, an alert may be triggered because that is considered outside of the norm. Detecting deviations from the movement trends of an individual also reveals adherence to a social expectation of regularity or routine.

Norms and values about how one should practice care are embedded within monitoring technologies and practices. Alerts only work if they are followed up by a call center or communicated to an emergency contact who decides how to respond. The sensor is useless without its attendant response service. Using the example of the bathroom enter/exit alert that is triggered after a preset time, the emergency contact could assume that the older adult is taking a long bath, or they could fear that the person has fallen and is in need of urgent help, and then must decide how to proceed. New roles created in this process, including the telecare operator and the emergency contact, have implications for how care is practiced and how responsibility is divided. A new passive monitoring system, for example, may replace a community check in system in senior housing whereby residents place a flag outside their door, informing a volunteer resident during his or her walk of the hallway that the person is up and about. Passive monitoring may be viewed as an improvement because a volunteer is no longer required and residents do not have to remember to place their flag. On the other hand, the non-technological system may have encouraged community awareness and knowledge about how residents were doing. It may also be more inclusive for those who do not adopt a passive monitoring system. Most concretely, responsibility is shifted from a fellow resident to a call center employee and emergency contact for each user.

Norms and values about how one should live as an older adult are embedded within monitoring technologies and practices. The emerging practice of passive monitoring is something that requires careful negotiation, between family members, possibly with social workers or home health aides, and around issues of independence, trust, autonomy, obtrusiveness, and responsibility. It has largely been assumed that privacy would be the biggest concern and barrier to passive monitoring use among older adults. However, the Ethical Technology in the Homes of Seniors (ETHOS) research team lead by a computer scientist, engineer and a gerontologist found that existing privacy frameworks inadequately captured concerns expressed by participants. To their surprise, older adults did not express concerns about seclusion, autonomy, spatial, property or data protection (Lorenzen-Huber et al. 2011). Older participants in focus group studies report that perceived need for passive monitoring may supersede privacy concerns, but discrepancies exist between elders' and caregivers' assessments of perceived need, health conditions and ideas about who would benefit from the technology (Courtney et al. 2008).

Older adults may use passive monitoring technologies to reduce their perception of caregiver burden (Lorenzen-Huber et al. 2012), reducing risks to others such as to family members who feel responsible for their well-being and ability to avoid having to move to assisted living or a skilled nursing facility, or they may refuse the technology because they perceive that it will cause burden. These negotiations occur in the context of family power relations that evolve over lifetimes, but the introduction of new forms of monitoring practices may involve the interplay of older adult-family-professional caregiver interactions as well as conflicting viewpoints. Little is known about how these dynamics and power relations play out in real life decision making. Researchers highlight the need for work that articulates privacy for this age cohort, as well as how autonomy, independence, and preferred levels of support are balanced in practice (Lorenzen-Huber et al. 2011; Shankar 2010). Sociologists who draw upon feminist STS, disability studies and health geography are investigating how passive monitoring practices impact care, its meaning, value, responsibilities and materiality. They recommend acknowledging remote monitoring's potential impact on care relations and looking to engagement with older adults and inclusive design to learn what people want in care and technologies to support it (Mort et al. 2011).

Technogeography draws on human geography to understand the intersection of technology, place and care. Through the lens of technogeography, the meaning of home and care are influenced by 'aging in place' technologies used in the home; the collection of movement and behavioral data in an elder's home reconfigures home and its meaning. Oudshoorn (2011) explains: "In contrast to the rhetoric on telecare, which emphasizes spatially unbounded care practices, telecare technologies still largely depend on locally grounded, situated care acts" (p. 24). One implication is that the home in which one is monitored is a legitimate and important place for the study of the technology as it provides the context in which users participate.

By studying how the user is imagined in the making of the technology itself, researchers can learn about social perceptions and expectations of old age. The design and development of health technologies are of interest to science and technology studies (STS) scholars who study the user representations, values, goals, and assumptions on which product design is based (Brown and Webster 2004; Lehoux 2008; Neven 2011). Neven's research on user representations of older adults in technology design reveals ageist assumptions, as well as the perception of older adults "as people who want to live at home independently for as long as possible" wherein home is assumed to be a good place where older adults want to remain and technologies with potential to cut costs by providing care at home are privileged over alternative interventions (Neven 2011, p. 130). These representations of older people influence the design of technologies (Neven 2011).

Another way in which technologies can tell us about dominant views of and expectations for older adults is by observing how the technology is designed to be used and who has control over its functions. A telling question is did the designers build in an option for the older adult to pause monitoring or otherwise control it? Is the option available to the emergency responder? Is the monitoring one-way or two-way; that is, does the older adult user also receive information they might find helpful, like when a loved one is home? Did designers consider that older adults may feel stigmatized by a device that appears to have a medical purpose? This is relevant for wearable devices as well as hardware installed in the home. What does this design convey about regard for older peoples' priorities surrounding identity and appearance? Do we expect conformity or compromises that are not expected of younger adults? What assumptions are perceptible in product designs about technology and culture? These are some examples of ways in which the technologies discussed here communicate expectations for their users.

Finally, to view older adults as technogenarians, as potentially capable of shaping technology's meaning and use (Brown and Webster 2004), those interested in improving access and experiences of technology use for older adults must take seriously the ways in which older adults choose to use it. Like technology research in general, research on passive monitoring has focused on the issue of adoption, leaving rejection, misunderstanding, resistance, supplementation, intermittent use, and creative use understudied. Researchers of the multi-disciplinary, multi-country project called the *Ethical Frameworks for Telecare Technologies for Older People at Home* (EFORTT) conclude that diverse forms of use should be respected as part of an ethical agenda that prevents these systems from becoming "totalizing and coercive" (EFORTT 2011, p. 16). Accessibility of technology also needs to be considered. Are those who want to use a particular technology able to afford, learn to use, or otherwise access it? The ideas that non-use matters helps push beyond the paternalistic notion of noncompliance and generational incompetency in technology use. Moreover, what it reveals when studied closely can inform design attentive to the needs of older adults.

Expanding the Scope of Inquiry to Improve Experiences

Technology touches multiple aspects of older adults' lives in ways that matter. The conditions that influence access and adoption of technology have broad research, practice, and policy implications. How one builds and maintains social ties, accesses appropriate medical care or information regarding one's health or learning interests has practical consequences for one's well-being. Technology is in many ways social, causing change, engaging diverse stakeholders, raising ethical concerns, and embodying values through its design and use. In this light, it becomes clear that these values should be made explicit through research and policy conversations so that technological systems can be better matched with desired outcomes and the desirability of the outcomes can be discussed openly (Lehoux 2006). Technology is not only a medium through which communication flows, but can also be read for clues about how aging and old age are understood, how care practices are changing, and what possibilities and expectations are in place for its intended users.

This chapter has highlighted the capacity technology has to enable and enhance connectedness, and improve access and efficiency of medical care. The chapter also introduced emerging issues in technology for older adults, particularly related to surveillance and implications for care. It should be understood that understanding technology for older adults as a single entity or phenomenon is an oversimplification and mistake. Indeed, a single disciplinary lens or quantifiable outcomes approach may be inadequate for the task of describing and improving technology-reliant practices with a diverse older adult population. Technology use involves multiple stakeholders and its outcomes are influenced by variables that are non-technological (Lehoux 2006). These practices engage intersections of age and dis/ability, gender, socioeconomic status, and the understudied intersections of technology, age and culture—all areas that can be viewed through a single disciplinary lens, but truly illuminated through a multidisciplinary approach.

The emerging practice of passive monitoring has been described, including the multiple approaches to its increased understanding, from fields of science and technology studies, sociology, gender studies, nursing, engineering, and techno-geography. The implications for further research in gerontechnology and social gerontology that have been noted include the STS tenet that nonuse matters, the idea that the home is an important research site for the study of technology-based home services, and that understanding user representations in design processes is a worthy endeavor that potentially will result in more reflexive, age-aware design.

Emerging practices in technology use could be improved by providing practitioners the space to openly discuss dilemmas and conflicts they encounter in implementing health and care technologies. These issues must also be communicated to researchers and policy makers. Opportunities could be created by policy makers to engage stakeholders in an open dialogue about the consequences of passive monitoring systems, including the ethical and social aspects and implications of their use. Older adults, including those affected directly by specific policy decisions, must be consulted on a large scale. To promote transparency and dialogue in such consultations, objectives should be stated clearly without relying on undefined terms. Policy makers must also consider barriers to use of a variety of technologies, including cost and availability of broadband. Questions about how specific technologies affect opportunities for independence, safety, identity, relationships, connectedness and care belong on the table together with the challenges of expanding access to those who might benefit. Ultimately the question to be asked of each technology of interest is what new ways of living as an older adult does it render and which does it close? The growing field of gerontechnology and affiliated disciplines are making headway, yet there is still much to learn.

References

AARP. (April 2005). Beyond 50.05 survey. Retrieved from http://assets.aarp.org/rgcenter/il/beyond_50_05_survey.pdf

Baker, J., & Bass, G. (2003). *Assistive technology and older adults: The journey through caregiving*. Manual for The North Dakota Family Caregiver Project, North Dakota State University.

Bowes, A., Dawson, A., & Bell, D. (2012). Ethical implications of lifestyle monitoring data in ageing research. *Information, Communication and Society, 15*(1), 5–22.

Broderick, A., & Steinmetz, V. (2013, January 30). Centura health at home: Home telehealth as the standard of care. *Case Study. The Commonwealth Fund*. Retrieved from http://www.commonwealthfund.org/Publications/Case-Studies/2013/Jan/Telehealth-Synthesis.aspx

Brodsky, J., Habib, J., Hirschfeld, M., & Siegel, B. (2002). Care of the frail elderly in developed and developing countries: The experience and the challenges. *Aging Clinical and Experimental Research, 14*(14), 279–286.

Brown, N., & Webster, A. (2004). *New medical technologies and society: Reordering life*. Cambridge: Polity.

Brownsell, S., Bradley, D., Blackburn, S., Cardinaux, F., & Hawley, M. S. (2011). A systematic review of lifestyle monitoring technologies. *Journal of Telemedicine and Telecare, 17*, 185–189.

Burchardt, T. (2004). Capabilities and disability: The capabilities framework and the social model of disability. *Disability and Society, 19*(7), 735–751.

Charness, N., Kelley, C., Bosman, E., & Mottram, M. (2001). Word-processing training and retraining: Effects of adult age, experience, and interface. *Psychology and Aging, 16*, 110–127.

Cohen, L. (1998). *No aging in India: Alzheimer's, the bad family, and other modern things*. Berkeley: University of California Press.

Courtney, K.L, Demiris, G., Rantz, M., Skubic, M. (2008). Needing smart home technologies: The perspectives of older adults in continuing care retirement communities. *Informatics in Primary Care, 16*, 195–201. Retrieved from http://www.radcliffe-oxford.com/

CTA. (2013a). Winter issue brief. Strategies for incorporation of telehealth-based care coordination and management solutions into programs to integrate care for dual eligibles. Retrieved from http://www.agetechwest.com/files/CTA%20Dual_Eligibles_Brief_2012.2.3.pdf

CTA. (2013b). The new era of connected aging: A framework for understanding technologies that support older adults in aging in place. Center for technology and aging issue brief. Retrieved from http://www.techandaging.org/ConnectedAgingFramework.pdf?utm_source=Master+List&utm_campaign=976104dcb4May_CABrief5_6_2013&utm_medium=email&utm_term=0_e60c8920ea-976104dcb4-16699438

Czaja, S. J., & Lee, C. C. (2007). The impact of aging on access to technology. *Universal Access in the Information Society, 5*(4), 341–349.

Darkins, A., Ryan, P., Kobb, R., Foster, L., Edmonson, E., Wakefield, B., et al. (2008). Care coordination/home telehealth: The systematic implementation of health informatics, home telehealth, and disease management to support the care of veteran patients with chronic conditions. *Telemedicine Journal and E-Health, 14*(10), 1118–1126.

Demiris, G. (2009). *Independence and shared decision making: The role of smart home technology in empowering older adults*. Engineering in Medicine and Biology Society, EMBC 2009. Annual International Conference of the IEEE (pp. 6432–6436).

Demiris, G. (2010). Information technology and systems in home health care. In S. Olson (Ed.), *The role of human factors in home health care: Workshop summary* (pp. 173–200). Washington: The National Academies Press.

Demiris, G., & Hensel, B. K. (2008). Technologies for an aging society: A systematic review of "smart home" applications. *IMIA Yearbook of Medical Informatics, 47*, 30–39.

Demiris, G., Skubic, M., Rantz, M., Keller, J., Aud, M., Hensel, B., & He, Z. (2006). Smart home sensors for the elderly: A model for participatory formative evaluation. *Proceedings of IEEE EMBS International Special Topic Conference on Information Technology in Biomedicine*, Ioannina, Greece, October 26–28 (pp 1–4).

Department of Health. (2011). Whole system demonstrator programme headline findings. Retrieved from http://www.dh.gov.uk/en/Publicationsandstatistics/Publications/PublicationsPolicyAndGuidance/DH_131684

Dring, D. (2013). *Personal communication*

EFORTT. (2011). Deliverable 7: Final research report of ethical frameworks for telecare technologies for older people at home. Submitted to the European Commission, available from Maggie Mort, Department of Sociology, Lancaster University.

Federal Communications Commission (FCC). (2013). Connecting America: The national broadband plan. Retrieved from http://www.broadband.gov/download-plan/

Fisk, A. D., Rogers, W. A., Charness, N., Czaja, S. J., Sharit, J., & Gables, C. (2009). *Designing for older adults: Principles and creative human factors approaches* (2nd ed.). Boca Raton: CRC Press.

Gardner, P., Netherland, J., & Kamber, T. (2012). Getting turned on: Using ICT training to promote active ageing in New York City. *Journal of Community Informatics, 8*(1), 1–16.

Gibson, H. (2000). *Loneliness in later life*. New York: St. Martin's Press.

Goins, R. T., Williams, D. A., Carter, M. W., Spencer, S. M., & Solovieva, T. (2005). Perceived barriers to health care access among rural older adults: A qualitative study. *Journal of Rural Health, 21*, 206–213.

Grenier, A. M. (2003). Unhinging the assumptions within independence: Toward a broader conceptualization of diversity and difference in home care. *Canadian Review of Social Policy, 51*, 29–48.

Herman Bouma Fund for Gerontechnology. (2013). What is gerontechnology? Retrieved from http://gerontechnologie.nl/what-is-gerontechnology/

International Society for Gerontechnology. (2013). Home website: http://gerontechnology.info/index.php/journal/pages/view/isghome

Jaeger, B. (2004). Trapped in the digital divide? Old people in the information society. *Science Studies, 17*(2), 5–22.

Jaeger, B. (2012). *New frontiers in the digital divide: Revisiting policy for digital inclusion.* Paper presented at The European Group for Public Administration (EGPA) Annual Conference, Bergan, Norway, September 5–8. Retrieved from https://www.scss.tcd.ie/disciplines/information_systems/egpa/bergen.htm

Kane, R. L. (2009). Empowerment of seniors and paradigm change: A commentary on Haber's empowerment paradigm. *Journal of Applied Gerontology, 28*, 298–303.

Katz, S., & Marshall, B. (2004). Is the functional 'normal'? Aging, sexuality and the bio-marking of successful living. *History of the Human Sciences, 17*(1), 53–75.

Kodner, D. L. (2003). Consumer-directed services: Lessons and implications for integrated systems of care. *International Journal of Integrated Care, 3*(12), e12.

Krogh, K. (2004). Redefining home care for women with disabilities: A call for citizenship. In K. Grant, C. Amaratunga, P. Armstrong, M. Boscoe, A. Pederson, & K. Willson (Eds.), *Caring for/caring about.* Aurora: Garamond Press.

Ladner, R. E. (2010). Accessible technology and models of disability. In M. M. K. Oishi, I. M. Mitchel, & M. V. H. F. Loos (Eds.), *Design and use of assistive technology: Social, technical, ethical, and economic challenges.* New York: Springer Publishing.

Lehoux, P. (2006). *The problem of healthcare technology: Policy implications for modern health care systems.* New York: Routledge.

Lehoux, P. (2008). The duality of health technology in chronic illness: How designers envision our future. *Chronic Illness, 4*, 85–97.

Loe, M. (2010). Doing it my way: Old women, technology and wellbeing. In K. Joyce & M. Loe (Eds.), *Technogenarians: Studying health and illness through an ageing, science, and technology lens.* Hoboken: Wiley.

Lorenzen-Huber, L., Boutain, M., Camp, L. J., Shankar, K., & Connelly, K. H. (2011). Privacy, technology, and aging: A proposed framework. *Ageing International, 36*, 232–252.

Lorenzen-Huber, L., Shankar, K., Caine, K., Connelly, K., Camp, J., Walker, B., et al. (2012). How in-home technologies mediate caregiving relationships in later life. *International Journal of Human-Computer Interaction, 29*, 441–455.

Madden, M. (2010). Older adults and social media. Pew Internet. Retrieved from http://pewinternet.org/Reports/2010/Older-Adults-and-Social-Media.aspx

Martin, S. (2010). Research into the future housing and support needs of older people. Electronic assistive technology—supporting older people within local communities. Final Report. Retrieved from http://www.nihe.gov.uk/electronic_assistive_technology__supporting_older_people_within_local_communities__published_october_2010_.pdf

Morris, J. (2001). Impairment and disability: Constructing an ethics of care which promotes human rights. *Hypatia, 16*(4), 1–16.

Mort, M., Roberts, C., & Milligan, C. (2011). Telecare and older people: Re-ordering social relations. In R. von Schomberg (Ed.), *Towards responsible research and innovation in the information and communication technologies and society technologies fields* (pp. 150–164). The European Commission's Science in Society Initiative EFORTT Project. Retrieved from http://www.synbioproject.org/process/assets/files/6613/_draft/mep-rapport-2011_en.pdf

Neven, L. (2011). Representations of the old and ageing in the design of the new and emerging: Assessing the design of Ambient Intelligence technologies for older people. Published dissertation, University of Twente.

Older Adults Technology Services (2013). http://www.oats.org

Olson, K. E., O'Brien, M. A., Rogers, W. A., & Charness, N. (2011). Diffusion of technology: Frequency of use for younger and older adults. *Ageing International, 36*, 123–145.

Oudshoorn, N. (2011). *Telecare technologies and the transformation of healthcare*. New York: Palgrave Macmillan.
Pecchioni, L. L., Ota, H., & Sparks, L. (2004). Cultural issues in communication and aging. In J. F. Nussbaum & J. Coupland (Eds.), *Handbook of communication and aging* (2nd ed., pp. 115–138). Mahwah: Erlbaum.
Perissinotto, C., Cenzer, S. I., & Covinsky, K. (2012). Loneliness of older persons: A predictor of functional decline and death. *Archives of Internal Medicine, 172*(14), 1078–1084.
Rubinstein, R., Kilbride, J., & Nagy, S. (1992). *Elders living alone: Frailty and the perception of choice*. New York: Aldine De Gruyer.
Ruggiano, N. (2012). Consumer direction in long-term care policy: Overcoming barriers to promoting older adults' opportunities for self-direction. *Journal of Gerontological Social Work, 55*(2), 146–159.
Scala, M. A., & Nerney, T. (2000). People first: The consumers in consumer direction. *Quarterly Journal of the American Society on Aging, 24*(3), 55–59.
Secker, J., Hill, R., Villeneau, L., & Parkman, S. (2003). Promoting independence: But promoting what and how? *Ageing and Society, 23*, 375–391.
Shankar, K. (2010). Pervasive computing and an aging populace: Methodological challenges for understanding privacy implications. *Journal of Information, Communication and Ethics in Society, 8*(3), 236–248.
Skubic, M., Alexander, G., Popescu, M., Rantz, M., & Keller, J. (2009). A smart home application to eldercare: Current status and lessons learned. *Technology and Health Care, 17*, 183–201.
Theeke, L. A. (2010). Sociodemographic and health-related risks for loneliness and outcome differences by loneliness status in a sample of older US adults. *Research in Gerontological Nursing, 3*(2), 113–125.
Torres-Gil, F., & Hofland, B. (2012). Vulnerable populations. In H. Cisneros, M. Dyer-Chamberlain, & J. Hickie (Eds.), *Independent for life: Homes and neighborhoods for an aging America* (pp. 221–232). Austin: University of Texas Press.
Vasunilashorn, S., Steinman, B.A., Liebig, P.S., & Pynoos, J. (2012). Aging in place: Evolution of a research topic whose time has come. *Journal of Aging Research.*
Wahl, H. W., Iwarsson, S., & Oswald, F. (2012). Aging well and the environment: Toward an integrative model and research agenda for the future. *The Gerontologist, 52*(3), 306–316.
Webster, F. (2002). *Theories of the information society* (2nd ed.). London: Routledge.
Zickuhr, K., & Madden, M. (2012, June 6). Older adults and internet use. Pew internet. Retrieved from http://www.pewinternet.org/Reports/2012/Older-adults-and-internet-use/Summary-of-findings.aspx
Zickuhr, K., & Smith, A. (2012, April 13). Digital differences. Pew internet. Retrieved from http://pewinternet.org/Reports/2012/Digital-differences/Overview.aspx

Health Promotion for African American Elders: Church is the Likely Place

Cheryl Waites, Angela Kaiser and Fayetta Martin

Abstract African Americans living in neighborhoods with limited access to health promotion programs, services or amenities—safe places to walk and exercise, health care facilities, senior centers, food security and affordable healthy foods, and transportation—are at risk of having poor health outcomes as they age (Balfour and Kaplan 2002; Bouchard et al. 2007; Ross and Mirowsky 2001). The Black church and its historic roots in African American communities may be a source of support (Bopp et al. 2006; Watson et al. 2003). Historically, the Black church has provided social and spiritual support networks to assist with living in rural and urban communities. Many African American elders attend church services regularly and benefit from a sense of community (Levin and Taylor 1993; Trinitapoli 2005). Therefore, this institution seems a likely place to provide health promotion services in many neighborhoods. This chapter discusses the African American church as a good location and partner for health promotion services. A framework for understanding healthy aging services is presented in a study of 38 African American churches in the Detroit, Michigan, metropolitan area, a case example to illustrate the potential of the Black church as a resource to promote healthy aging. Ideas on strategies to expand healthy aging programs and services for African American elders are discussed throughout the chapter.

The Black church plays a "unique role in advancing the physical, emotional, and spiritual well-being of its members" (Watson et al. 2003, p. 198).

C. Waites (✉) · F. Martin
Wayne State University, Detroit, MI, USA
e-mail: dv7029@wayne.edu

A. Kaiser
Oakland University, Rochester, MI, USA

H. F. O. Vakalahi et al. (eds.), *The Collective Spirit of Aging Across Cultures*, International Perspectives on Aging 9, DOI: 10.1007/978-94-017-8594-5_11,

Although medical advances have increased the possibility of a longer and healthier life, African American elders experience a disproportionate occurrence of chronic illness when compared with other racial and ethnic groups (James 2006; Satcher et al. 2005). The intersection of race/ethnicity, age, and wealth seem to confound health outcomes for the elders. In addition, African Americans living in neighborhoods with limited access to health promotion programs, services or amenities—safe places to walk and exercise, health care facilities, senior centers, food security and affordable healthy foods, and transportation—are at risk of having poor health outcomes as they age (Balfour and Kaplan 2002; Bouchard et al. 2007; Ross and Mirowsky 2001). It is vitally important to address these issues and identify strategies for promoting healthy aging for African American elders across contexts. The Black church and its historic roots in African American communities may be a source of support (Bopp et al. 2006; Watson et al. 2003).

The health condition of older African Americans continues to be a major issue. As one of the fastest growing segments of the older adult population (U.S. Administration on Aging, U.S. Department of Health and Human Services 2010), African American elders are more likely to die prematurely or suffer from poor health and chronic health conditions as they age. Most African American older persons have at least one chronic condition and many have multiple ailments. The most frequently occurring conditions among African American elders in 2005–2007 were hypertension (84 %), diagnosed arthritis (53 %), all types of heart disease (27 %), sinusitis (15 %), diabetes (29 %), and cancer (13 %). The comparable figures for all older persons were hypertension (71 %), diagnosed arthritis (49 %), all types of heart disease (31 %), sinusitis (14 %), diabetes (18 %), and cancer (22 %). This disparity presents a challenge to families and health care systems (U.S. Administration on Aging, U.S. Department of Health and Human Services 2010).

Historically, the Black church has provided social and spiritual support networks to assist with living in rural and urban communities. The Black church is comprised of churches with predominately African American membership. The Black church, as a social institution, has been serving the African American community and has provided spiritual, emotional, and social support to its congregates and the surrounding community. Many African American elders attend church services regularly and benefit from a sense of community (Levin and Taylor 1993; Trinitapoli 2005). Therefore, this institution seems a likely place to provide health promotion services in many neighborhoods. This chapter discusses the African American church as a good location and partner for health promotion services. A framework for understanding healthy aging services is presented in a study of 38 African American churches in the Detroit, Michigan, metropolitan area, a case example to illustrate the potential of the Black church as a resource to promote healthy aging. Ideas on strategies to expand healthy aging programs and services for African American elders are discussed throughout the chapter and presented in Tables 1, 2 and 3.

Table 1 Dimension of wellness

Dimensions of wellness	Program or activities—predetermined codes
Intellectual	Computer classes, courses, workshops, seminars, heath information materials
	Lectures and educational presentations
	Trips to museums, libraries, lectures, historical sites
	Cognitive oriented games, reading activities
Physical/functional	Nutrition education
	Health fairs, health screenings, health related clinics/services
	Flu shots, immunizations
	Walking clubs, yoga classes, chair aerobics, water aerobics, exercise classes
	Sports—baseball, basketball, golf, swimming, etc.
Emotional	Support groups e.g. grief and loss, caregiving, grandparent raising grandchildren, coping with chronic illness
	Seminars on emotional wellness or mental health issues
	Mental health services
	Pastoral counseling services
Vocational	Volunteer programs/activities
	Mentoring programs/activities
	Job training for elders
	Employment referral services
	Leadership development
Spiritual	Senior ministries
	Bible study, senior prayer meeting
	Special church services or programs (i.e., senior sunday)
	Other ministries
Social	Celebrations (holiday, birthdays, parties)
	Concerts, dances
	Movies, plays, etc.
	Day trips—casino, museums
	Choir or singing groups and activities
	Bingo, card games, pool, etc.
	Travel groups—longer trips
Community engagement/ civic engagement	Advocacy and organizing around community issues
	Voter registration drives
	Political activity
	Volunteering in community organizations
	Serving on boards
	Community outreach activities
	Collaborating with other organizations

Black Churches: A History of Service

The Black church, as a social institution in the United States, has a history of serving the African American community in a myriad of ways (Billingsley and Morrison-Rodriguez 1998; Waites 2009). The collective spirit of the Black church

Table 2 Interviews with pastor of church representative

Church size	Church denomination
Mega	Unity
Mega	Non-denominational
Mega	Baptist
Mega	Non-denominational
Large	Baptist
Large	Methodist
Medium	Baptist
Medium	Methodist
Medium	Baptist
Small	Baptist
Small	Methodist episcopal

starting from the time of slavery, through the Civil Rights Movement era, and continuing in the present, has played a vital role in providing economic guidance and support, social activities and interaction, community services, political messages and movements, education, and support for physical and mental well-being (Brown and Gary 1991; Frazier 1963; Lincoln and Mamiya 1990; Morris 1984; Nelsen and Nelsen 1975; Pattillo-McCoy 1998). In terms of promoting health and well-being, Black churches have been directly addressing the unmet needs of African American communities as early as the 1920s, offering food programs for the unemployed, free health clinics, recreational activities, and child care programs (Mays and Nicholson 1933).

Even more significant in establishing a connection between the Black church and the health of African Americans was the National Negro Health Movement of 1915–1951. This movement emerged in response to major health disparities as well as a lack of quality health services and providers in African American communities (Quinn and Thomas 2001). Because the church already played such a significant role in the lives of African American people, this movement utilized churches to mobilize community members around the issue of improving access to community health services for African Americans (Quinn and Thomas 2001). The Black church has continued to serve as a vehicle for linking elder community members with formal and informal systems of care, particularly for community elders who are harder to reach through other mechanisms (Eng et al. 1985; Taylor and Chatters 1986). Programs to support healthy aging such as senior apartments, senior appreciation day, transportation, senior luncheons, meals on wheels, and health screenings services have been affiliated with the Black church over the past 90 years.

There are several factors that make the Black church an ideal place for health and wellness programs. As discussed above, the Black church has taken an active role in promoting civic engagement to address community needs and reduce disparities, as well as connecting congregants to information and services since the early 1900s (Mays and Nicholson 1933; Quinn and Thomas 2001). Churches assumed this role early on due to health disparities and the lack of appropriate services in Black communities. While access to health-related services has improved, the need remains and African American elders continue to rely on churches to fill service gaps

Table 3 Summary of health promotion activities and program inventory

Dimensions	Programs/activities for elders	Church size and number of program/activities				
		Mega n = 13	Large n = 8	Medium n = 10	Small n = 7	Total N = 38
Physical	Yoga, aerobics, chair exercises[a]	18	10	8	4	40
	Weight watchers, weight management					
	Walking, running[a]					
	Smoking cessation					
	Sports—golf, bowling, etc.[a]					
	Health screenings, free exams, clinics					
	Medicare prescription sign-up assistance					
	Health ministry, church nurses[a]					
	Healthy diet/eating/cooking					
	Meals on wheels, food panties, reduced cost food program					
	Senior apartments					
Intellectual	Seminars, workshops, adult well-being classes[a]	11	6	4	2	23
	Driver refresher course[a]					
	Health seminars, health fairs[a]					
	Adult literacy					
	Black health empowerment					
	Nutrition information					
	Art of aging conference					
Social	Luncheons, dinners, coffee hours[a]	15	6	2	4	27
	Field trips, fellowship activities[a]					
	Women's or men's day programs[a]					
	Leisure/travel group[a]					
	Celebrations [holiday, birthdays, parties][a]					

(continued)

Table 3 (continued)

Dimensions	Programs/activities for elders	Church size and number of program/activities				
		Mega n = 13	Large n = 8	Medium n = 10	Small n = 7	Total N = 38
Emotional	Bereavement support groups	7	1	1	1	10
	Caregiver's ministry network[a]					
	Grandparents as parents group					
	Nursing home visiting					
	Pastoral or lay counseling, social services[a]					
Spiritual	Visitation, convalescent, nursing home ministries	15	7	3	2	27
	Senior's ministry [promotes spiritual growth]					
	Bible study[a]					
	Senior choir					
	Healing services					
Civic engagement	Charity soup kitchen	10	3	1	3	17
	Empty bowls luncheon					
	Political awareness and voting assistance					
	Voters registration					
	Linked to outside groups that advocates for better care for seniors					
	Healthcare reform meeting					
	Community breakfast on sundays					
Vocational	Professional developing, seminars, workshops[a]	2	1	1	0	4
	Job training/placement[a]					

[a] All identified activities and programs geared towards for older adults usually 65 and older

(James 2006; Satcher et al. 2005). History of acting as the primary provider for physical, social, and emotional needs in communities has had a profound impact on members who rely on churches for these services.

Community trust has also played a role in many churches being an ideal place for health and wellness services for the elders. For many church-goers across denominations, clergy are often the first and only professionals they encounter (Taylor et al. 2000). In most cases, congregants may have already established a personal relationship with the pastor and they feel more comfortable approaching the pastor regarding physical or emotional challenges that impact their overall well-being. Many elders also turn to other trusted congregants for socialization and support (Levin and Taylor 1993; Taylor and Chatters 1986; Walls and Zarit 1991). Still some church members may not be able to see a professional for emotional issues due to financial or geographical limitations (Taylor et al. 2000), as well as past experiences with discriminatory practices. Relationships with other church members as well as with the pastors may provide the Black church with a level of built-in trust that may facilitate improved usage of services.

The fact that many elders are already engaged in church activities also makes it an ideal place to offer services for health and well-being. Many elders are limited in terms of accessing transportation or participating in activities that are financially taxing. African American elders may be less likely to seek support in formal institutions outside of the church environment (Walls and Zarit 1991). Providing services in a venue that is both familiar and accessible may increase the likelihood of participation in health and wellness programs for African American elders.

Many churches including Black churches continue to address needs of health and wellness using a broad range of services and activities. They are providing health and wellness services to church attendees and community members (Thomas et al. 1994; Stineman et al. 2011) as well as establishing community outreach networks. Older congregants and community members are often made aware of the programs and activities available to them in their communities through church bulletins, radio ads, and by word of mouth. When programs or activities are church-affiliated, church and community members are more likely to take advantage of these activities. The Black church in particular embodies an inclusive spirit, engaging in providing services for health and wellness among African American elders.

Exploring Health Promotion Activities and Programs: A Case Example

A Wellness Framework

A few studies have explored African American elders' perceptions regarding what constitutes wellness and healthy lifestyles. In sum, a holistic view of health that includes caring for the mind and body and the significance of spirituality and religion was expressed throughout these studies (Waites 2012). With this in mind, a holistic wellness framework is used as a means for understanding the framing and development of healthy aging programs and activities for African American elders.

The concept of wellness has been used to describe an active process through which people become aware of, and make choices toward, a positively affirming and healthy lifestyle (Hettler 2010). Six dimensions of wellness have been delineated by the National Wellness Institute to describe a holistic wellness system. The six dimensions aptly fit with research regarding African American elders' holistic perception of healthy aging. These dimensions are intellectual, physical/functional, emotional, spiritual, social, and vocational wellness (see Table 1). Intellectual wellness refers to cognitive exercises and processes to not only maintain but also improve mental functioning. Physical wellness focuses on exercising the body and maintaining or improving functional capacity. Emotional wellness refers to the acceptance of one's emotions and feelings, as well as the ability to handle stressful situations. Vocational wellness covers a wide range of productivity issues, from maintaining one's physical abilities to basic skills such as writing, and continuing to participate in hobbies. Spiritual wellness involves the desire to explore life issues and meanings. This might include Bible study, participation in ministries, and church attendance. Social wellness requires an ability to relate to other people in a healthy, productive way. A seventh dimension, civic engagement, was added, which refers to a personal or collective action designed to address a public concern or an unmet human, educational, health care, environmental, or public issue or need. The wellness framework is used to describe health and wellness and to explore the health promotion activities and programs for older adults in Black churches in the Detroit metropolitan area.

The Black Church

The Black churches' involvement in health promotion activities for the elders as well as their commitment to eliminating health disparities and promoting healthy aging has been documented. According to Samuel and Sanders (1991), many churches have programs and services that fit nicely into a wellness (health promotion) framework. Knapp (2001) also found that the size and budget of the church matters; larger churches with more money are more likely to have programs specifically for seniors. Similarly, Looney and Haber (2001) found that the more resourceful the church, the more likely the response to health programs would be positive. They defined a resourceful church as one that has a large number of members, adequate space, and a membership that, on average, has a higher socioeconomic status. The case example presented in this chapter explores health promotion for African American older adults using a wellness framework to conceptualize programming and activities in Black churches.

Research Design

A qualitative approach was employed to explore health promotion activities and programs in 38 Black churches serving the Detroit Metropolitan area. A content analysis of health promotion activities was conducted. Follow-up interviews with church leaders were also carried out to provide member checking and confirm health promotion strategies.

Sampling Strategy

Churches located in the Detroit Metropolitan area's Eastside, Westside, and the city of Southfield were included in this study. A purposive sampling scheme was used and churches were selected based on the following criteria: (1) church congregation size (small, medium, large and mega), (2) a predominantly African American membership, and, (3) denomination. Church size distinction was determined through self-identification, as well as through church websites. Church sizes were categorized as Small = 0–499, Medium = 500–999, Large = 1,000–1,999, Mega = 2,000 (Watson et al. 2003) (see Table 2). A research assistant familiar with the region compiled a list of churches with predominantly African American congregations. Church location, size, and congregation composition were identified and verified with information from church websites, and follow-up phone contacts with church officials. The sample was structured to include both churches that did or did not report health promotion activities. A purposive sample of pastors and or church leaders was interviewed following the content analysis as a method for member checking and to confirm the identified health promotion strategies.

Data Collection and Instruments

Once the researchers identified churches, a tally sheet was completed for each church to provide an inventory of the data. These tally sheets captured descriptive information regarding church size, location, denomination and health promotion activities, obtained through secondary data sources such as church websites, church bulletins, flyers, calendars, and articles, as well as grants and collaborative projects that were posted on the internet. When written information was not available, churches were contacted by phone and asked if they provided health promotion programs for older adults. Researchers focused on activities and programs that appeared to promote healthy aging for older adults (i.e., physical activity programs; nutrition and healthy diet programs; and social support activities for older adults or caregivers). The seven dimensions of wellness comprised the coding scheme. Activities relevant to each dimension were recorded (see Table 1). The data collection ended when a saturation point was reached, where searches continued to identify the same programs and services for each church.

Data was also collected through individual interviews with church leaders which consisted of several steps. First, prior to the interviews, investigators obtained approval of the Institutional Review Board (IRB approval obtained from Wayne State University). Second, researchers identified 11 churches using sampling criteria of church size and the presence or absence of health promotion activities. Interviews were scheduled with church leaders. Researchers used a semi-structured interview guide with open-ended questions to conduct face-to-face interviews with pastors and church leaders who were active in their respective churches and knowledgeable about the older adult population and healthy aging. Prior to being interviewed, participants signed a consent form that explained the purpose of the study and how the information would be used. Two researchers, one African American and one Caucasian, conducted the interviews. These interviews were audio taped and notes were taken.

Data Analysis

A content analysis of the secondary data was conducted using predetermined theoretical codes (see Table 1 for code list). Furthermore, interviews were transcribed and analyzed and responses were compared to the secondary data and other responses using a constant comparative method. A summary of all data outlining programs, information sources, special projects, church denomination, size and apparent interest in health promotion for older adults was prepared by the study director (PI) with input from research team members. Thereafter, an inventory detailing health promotion activities and programming for all 38 churches was created (see Table 3).

To address trustworthiness of the results in terms of credibility, transferability, audibility and conformability (Padgett 2004) a "community gerontology expert" reviewed the inventory, consolidated themes and theories that emerged from the study. The expert was asked to review the link between the inventory, interviews and her knowledge regarding elders in the community and whether the study reached saturation in terms of depth of the information gathered. The expert was also asked to provide feedback regarding the soundness of the study procedures and to review the findings and recommendations to insure that they were clearly linked to the data. This expert was very familiar with the concerns and needs of older adults, particularly African American elders, in the metropolitan area. Comments and suggestions from the community expert were discussed and integrated into the final report.

Findings

The findings of this study showed that the participating 38 Black churches in the Detroit metropolitan area were providing access to health and wellness programs and activities. The number and types of programs and activities varied depending on the available resources and church size. Larger churches, for the most part, were engaged in health promotion activities for its members. Churches, which reported younger memberships, appeared to offer few programs that specifically addressed the needs of its elder members. There was also variation regarding attention paid to some dimensions of wellness. Findings showed that the spiritual needs of older adults were addressed with church attendance and participation in organized church activities. In addition, most churches provided physical and socially oriented health and wellness activities and programs; the larger churches also provided informational sessions, lectures and workshops. Often the intellectual dimension was facilitated through collaboration with community organizations and university programs. Other dimensions of wellness–emotional, civic engagement, vocational—were less often addressed through planned programs or church sponsored activities for older adults.

Dimensions of Wellness

Physical activities and programs. Most churches provided physical health promotion or wellness activities and programs. Mega and larger churches provided more health promotion activities and programs than medium and smaller churches. Physical activities included exercise, dance, yoga, golf, and other activities included health fairs, food and nutrition programs, and health screenings. Many churches have developed fair weather walking groups and are using convenient and safe places to walk such as parks, local high school tracks, riverfront walking areas, shopping malls, fitness centers, and local senior centers. These options are convenient for those with adequate transportation and available 7 days a week. Many senior centers with wellness programs have partnered with churches.

Intellectual activities and programs. The majority of church supported activities included information sessions or seminars on health-related topics, computer classes and adult literacy programs. Many churches collaborated with the universities and other community organizations to provide services to older adults. Through a Wayne State University and University of Michigan partnership, as well as community outreach, the Healthier Black Elders Center puts on a major event, the Healthier Black Elders Reception, each year. Hundreds of elders attend this free one-day event which provides educational panel discussions, with question and answer sessions, health screenings, a nutritious lunch, line dancing and chair exercises. This event cuts across several dimensions of wellness. Additionally, the Healthier Black Elders Center provides a Lunch and Learn series at many local churches from December to June each year. The Art of Aging conference, also offered by the Wayne State University Institute of Gerontology, is another example of a community and university partnership. This is a conference organized by seniors for seniors to promote successful aging. The Association of Black Cardiologists (ABC) has conducted presentations on hypertension at one of the larger churches. They have also provided information on other identified illnesses such as diabetes, heart disease, high cholesterol, obesity, physical inactivity and tobacco use. ABC sponsors an annual Grandparents 2 K/5 K Fun Walk in many inner cities to support intergenerational bonds and activities.

Social activities and programs. Social programs were also offered by all but the smallest churches. Activities included senior appreciation days, senior luncheons, and group activities such as knitting or playing cards. Additionally, some churches offer day trips to shopping malls, farmer's markets, and the movies, as well as travel clubs. Other churches provide activities such as coffee hours, women's day programs, praise fests, and social outings for mothers' young and old.

Emotionally supportive activities and programs. The larger churches conducted programs that addressed emotional health. The majority of the activities identified were support groups, such as bereavement groups, caregivers support groups and "grandparents as parents" groups. In most churches, the pastor or other church leaders offered counseling services.

Spiritual activities and program. All churches provided worship services. Most churches had programs where members visited nursing homes to bring worship to older members and also conducted Bible study groups. Smaller churches did not offer specific spiritual activities for older adults outside of regular church services.

Civic engagement activities and program. Mega churches were more likely to report civic engagement activities. Most of the activities identified in the interviews for this dimension were related to volunteerism in the community. One mega church highlighted their association with outside groups that advocate for better care for seniors. A few responses were related to political engagement, but the majority focused on volunteering at soup kitchens or special events. Many seniors volunteered to work at the voting polls, helped get the word out regarding healthcare reform, or served at community breakfasts on Sundays.

Vocational activities and program. Few churches offered vocational programs. One of the larger churches provided job training and placement services for all its members. Another church provided professional development, seminars and workshops, and others offered computer classes.

Overall, this study demonstrated that Black churches remain involved in promoting the health and well-being of their older adult members and their surrounding communities. The church leaders who were interviewed expressed concerns regarding the health of their elder church members. Churches providing limited healthy-aging programs indicated that the lack of church resources was a deterrent. In response, several churches have collaborated with aging network organizations such as Senior Centers, Detroit Area Agency on Aging, Healthier Black Elders Center, and The Association of Black Cardiologists to gain access to programs and activities for their older adult members. All indicated that their churches functioned as a source of social support for their congregants and for the surrounding community.

Church is a Likely Place

> At the heart of all that civilization has meant and developed is 'community' – the mutually cooperative and voluntary venture of man to assume a semblance of responsibility for his brother.
>
> Martin Luther King Jr.

With a long history of being the primary source of support in African American communities, it is not surprising that the Black church continues to be an epicenter of social and health-related activities. Churches have good outreach potential with their built-in membership, and in particular, for their older adult members who often rely on churches for social interaction and for supportive services (Felton and Berry 1992; Levin and Taylor 1993). The strong presence of Black churches in African American communities makes them ideal sources for programs promoting the health and well-being of community members.

The Black church is a likely place to do outreach and to launch health promotion programming. Churches in this study were aware of the health disparity challenges facing African Americans and the impact of chronic illness on the quality of life of elders. They were engaged in providing some level of health promotion activities ranging from spiritual support and physical activities to, in some cases, intellectual and vocational activities. African American elders want better access to health promotion amenities and services. There is a need to expand church programs in order to meet the needs of our elder communities. Policies can allot funding for such expansion. This is especially true for smaller churches with limited financial resources.

The network of aging agencies, with missions to address health disparities and to promote healthy aging, must work with churches to cultivate working partnerships and to tailor compatible services. Although some exchange is in place, there is a need to further engage the Black Church, which has historically functioned as a natural support system in African American communities. The church is trusted by its members and has the potential to expand health promotion efforts beyond their congregation and into their surrounding communities. Partnerships with the network of aging agencies and churches can contribute to the advancement of health promotion and healthy aging for African American elders. This partnership would benefit from a comprehensive strategy-driven approach that emphasizes collaboration, information sharing, participatory planning, and resource building. Recruiting volunteers, who might provide professional services, in-kind donations in the form of transportation and space, and community outreach, though not new ideas, when bundled with a comprehensive plan could expand the reach of the services and create healthy aging options for African America elders. Through a planned alliance with Black churches, resources can be leveraged to build the capacity of all churches and the communities where African American elders reside.

References

Balfour, J. L., & Kaplan, G. (2002). Neighborhood environment and loss of physical function in older adults: Evidence from the alameda county study. *American Journal of Epidemiology, 155*(6), 507–515.

Barry, C. A., Britten, N., Barber, N., Bradley, C., & Stevenson, F. (1999). Using reflexivity to optimize teamwork in qualitative research. *Qualitative Health Research, 9*(1), 26–44. doi:10.1177/104973299129121677.

Billingsley, A., & Morrison-Rodriguez, B. (1998). The black family in the 21st century and the church as a social action system: A macro perspective. *Human Behavior and the Social Environment, 1*(2/3), 31–47.

Bopp, M., Lattimore, D., Wilcox, S., Laken, M., McClorin, L., Swinton, R., Gethers, O., & Bryant, D. (2006). Understanding physical activity participation in members of an African American church: A qualitative study. *Health Education Research, 22*(6), 815–826. Oxford: Oxford University Press.

Bouchard, C., Blair, S. N., & Haskell, W. L. (2007). Physical Activity and Health. Why study physical activity and health. *Physical Activity and Health*, 3–19.

Brown, D. R., & Gary, L. E. (1991). Religious socialization and educational attainment among African Americans: An empirical assessment. *The Journal of Negro Education, 60*, 411–426.

Eng, E., Hatch, J., & Callan, A. (1985). Institutionalized social support through the church and into the community. *Health Education Quarterly, 12*, 81–92.

Felton, B. J., & Berry, C. A. (1992). Do the sources of the urban elderly's social support determine its psychological consequences? *Psychology and Aging, 7*, 89–97.

Frazier, E. F. (1963). *The Negro church in America*. New York: Shocken.

Hettler, the National Wellness Institute. (2010). *National Wellness Institute*. http://www.nationalwellness.org/index.php?id_tier=2&id_c=25

James, E. H. (2006). Race disparities in health among older adults: Examining the role of productive engagement. *Health and Social Work, 31*(4), 275.

Knapp, J. L. (2001). The impact of congregation-related variables on programs for senior adult members. *Journal of Applied Gerontology, 1*, 24–38.

Levin, J. S., & Taylor, R. J. (1993). Gender and age differences in religiosity among Black Americans. *The Gerontologist, 33*(1), 16.

Lietz, C. A., Langer, C. L., & Furman, R. (2006). Establishing trustworthiness in qualitative research in social work: Implications from a study regarding spirituality. *Qualitative Social Work, 5*(4), 441–458. doi:10.1177/1473325006070288.

Lincoln, C., & Mamiya, L. (1990). *The Black church in the African American experience*. Durham: Duke University Press.

Looney, C., & Haber, D. (2001). Interest in hosting an exercise program for older adults at African-American churches. *Journal of Religious Gerontology, 13*, 19–29.

Mays, B. E., & Nicholson, J. W. (1933). *The Negro's Church*. New York: Russell and Russell.

Morris, A. (1984). *The origins of the civil rights movement: Black communities organizing for change*. New York: Free Press.

Nelsen, H. M., & Nelsen, A. K. (1975). *Black church in the sixties*. Lexington: University of Kentucky Press.

Padgett, D. K. (1998). *Qualitative methods in social work research: Challenges and rewards*. Thousand Oaks: Sage.

Padgett, D. K. (Ed.). (2004). *Introduction: Finding a middle ground in qualitative research*. Belmont: Thomson/Brooks and Cole.

Pattillo-McCoy, M. (1998). Church culture as a strategy of action in the black community. *American Sociological Review, 63*, 767–784.

Quinn, S. C., & Thomas, S. B. (2001). The National Negro Health Week 1915–1951: A descriptive account. *Minority Health Today, 2*, 44–49.

Ross, C. E., & Mirowsky, J. (2001). Neighborhood disadvantage, disorder, and health. *Journal of Health and Social Behavior, 42*, 258–276.

Samuel, M., & Sanders, G. F. (1991). The role of churches in the supports and contributions of elderly persons. *Activities, Adaptation and Aging, 16*, 67–79.

Satcher, D., Fryer, G. E., McCann, J., Troutman, A., Woolf, S. H., & Rust, G. (2005). What if we were equal? A comparison of the black-white mortality gap in 1960 and 2000. *Health Affairs, 24*(2), 459–464. doi:10.1377/hlthaff.24.2.459.

Stineman, M. G., Strumpf, N., Kurichi, J. E., Charles, J., Grisso, J. A., & Jayadevappa, R. (2011). Attempts to reach the oldest and frailest: Recruitment, adherence, and retention of urban elderly persons to a falls reduction exercise program. *The Gerontologist, 51*(suppl 1), S59–S72.

Taylor, R. J., & Chatters, L. M. (1986). Patterns of informal support to elderly black adults: Family, friends, and church members. *Social Work, 31*(6), 432–438.

Taylor, R. J., Ellison, C. G., Chatters, L. M., Levin, J. S., & Lincoln, K. D. (2000). Mental health services in faith communities: The role of clergy in Black churches. *Social Work, 45*(1), 73–87.

Taylor, R. J., Thornton, M. C., & Chatters, L. M. (1987). Black American's perceptions of the socio-historical role of the church. *Journal of Black Studies, 18*, 123–138.

Thomas, S. B., Quinn, S. C., Billingsley, A., & Caldwell, C. (1994). The characteristics of Northern Black churches with community health outreach programs. *American Journal of Public Health, 84*(4), 575–579.

Trinitapoli, J. (2005). Congregation-based services for the elders: An examination of patterns and correlates. *Research on Aging, 27*(2), 241.

U.S. Administration on Aging, U.S. Department of Health and Human Services. (2010). *Profile of older Americans 2010*. Retrieved May 20, 2013, from http://www.aoa.gov/aoaroot/aging_statistics/profile/2010/docs/2010profile.pdf

Waites, C. (2009). Building on strengths: Intergenerational practice with African American families. *Social Work, 54*(3), 278.

Waites, C. (2012). Examining the perceptions, preferences, and practices that influence healthy aging for African American older adults: An ecological perspective. *Journal of Applied Gerontology,*. doi:10.1177/0733464812446020.

Walls, C. T., & Zarit, S. H. (1991). Informal support from Black churches and the well-being of elderly blacks. *The Gerontologist, 31*, 490–495.

Watson, D. W., Bisesi, L., Tanamly, S., Sim, T., Branch, C. A., & Williams, E, I. I. I. (2003). The role of small and medium-sized African American churches in promoting healthy life styles. *Journal of Religion and Health, 42*, 191–200.

City Life: What a Wonderful Way of Life—Aging in the Urban Environment

Kimberly Yancey, Tamika Baldwin, Ama R. Saran and Halaevalu F. Ofahengaue Vakalahi

Abstract This chapter attempts to discuss the complexities of aging in an urban context with specific attention to a few salient issues that intersects across cultural groups in urban communities. Older adults in urban environments are examined from People in Place of what it is like growing up in an urban environment. Challenges for Urban Elders dealing with urban issues of substance abuse, mental health, HIV/AIDS, and general health are addressed. Existing strategies and approaches to dealing with issues of urban environments for older adults are considered. Intersectionality, Systems/Ecological, and the Strengths Perspective are the theories guiding the discussion of both the positive and challenges of older adults aging in an urban environment. Through the lens of these theories, the multi-facets of urban context and older adults will be better understood.

K. Yancey (✉) · T. Baldwin · H. F. O. Vakalahi
Morgan State University, Baltimore, USA
e-mail: Kimberly.Yancey@morgan.edu

A. R. Saran
Social Research Strategies and Solutions, The Bryant Group, Indianapolis, USA

H. F. O. Vakalahi et al. (eds.), *The Collective Spirit of Aging Across Cultures*,
International Perspectives on Aging 9, DOI: 10.1007/978-94-017-8594-5_12,

Daybreak: The Clare Court Mural Project by Allison L. Duggan.
Courtesy of Annette Saunders, President of GPTSA

The Clare Court intergenerational housing wall Mural was created by grandparents, grandchildren, and Nuns (2011). The mural was created to reflect the collectivity and inclusivity of urban environments, the acceptance of people from all walks of life, diverse, yet accepting of each other, white, black, yellow other hands held together… paint me a dream of daybreak in Clare Court, hand in hand.

Let Them Know

Hey, thought that was you. You look good.
Come in. Sit down.
Here, take the tallest, broadest wide-hipped, cushiest, purple suede chair.
Yeah, that one.
Crowned with that lush velvety, long, fringed lavender throw.
That's it, slide into the sunniest spot, right here by the high flying window with the 10 foot palm pushing against white wood window panes.
Probably wishing to join the Japanese maple and pink-red tulip trees in spring's first bloom, but more than that, *outside*.
Talk to me, settle into the company of grace, good times, sorrows and sensibility, triumph, terrors (night and day), greetings, goodbyes, leave-takings and longings.
Fire up fat, chunky, white, slow-melting honey-comb candles clearing space like wasabi wads clearing cold-crammed sinuses.
Serve up tea-for-the fever: chamomile, lemon, Constant Comfort'
Feet up, head back, hips settled?
Talk about traveling, the journeys to your self
Let them know we are not merely memory at 50, 60,70, 80, 90 or 100
We are magic mindfulness, redolent of ritual, grounded in grace
Let them know adventure still claims our lives,
Talk to them of delights newly discovered and old ones kept close; still warm.

Settle in some more relishing the relating
Like soft butter on crisp bread, sugar sliding into the cup of the spoon, honey hanging poised for pouring onto ready lips.

Let them know:

Life has yet to be done with us, the silly and strong, the weak and wailing, the still sexual and sensuous self.
Let them know the delicious ways wine still wets our souls.
Curl into our dreams, delve into our desires—loitering, but not languishing,
Still and always, our tickets to ride onto urban cityscapes
Freighted with those sisters, Unseen and Unexpected, even for aging eyes and lives.

Let them know:

I am not done, nor are you.
I am not through, nor do WE ever intend to be.
Not as long as cellular memory celebrates our being.

Let them know:

Insist on telling, relish the relating:
We are here, still whole as possible through decades of life declaration.

Let them know:

This is certainly ME. Let them know. This sure as hell is YOU.
(Ama R. Saran, Ph.D. ©2013)

Aging in an urban environment affords both positive and challenging lived experiences. For some, the urban community is a place where lack of access, resources, networks, support, connection, economics, resilience, and isolation are anxiety provoking issues, whereas for others, it is just home—a wonderful place of refuge filled with people they love. The urban community offers a very unique context for aging across populations from the recent immigrants to those who are native to that particular urban community. It also offers endless possibilities for interdisciplinary work across professions and organizations that serve the aging population. Defining and describing aging in the urban environment is a daunting task because it is a complex and multi-dimensional context comprised of people from across cultures with unique histories, perspectives, values, beliefs, and practices. This chapter will attempt to discuss these complexities of aging in an urban context with specific attention to a few salient issues that intersects across cultural groups in urban communities and calls for interdisciplinary works. A strengths-based perspective of the urban environment for urban elders is emphasized.

At the Intersection of Aging and the Urban Environment

Theoretical Frameworks

The unique and diverse aspects of aging in the urban context call for a conceptual framework that is collective, inclusive, and intersectional. Guiding this discussion are relevant aspects of intersectionality theory, Systems/Ecological Theory, Strengths Perspective and the concept of resilience. These perspectives reflect the multiple dimensions of urban context and the complexities of the lived experiences of the elders in terms of the intersecting demands of urban living, the need to navigate systems at all levels from the micro to macro, and the importance of focusing on individual, familial, and communal strengths as a source of resilience for these urban elders.

The theory of intersectionality examines the interaction among social, cultural, and biological dimensions of systems which often negatively contributes to social inequality, racism, sexism, ageism, and how these dimensions interrelate to create a system of oppression that is intersected and often difficult to break, particularly in urban communities among older adults (McCall 2005). Understanding these intersections and the resulting relationships are especially critical for creating availability and accessibility to resources needed to sustain the urban elders. These interactions and relationships may be protective factors that lead to recognition of needs, establishment of policies that protect, and channeling of needed resources into the urban communities of older adults.

Similarly, Systems Theory/Ecological Theory (Germain and Gitterman 1980), discusses the various sub-systems that influence an individual or the older adult on the micro, mezzo and macro levels. These systems include family, community, neighborhood, organizations/institutions, and other subsystems of the environment that interact and relate to an elder's attributes (i.e., race, gender, age, class) and consequently influencing the outcomes of health and well-being of these elders. Because micro, mezzo, and macro systems are interdependent, the effects of these systems are reciprocal. For older adults in urban communities, the goal is for improving or enhancing systems connections that will produce positive outcomes.

Of major importance to how we conceptualize aging in an urban environment are the issues of resilience and the strengths perspective (Hill 1972) which emphasizes diversity and difference by culture, race, gender, and other dimensions as strengths and not deficiencies. The resilience of older adults in urban environment is linked to family and community inclusiveness and interdependency. To fully understand the urban environment begs the importance of a resilience and strengths perspective. According to the strengths-based perspective (Hill 1972; Saleebey 1996), focusing on the strengths of individuals, families, and communities will generate the best and most relevant solutions to problems impacting those individuals, families, and communities. The elders in the urban communities are the best resource for identifying solutions to challenges of their lived experiences. The strength of community in urban context relative to its contributions to the lives of older adults is well known. Through the lens of these theories, the multi-facets of urban context and older adults will be better understood. Accurate understanding and information will thereafter inform culturally relevant services and programs for those aging in the urban environment.

People in Place

> Growing up as a child in the Detroit area, in a city called Highland Park, my family, including several generations, were no strangers to the ramifications of living in the city entailed. My grandparents, great aunts and uncles, and other elder relatives have lived in urban communities for decades. Coming from the South, many of my family, like a lot of people of color, moved to places like Detroit and Chicago during the Great Migration of the 20th century. Several of them still live in the city and have lived with their children and families for generations. They view the city life as their way of life. As a product of intergenerational relationships, I have learned to respect my elders as keepers of our histories, to never take things for granted, and to be grateful for the little things in life, especially the love of family which is anchored on the strength of the elders.
>
> Tamika Baldwin, personal communication

> Aunt Joyce, the matriarch of the family who is seventy years old, first describes herself as blessed with no aches and pains, sharp, smart, and still cute. Aunt Joyce explains that it does not matter where you live and living in an urban community for seventy years has made her become a stronger person. Aunt Joyce points out that it is the respect that gives you the strength to enjoy your community no matter where it is. That you respect the inside and outside of your home and you require that from family, friends, neighbors and those passing by. Aunt Joyce describes that one day she and her eldest sister Doris who is seventy four years old was sitting on the front porch one evening and a gentleman threw an empty beer can in front of the stoop after she just finished sweeping. Doris said to the gentleman, pick that can up and put it in the trash. Respectfully, the gentleman picked up the can and apologized and promised that he would put in a trash can. Later that evening the man passed through again and said to both Aunt Joyce and her sister Doris, I put the can in the trashcan just like you told me and you don't have to worry about that again. Aunt Joyce further explains that you may not have the finest clothes or material things, but obtaining education is instilled in those living in an urban community.
>
> Kimberly Yancey, personal communication

As reflected in these personal accounts, there is a special innate quality of elders in urban communities in terms of the positive effects of their wisdom and experiences on people in places such as urban communities. As in these accounts, these elders have been relied upon to strengthen relationships both on the family and community levels from one generation to another.

According to the U.S. Census 2010, there were approximately 3,573 urban areas in the United States (U.S. Census Bureau 2010). Unlike rural areas, urban areas are defined as densely developed territory, which incorporates commercial, residential, and other non-residential urban land uses. An urban area is comprised of a minimum of 2,500 people of which 1,500 people reside outside of institutional group quarters. There are two types of urban areas: an urbanized area made up of at least 50,000 people, and an urban cluster made up of 2,500–50,000 people (U.S. Census Bureau 2010).

In the U.S., the number of adults ages 65 and older is on the rise (Baskind and Briar-Lawson 2005; U.S. Census Bureau 2010). The U.S. Census projects that the elder population will more than double from about 35 to 72 million representing nearly one-fifth of the entire population, by year 2030 (Bureau of Labor Statistics 2008; Slack and Jensen 2008). These elders come from all walks of life, cultures, and environments, with varying perspectives, and many live or will live in urban communities. According to Masotti et al. (2006), as there is an increase trend in aging, there is also an increase global trend toward urbanization, with the percentage of urban dwellers, most of whom are older adults, increasing from 47 % in 2000 to 60 % by year 2030.

Various terms have often been associated and used interchangeably with the word 'urban', including city, metro/metropolitan area, low-income area, neighborhood, the hood, ghetto, and the projects. The term 'urban' typically, and perhaps unfairly, has been associated deeply with negative connotations such as high rates of poverty, crime infestation, rampant homicide, disproportionate high school dropouts, racial residential segregation, unemployment, and low rates of community participation (Robert and Ruel 2006). The authors of this chapter beg to differ. Oswald et al. (2007) argue that environmental factors in urban communities play both negative and positive roles in health-related outcomes of urban residents, indicating that environmental-related factors can either contribute positively to older adults' independence in daily activities and subjective well-being, or negatively in terms of dependency and ill-being.

An environmental factor that is crucial to the functioning of older adults in urban communities is the built environment (BE), defined as one's physical surroundings including homes, parks and recreation areas, business areas, and transportation systems (Shendell et al. 2011). BE includes spaces, buildings, and products created or modified by individuals and groups of people as well as land use planning and other policies affecting communities (Clarke et al. 2008). Thus, some of the benefits of the urban areas that senior citizens enjoy are accessible public transportation, health facilities, shops, and other amenities, in addition to more opportunities for stimulation through social interaction and involvement in the community (Burton et al. 2011).

The link between BE and the health of older adults especially in promoting increased physical activity and social interaction has been established (Brown et al. 2009). BE helps determine whether or not an elder, who is physically capable of walking and biking, will actually walk or bike around the community (Li et al. 2005). Older adults are intensely dependent on their local environments to accomplish their daily activities (Clarke and George 2005). For instance, in naturally occurring retirement communities (NORCs), which are often in urban areas, there are more civic and social opportunities, like entertainment avenues, volunteer charities, and community and neighborhood organizations, with which the elders engage in their daily activities (Quinn 2008).

City life offers many positive experiences for elders from across cultures. Data shows that urban elders have lower rates of functional limitations, cognitive impairment, and chronic health conditions, compared with their rural counterparts

(Chumbler et al. 2001). Additionally, urban elders are more likely to have higher incomes than rural elders (Chumbler et al. 2001). Urban elders also tend to be keenly aware of their surroundings and responding to or managing risky situations. Thus, contrary to popular negative views, urban communities have a deeply collective orientation, a rich fictive kinship, extended access to public transportation, recreational amenities, and other support systems that combats the isolation of rural life, which are major strengths of urban communities (Burton et al. 2011). Older adults living in urban communities benefit greatly from this sense of collectivity and inclusivity in sharing material resources, family, and moral supports. Thus, maintenance of resources and structures are critical for continued positive outcomes for these elders.

Challenges for Urban Elders

Professionals alone have not been able to effectively deal with some age-old challenges affecting urban elders. These same challenges may require new approaches generated by the elders themselves in partnership with professionals from across disciplines and people from diverse cultures. With the growth of diverse older adult populations in urban communities, there is an urgent need for more multi-dimensional and innovative supports to deal with the following age-old challenges.

Substance Abuse. Alcohol use and abuse is a challenge among older adults living in urban communities which Ferrell and Sorocco (2006) referred to as the "invisible epidemic." Forty percent of all adults age 65 and older consume alcohol (National Institution of Alcohol Abuse and Alcoholism 2012). A review of the related literature indicates that few studies have comprehensively examined the use and effect of alcohol consumption among older adults, particularly in urban communities (Aira et al. 2005; Beck et al. 2001; Castellon et al. 2001; Chin et al. 1999; Ganry et al. 2000). Oftentimes risky behavior such as drinking is not associated with older adults. However, older adults are more at-risk for the effects of alcohol consumption given the natural ailments that comes with aging and that as people age their tolerance for alcohol decreases (Angermeyer et al. 2009). Interestingly, the medical profession and media are fully aware of the affects and dangers of teenage drinking, especially as it relates to driving. Do older adults not drive? Ross (2000) points out that, "clinicians should maintain the same or a greater degree of vigilance for alcohol use and abuse in those over age 65 as in those under age 65" (p. 71).

Alcohol consumption among older adults is also alarming because it is often taken with a prescription or over the counter (OTC) medication for age-related health ailments. Aira et al. (2005) found that "the risk of adverse effects of alcohol use was mostly related to medication-alcohol interactions rather than alcohol consumption" (p. 682). A study by Johnson (1997) on use of alcohol and OTC medications as sleep aids among urban women age 85 and older, found that the use of alcohol and OTC medications were linked to the inability to sleep and that the inability to sleep was related to the fear of living in urban areas.

Abundant availability and easy access to alcohol also contribute to alcohol use and abuse among the urban aging population. In a study by Johnson (1997), a 90 year old woman who drank everyday stated, "having a few drinks is so easy—no fuss, no muss. Just go to the store and bring it home-couldn't be easier" (p. 186). Convenience and inexpensive cost of alcohol contributes to the perception that it is not harmful. This misconception could complicate existing health conditions of older adults with diabetes, high blood pressure, congestive heart failure, liver problems, osteoporosis, memory problems, and mood disorders (National Institute on Alcohol Abuse and Alcoholism 2012).

Additionally, socioeconomic factors contribute to alcohol use and abuse among older adults living in urban communities (De La Rosa et al. 2007). For instance, Angermeyer et al. (2009) found that older adults who had higher economic status consumed more alcohol. A study by Livingston and King (1993) of the prevalence of alcohol use and psychiatric problems among older women living in inner-city in the UK, found that participants had higher socioeconomic status. Ferrell and Sorocco (2006) further explained that there is a bias among health professionals in assessing for alcohol consumption among older adults that is linked to socioeconomic status, in that, "clinicians are less likely to screen for alcohol use problems among older individuals, women, the educated, and those with higher economic status" (p. 455).

Assessment. Given the major challenge of substance abuse among older adults and the possibility that alcohol disorders "can be mistaken for medical or psychiatric conditions otherwise common among the elderly" (Angermeyer et al. 2009, p. 1377), there is a need not only for access to appropriate health care, but also a need for a reliable and valid standard assessment for substance abuse as part of regular health care. An example of an assessment that clinicians and researchers have utilized to assess alcohol use among older adults is the CAGE. Culberson (2006) explains Cut down, Annoyed, Guilty, Eye-opener (CAGE) consisting of four questions:

1. Have you ever tried to Cut down on your drinking?
2. Have people Annoyed you by criticizing your drinking?
3. Have you ever felt bad or Guilty about your drinking?
4. Have you ever had a drink first thing in the morning to settle your nerves or to get rid of a hangover? (Eye-Opener)

Using CAGE, Montgomery et al. (2009) found that men scored higher than women in alcohol consumption and that this finding contributes to the validity of the CAGE in assessing women accurately. However, Brousse et al. (2009) challenged the use of CAGE in assessing alcohol consumption and older adult population, urging priority on the development of new measures that focuses on elder-specific factors. The primary challenge of the CAGE is that it is not population specific in assessing alcohol use and abuse among older adults (Culberson 2006); thus, there is an urgent need for development of urban elder-specific assessments for accurate evaluation of alcohol abuse among older adults in urban settings.

Mental Health. Addressing mental health issues of older adults in urban communities can be challenging because of its complexities. Urban communities struggle with issues of poverty and debilitating housing environments which may result in older adults retreating within their homes thus leading to social isolation (Massey and Denton 1993) that may contribute to depression and stressful experiences living in an urban community (Zheng et al. 2011). On the other hand, Wight et al. (2007) found that individual characteristics contributed to depressive symptoms rather than the urban community environment. Two thirds of the participants in this study were women which perhaps speak to the difference in depression among older male and female adults living in urban communities. As such Zunzunegui et al. (2007) found the difference was in socioeconomic factors during the lifespan. In other words, older women were disadvantaged compared to older men due to frequent social and economic challenges, as well as lifetime stressors. Thus, older adult women in urban communities may be at greater risk for depression.

The willingness and ability of older adults to seek out mental health services in urban communities is also a challenge. Urban families are historically known not to seek out public mental health treatment because of the attached stigma. Norris and Karlin (2006) found that older adults are less likely to seek mental health treatment compared to young adults in urban areas; however older adults were also less distressed. Uphold et al. (2005) further found that even with an HIV/AIDS diagnosis, older adults in urban areas are less likely to experience depression than older adults living in rural areas. Perhaps, although poverty, debilitating housing, socioeconomic challenges and HIV/AIDS can plague an urban community, older adults have kin, family, and community strengths that support their mental health.

HIV/AIDS. Despite stereotypes, many older adults are sexually active (Illa et al. 2010; Jacobson 2011). According to Jacobson (2011), the majority (75 %) of older adults ages 57–64, more than half of those ages 65–74, and more than a quarter of people 75 and older, are sexually active. Given these statistics, older adults are at high risk for contracting HIV/AIDS, with ethnic minorities being affected disproportionately by the disease (Ward et al. 2004). Of U.S. HIV/AIDS diagnoses in 2007, 12.5 % were age 50 and older, which is expected to increase to 50 % of all HIV/AIDS cases by year 2015 (Heckman et al. 2011). Due to the number of underreported cases and those who have 'aged in' with the illness, these statistics may be higher (Altschuler et al. 2004). Perception of risk is affected by different factors such as age, gender, ethnicity, religion, education, worries, knowledge about the disease, number of sexual partners, condom use, and environment (Ward et al. 2004). At higher risk are urban, low-income, senior-housing residents who may be greatly exposed to criminal activity, drug trafficking, and commercial sex trades (Ward et al. 2004). However, urban residents are less likely to be depressed than rural residents in relation to HIV infection (Uphold et al. 2005). Nevertheless, the effects of stigma related to HIV/AIDS have profound effects across populations (Sayles et al. 2007).

Symptoms of HIV/AIDS such as fatigue, memory loss, general aches and pains, headache, loss of appetite and weight, are also often mistaken among older adults and health care providers as solely signs of aging or existing chronic illnesses (Illa et al. 2010; Jacobson 2011). Given the belief that vulnerability to sexually transmitted diseases (STDs) declines with age and reduced sexual activity, it is imperative that education interventions raise awareness about behaviors beyond perceptions, which place elders at risk. Likewise, prevention must take place at the intrapersonal, interpersonal, institutional, community, and policy levels of urban communities (Jacobson 2011; Ward et al. 2004). There should be an integration of the Ryan White CARE Act, the Older Americans Act, and the Health Care Affordable Act in the implementation of services for HIV/AIDS infected and affected seniors in urban communities (Emlet and Poindexter 2004). Also, collaborating with religious institutions in the urban community is crucial in the fight against HIV/AIDS (Derose et al. 2010).

Nutrition. Research suggests that an adequate amount of fruits and vegetables consumed daily helps to lessen the risk for acquiring such illnesses as cancer, coronary heart disease (CHD), and stroke (Li et al. 2012). The World Health Organization recommends a daily amount of 400 g of fruits and vegetables at a minimum, a standard which most countries do not meet (Li et al. 2012). Availability and accessibility to nutrition is influenced by age, gender, educational attainment, socioeconomic status, knowledge, attitudinal, and psychological factors, as well as geographical contexts (Fox et al. 2000). In many countries, including the U.S., there are few programs that focus specifically on nutrition services for the elders (Charlton and Rose 2001). As Perez-Escamilla and Putnik (2007) stated in regards to Latino immigrants in the U.S., one's original environment is also a powerful force in shaping lifestyle behaviors after leaving one's home country. Although it is reiterated time and again to exercise, eat right, and monitor one's weight, this behavioral approach does not take into account community cultural practices and external conditions (Eisenhauer 2001).

Urban food retailing, supermarket redlining, and limited access to healthy foods and recreational opportunities are issues with which many elders living in low-income urban neighborhoods struggle (Eisenhauer 2001). Income not necessarily culture has also been found to be more closely associated with nutritional deficiencies among urban elders. In order to better understand the health statuses of elders living in these communities, a review of the urban history and urban context, linked with issues of class, race, and racism, must all be taken into consideration, in addition to individual behavioral factors (Eisenhauer 2001). Food insecurity, which is defined as the limited or uncertain access to or availability of culturally relevant, nutritionally adequate, and safe foods, (Ramsey et al. 2011) is also a major nutrition issue among urban elders. Wolfe et al. (2003) indicated that the lack of finances is a primary cause of food insecurity for older adults. However, even when seniors have adequate financial resources, factors such as transportation, functional limitations and impairments, and health issues are risk factors for food insecurity. It is crucial that programs such as Meals on Wheels work in collaboration with community organizations and faith communities to provide the nutrition services needed by urban elders (Buys et al. 2010). Food insecurity may also be a source of risk factor for HIV/AIDS infection if it is associated with power differences in intimate relationships, sexual coercion, and trading sex for food, relevant across ages (Kalichman et al. 2011).

Some Relevant Approaches

Working with older adults from diverse cultures in urban communities requires acknowledgement and consideration of alternative ways of knowing and dealing with issues and circumstances that may or may not fit the "norm." Given the importance of considering gender, race, and socioeconomic differences among the elders, health, mental health, social service providers, and communities must seek to better understand the unique traits and abilities of older adults in urban communities (Briggs et al. 2011; Hanson and Gutheil 2004). For example, Yeh et al. (2006) discussed the use of collectivistic coping strategies such as familial and intra-cultural coping, forbearance, relational universality, fatalism/spirituality, and indigenous healing methods, whereas Blanchard-Fields et al. (2007) found the use of problem-focused strategies by urban elders from different cultures. In addition, with an increasing number of aging inmates and increased number of older adults suffering from years of abuse and addiction, compounded with a gross lack of available and accessible age and culturally relevant interventions, health, mental health, and social service professionals in urban communities are being challenged to think outside the box for novel approaches (Clay 2010; Hanson and Gutheil 2004; Snyder et al. 2009).

As a basic premise for approaching the challenges of urban elders, empowerment offers an opportunity for validation and support (DeCoster and Dabelko 2008). Raising awareness about healthy living is a foundational strategy (Richard et al. 2005), and identifying external barriers and ways to problem solve and access information and services including self-management services and support groups are components of the quest for health promotion among urban elders. Of great importance are social-action and education as strategies for advocating for elders in urban communities.

A model for working with older adults in urban communities by Marshall and Altpeter (2005) provides an example from social work that can be utilized across helping professionals. The model includes eight strategies that are relevant for the works of professionals across fields of practice:

1. Develop an activity-friendly community;
2. Advocate for transportation-friendly communities and an adequate infrastructure to enhance mobility;
3. Develop healthy work environments;
4. Foster coordination and collaboration across jurisdictions, government levels, and agencies;

5. Create opportunities to promote healthy aging in clinical services;
6. Enhance community competence;
7. Support economic development for health;
8. Promote profession-community research and information partnerships.

Furthermore, motivational interviewing is a versatile approach that has been found effective in working with older adults from across cultures and contexts. According to Miller and Rollnick (2012), motivational interviewing is a collaborative process that focuses on identifying and strengthening factors that motivates commitment and behavioral change. It is client-centered and semi-directive, and based on five general principles which include:

1. Expression of empathy,
2. Development of discrepancy between the envisioned status and current status of the person in order to appreciate change,
3. Avoiding arguments in order to prevent power imbalances,
4. Accepting reluctance to change as a natural process, and
5. Supporting self-efficacy, autonomy, and self-confidence (Miller and Rollnick 2012).

Overall, motivational interviewing emphasizes that the motivation to change comes from within an elder, not imposed from the outside; that interpersonal interactions can contribute to an elder's readiness to change; and that partnership in the process is critical (Miller and Rollnick 2012).

Cummings et al. (2009) found that using motivational interviewing contributed to major improvements in physical activity, diet, cholesterol, blood pressure, glycemic control, and increased smoking cessation among older adults. Bennett et al. (2005) also found in a study with nurses that using motivational interviewing on the telephone and email was a positive method for promoting behavioral change with urban older adults. Likewise, Cooper (2012) concluded that the combination of motivational interviewing and cognitive-behavioral therapy was effective in working with older adults who abuse drugs and alcohol.

Additionally, Min (2005) proposes five approaches for effectively working with racially and ethnically diverse elders which are relevant to those living in urban communities:

1. Re-conceptualize race/ethnicity and diversity in practice;
2. Identify and develop a conceptual framework for working with racially and ethnically diverse elders;
3. Consider a multidisciplinary community-oriented and neighborhood-based approach;
4. Advance culturally competent gerontological social work with diverse elders; and
5. Strengthen gerontological social work education with an emphasis on cultural competence.

Technology. Information Technology (IT) has become the new norm as a form of communication, society's vehicle for staying socially connected. IT has branched into various fields of practice relevant to older adults living in urban communities, such as the healthcare community in terms of physicians using a tablet to enter information about a patient's current health status. In a study of the use of HIT with nursing home patients in urban communities, Boratgis et al. (2011) established that most of the nursing home patients were not challenged with the use of electronic devices in their care and that only 23 % of these patients indicated that the use of electronic devices interfered with patient and staff time. HIT have also benefited older urban adults living at home, particularly those who have limited mobility, yet are able to use the Internet to access health information. Balcazar-Suarez et al. (2003) found that older adults in urban communities who have access to the internet and therefore access to health information experienced increased empowerment and feelings of being informed and in control as well as greater appreciation for the role of IT in access.

Campbell (2013) likewise suggests that the internet has become a mechanism for ensuring that older urban adults are actively involved in their care, particularly in situations where health care providers are not readily available. The internet can help with general and specific health information and patients can receive an immediate response regarding a health concern. E-visits through email or webcam have been approved by the Medical Association as a means for communication between physicians and patients. In terms of frail older adults living in an urban environment, Finkelstein et al. (2011) affirmed the use of tele-health technologies to enable maintenance of independent living. Consequently, HIT is a viable resource for older adults living in an urban environment which can aid in maintaining social connections with health care providers. It is the new norm for obtaining immediate access to health care information and it is extremely advantageous for older adults in urban communities. Nevertheless, while HIT can be valuable, use of HIT can also be challenging for older patients who often rely on a doctor's visit as an opportunity for social engagement with another person. Thus, although older adults in urban communities have begun to take advantage of these technological advances to maintain family and social connections as well as access information and services online, IT may still present challenges of personal nature.

Spirituality and Faith-based Organization. The urban community is enriched with faith-organizations through which many elders practice their spirituality. Religion and spirituality are core factors in the health and well-being of many older adults in urban communities thus, incorporating these factors into practice is essential (Hodge 2005). Some specific helping activities for urban elders include: prayer, meditation, and visualization, clergy referrals and collaboration, rituals, inspirational readings, spiritual narratives and dialogues, religious language, and touching client for helping purposes (Canda et al. 2004). Appropriately discussing an elder's spiritual and religious beliefs can help build rapport and form a trusting client-worker relationship (Damianakis 2001).

Among older adults from across ethnic cultures in urban communities, spirituality is used for multiple purposes including coping with life, death, tragedies, and as a source of hope, meaning, inner strength, health and well-being (Cascio 1998; Gotterer 2001; Hodge and Limb 2010). A crucial point in the evolution of urban communities in the U.S. is the historical significance of spirituality and religiosity and the role of the elders in the survival of people from across ethnic cultures including immigrants. For instance, an iconic feature of many heavily populated urban communities, historically, is the Black churches which served as a refuge and source of spiritual empowerment for African Americans. Led by the elders in the community, Black churches have been recognized as the oldest and most influential institution by and for Black people (Adksion-Bradley et al. 2005) attending to the physical, social, psychological, and religious needs of Black people. The Black churches not only concerned itself with matters of the soul but also help people cope with the injustice of society. It was viewed as a response to the needs of Black people requiring social services after slavery (Martin and Martin 2002).

Fictive Kin. The reciprocal benefits of intergenerational relationships between the youngians and their elders are magnified in settings such as the urban environment. Across cultures of immigrants as well as long-term dwellers in urban communities, a sense of social, communal or kinfolk support exists between and across generations. The resiliency of urban African American, Latino American, Native American, Asian American and Pacific Islander families with all of its unique dynamics relies heavily on the support of the elders in the community. Before modern day strategies such as therapy and support groups, the elders from different cultures living in urban communities were the primary sources of social and kin support for families. For example, nearly all families in the urban community had a "*big momma*" who was the *rock* that held the family together. Her wisdom in counsel is readily received. One can find this female elder sitting on the porch, peeping out the window, or at church. She knew how to pray and often time nobody knew she was praying until situations were better. At that point, this elder would reveal in her confident soothing voice prayers such as "The Lord is my Shepherd and I shall not want." Often time individuals did not know what the prayer meant, but this elder knew what needed to be said for each family member. When problems occur in the family, people would reference what this elder would do or say. Even in her passing, her legacy of using prayer to uplift the family remains the strategy for the family.

A variety of urban families including African American families have also relied on Aunt Gladys who is not blood related, but is respected as aunt. This aunt has the same authority and respect as blood relatives. Urban families can rely on this aunt for food, shelter, and overall support. She has the authority to discipline and the same family rights as any other family member. She holds the families together when its members are having problems. She often steps up and takes on parenting responsibilities when children come to the attention of the child welfare system. The Aunt Gladys today are legally known as fictive kin, although this formal term is not commonly used across families of color in urban communities. Historically, Maryland did not recognize fictive kin as a bond; however, because so many Aunt Gladys wanted to care for abused or neglected children, Maryland changed the law and recognized the strong bond of fictive kin (Family Law Article §5-507(b) Annotated Code of Maryland 2010).

Organizational Example: Baltimore County, Department of Aging
Baltimore, Maryland

The Baltimore County Department of Aging, Maryland, takes great pride in ensuring healthy and happy older adults in their predominantly urban environments. Healthy aging in the urban environment is the commitment of Baltimore County Department of Aging as reflected in a comprehensive service plan that includes the County's Community Resources, Maryland Access Point (MAP) website, and Aging and Disability Resource Center (ADRC), to name a few of the services and benefits available to older adults. With a vision to ensure that all residents in Baltimore realize their full health potential, Healthy Baltimore 2015 commits to advocacy and partnership for the promotion of access, equity, and creation of healthy individuals and communities.

Arnold Appel, Director
Department of Aging, Baltimore County, Maryland

Advocacy Example: In the Spotlight

Mrs. Annette Saunders is a native Baltimorean and product of Baltimore City Public Schools. She has a Master's in Human Services from Sojourner Douglass College. Mrs. Saunders is an activist for parent education and community engagement on behalf of grandparents raising grandchildren. She is the founding president of Grandfamilies PTSA (GPTSA), the "first in the nation" PTA serving the needs of grandparents raising grandchildren. In 2011, she received the Generations United Grandfamilies Award. GPTSA is a community based entity affiliated with Maryland PTA and National PTA. Mrs. Saunders is also a member of the National PTA Urban Family Engagement Network, Baltimore City Local Engagement Leadership Team, and Baltimore City Public Schools' Parent and Community Advisory Board (PCAB). She is a consultant with Baltimore City Health Department's Office on Aging and CARE Services and the University of Maryland's School of Social Work.

In celebration of the 2012 National Grandparents Day, Mrs. Saunders developed and coordinated a week of activities for grandparents across Baltimore City. The week concluded with the 1st Annual GRAND CARE Conference in Baltimore City and a respite day trip for grandparents to The National Constitution Center in Philadelphia, PA. Mrs. Saunders also served as a panelist on the Grant Makers for Children, Youth and Families' briefing on the subject "Second Generation Parenting: Raising Grandchildren and/or Adult Children with Disabilities," held at the Dirksen Senate Building in Washington DC. Mrs. Saunders has a special commitment to grandparents living in urban communities. She promotes the importance of celebrating urban families in this dispensation as the family begins to be realigned and restored.

As a product of the urban community, she embraces the strengths of urban environment and its extraordinary people, in relation to endurance, reciprocity, adaptability, relationships and obligations. She is a firm believer that grandparents have the special gift of picking up "the fruits" (grandchildren) and making phenomenal creations with it. Contrary to popular perspective of brilliance as high IQ, educational achievements, financial increases, Mrs. Saunders believe in the brilliance of the everyday people in the urban environment who are able to "feed 8 children with 5 potatoes." At the end of the day, the most precious possession to Mrs. Saunders is spending time with her two marvelous grandsons, passing on a legacy of strength, honor, and being fearless about standing up for and with others.

Discussion

As highlighted throughout the chapter, older adults living in urban communities face multiple challenges in today's society, many of which are the usual age-old challenges, while others are emerging as a result of changing times. With the baby boomer generation growing older and living longer, the stereotypical view of what aging looks like must change to accommodate this population. Not only will older adults need to be seen in a different light, but also in a light that takes into consideration their unique lifestyles, cultures, and environments. Different approaches relevant to working with urban elders have been discussed for professionals across the disciplines. These methods are ever-evolving and must be reinvented for the next generation of urban elders.

Among the many implications for research is an urgent need for more in-depth understanding of ways to combat challenges encountered by African Americans and immigrants living in urban communities. For example, there is limited research that explores the impact of substance abuse on older adults living in urban communities. Moreover, an examination of older adults' mental health seeking behaviors in urban communities is extremely important particularly because aging is a time of loneliness and sometimes withdrawal from family, friends, and society. Developing new, effective, evidence-based strategies that will aid in the longevity of African Americans and immigrants in terms of aging successfully in urban communities is paramount.

In terms of policy implications, there is a need for health care providers to improve the policies on how older adults are assessed and screened for general health care and specifically substance abuse and HIV/AIDS. Likewise, there has been a new wave of apartment style buildings exclusively for older adults emerging in urban communities. It is important that these projects continue to receive funding through local and state governments given that debilitating housing conditions has an impact on the mental health of older adults in urban communities. Additionally, it is essential that mandated policies such as the Health Care Affordability Act aka "Obama care" do not fall by the way side regardless of what political party is in office. Older adults benefit from the preventive screenings and the affordability of obtaining over the counter prescriptions.

Lastly, academia is in a prominent position to contribute positively to the lives of the elders living in urban communities which are most often the communities that surround a university campus. Among the many implications for education is the importance of educating and training students to work with and advocate for older adults in the urban community relative to health promotion, equity and equality in treatment, access to resources and quality services, to name a few. The internship, field education and service learning aspects of most professional schools in a university provide an opportunity for students to learn first-hand the art and science of advocacy and providing culturally-sensitive services to an ever growing diverse communities of urban elders. Likewise, more interdisciplinary advocacy among public health, social work, criminal justice, medicine, nursing for example, is needed in order to respond to the multi-dimensional needs of urban elders.

References

Adkison-Bradley, C., Johnson, D., Lipford Sanders, J., Duncan, L., & Holcomb-McCoy, C. (2005). Forging a collaborative relationship between the Black church and the counseling profession. *Counseling and Values, 49*, 147–154.

Aira, M., Hartiainen, S., & Sulkava, R. (2005). Community prevalence of alcohol use and concomitant use of medication—a source of possible risk in the elderly aged 75 and older? *International Journal of Geriatric Psychiatry, 20*, 680–685.

Altschuler, J., Katz, A. D., & Tynan, M. (2004). Developing and implementing an HIV/AIDS educational curriculum for older adults. *The Gerontologist, 44*(1), 121–126.

Angermeyer, M. C., Bickel, E. M., Bussche, H. V., Cvetanovska-Pllashiniku, G., Eifflaender-Gorfer, S., Haller, F., et al. (2009). At-risk alcohol drinking in primary care patients aged 75 years and older. *International Journal of Geriatric Psychiatry, 24*, 1376–1385.

Balcazar-Suarez, Y., Cassey, M., Kinney, L., Masi, C. M., & Piotrowski, Z. H. (2003). Internet access and empowerment: A community-based health initiative. *Journal of Internal Medicine, 18*, 525–530.

Baskind, F. R., & Briar-Lawson, K. (2005). Advancing social work with older adults. *Families in Society, 86*(3), 424–430.

Beck, J. C., Fielding, J. E., Harwar, N., Higa, J., Morgenstern, H., & Moore, A. (2001). Are older hazardous and harmful drinkers less likely to participate in health-related behaviors and practices as compared with nonhazardous drinkers? *JAGS, 49*, 421–430.

Bennett, J. A., Perrin, N. A., Hanson, G., Bennett, D., Gaynor, W., Flaherty-Robb, M., et al. (2005). Healthy aging demonstration project: nurse coaching for behavior change in older adults. *Research in Nursing and Health, 28*, 187–197.

Blanchard-Fields, F., Mienaltowski, A., & Seay, R. (2007). Age differences in everyday problem-solving effectiveness: Older adults select more effective strategies for interpersonal problems. *Journals of Gerontology: Psychological Science, 62*, P61–P64.

Boratgis, G., Chen, E. K., Eimicke, J. P., Henderson, C. R., Lachs, M. S., Meador, R. H., et al. (2011). Effects of electronic information technology implementation on nursing home resident outcomes. *Journal of Aging and Health, 24*, 92–111.

Briggs, W. P., Magnus, V. A., Lassiter, P., Patterson, A., & Smith, L. (2011). Substance use, misuse, and abuse among older adults: Implications for clinical mental health counselors. *Journal of Mental Health Counseling, 33*(2), 112–127.

Brousse, G., Llorca, P. M., & Malet, L. (2009). CAGE unsound for alcohol misuse in seniors. *International Journal Geriatric Psychiatry, 24*, 432–435.

Brown, S. C., Mason, C. A., Lombard, J. L., Martinez, F., Plater-Zyberk, E., Spokane, A. R., et al. (2009). The relationship of built environment to perceived social support and psychological distress in Hispanic elders: The role of "eyes on the street". *The Journals of Gerontology, 64B*(2), 234–246.

Bureau of Labor Statistics. (2008). *Spotlight on statistics: Older workers*. Washington, DC: U. S. Department of Labor. Retrieved July 12, 2013 from http://stats.bls.gov/spotlight/2008/older_workers/pdf/older_workers_bls_spotlight.pdf

Burton, E. J., Mitchell, L., & Stride, C. B. (2011). Good places for ageing in place: Development of objective built environment measures for investigating links with older people's wellbeing. *BMC Public Health, 11*(1), 839. doi:10.1186/1471-2458-11-839.

Buys, D. R., Marler, M. L., Robinson, C. O., Hamlin, C. M., & Locher, J. L. (2010). Recruitment of volunteers for a home-delivered meals programme serving homebound older adults: A theoretically derived programme among faith communities. *Public Health Nutrition, 14*(8), 1473–1478. doi:10.1017/S1368980010002880.

Campbell, R. J. (2013). Internet–based health information seeking among low-income, minority seniors living in urban residential center. *Home Health Care Management and Practice, 21*, 195–202.

Canda, E. R., Nakashima, M., & Furman, L. D. (2004). Ethical considerations about spirituality in social work: Insights from a national qualitative survey. *Families in Society, 85*(1), 27–35.

Cascio, T. (1998). Incorporating spirituality into social work practice: A review of what to do. *Families in Society, 79*(5), 523–532.

Castellon, S. A., Daum, G., Dickerson-Fuhrman, E., Hinkin, C., Jaffe, J., & Jarvik, L. (2001). Screening for drug and alcohol among older adults using a modified version of the CAGE. *The American Journal of Addictions, 10*, 319–326.

Charlton, K. E., & Rose, D. (2001). Nutrition among older adults in Africa: The situation at the beginning of the millennium. *The Journal of Nutrition, 131*(9), 2424S–2428S.

Chin, M. H., Friedmann, P. D., Hayley, D. C., Jin, L., Karrison, T., Miller, A., et al. (1999). The effect of alcohol abuse on the health status of older adults seen in the emergency department. *Drug Alcohol Abuse, 25*(3), 529–542.

Chumbler, N. R., Cody, M., Booth, B. M., & Beck, C. K. (2001). Rural-urban differences in service use for memory-related problems in older adults. *The Journal of Behavioral Health Services and Research, 28*(2), 212–221.

Clarke, P., & George, L. K. (2005). The role of the built environment in the disablement process. *American Journal of Public Health, 95*(11), 1933–1939.

Clarke, P., Ailshire, J. A., Bader, M., Morenoff, J. D., & House, J. S. (2008). Mobility disability and the urban built environment. *American Journal of Epidemiology, 168*(5), 506–513. doi:10.1093/aje/kwn185.

Clay, S. W. (2010). Treatment of addiction in the elderly. *Aging Health, 6*(2), 177–189. doi:10.2217/ahe.10.14.

Cooper, L. (2012). Combined motivational interviewing and cognitive-behavioral therapy with older adult drug and alcohol abusers. *Health and Social Work, 37*(3), 173–179.

Culberson, J. W. (2006). Alcohol use in the elderly: Beyond the CAGE part 2: Screening instruments and treatment strategies. *Geriatrics, 61*(11), 20–26.

Cummings, S. M., Cooper, R. L., & Cassie, K. M. (2009). Motivational Interviewing to affect behavioral change in older adults. *Research on Social Work Practice, 19*, 195–204.

Damianakis, T. (2001). Postmodernism, spirituality, and the creative writing process: Implications for social work. *Families in Society, 82*(1), 23–34.

De La Rosa, M., Delva, J., Kim, S., & Rice, C. (2007). Prevalence of smoking and drinking among older adults in seven urban cities in Latin America and Caribbean. *Substance Use Misuse, 41*, 1455–1475.

Decoster, V. A., & Dabelkoo, H. I. (2008). Forty-four techniques for empowering older adults living with diabetes. *Health and Social Work, 33*(1), 77–80.

Derose, K. P., Mendel, P. J., Kanouse, D. E., Bluthenthal, R. N., Castaneda, L. W., Hawes-Dawson, J., et al. (2010). Learning about urban congregations and HIV/AIDS: Community-based foundations for developing congregational health interventions. *Journal of Urban Health, 87*(4), 617–630. doi:10.1007/s11524-010-9444-6.

Eisenhauer, E. (2001). In poor health: Supermarket redlining and urban nutrition. *GeoJournal, 53*(2), 125.

Emlet, C. A., & Poindexter, C. C. (2004). Unserved, unseen, and unheard: Integrating programs for HIV-infected and HIV-affected older adults. *Health and Social Work, 29*(2), 86–96.

Ferrell, S. W., & Sorocco, K. H. (2006). Alcohol use among older adults. *The Journal of General Psychology, 133*(4), 453–467.

Finkelstein, S. M., Potthoff, S., Ratner, E. R., Speedie, S. M., & Zhou, X. (2011). Perception, satisfaction and utilization of the VALUE home telehealth service. *Journal of Telemedicine and Telecare, 17*, 288–292.

Fox, T., Brummit, P. S., Ferguson-Wolf, M., Abernethy, M., et al. (2000). Position of the American Dietetic Association: Nutrition, aging, and the continuum of care. *Journal of the American Dietetic Association, 100*(5), 580–595.

Ganry, O., Joly, J. P., Queval, M. P., & Dubreuil, A. (2000). Prevalence of alcohol problems among elderly patients in a university hospital. *Addiction, 95*, 107–113. doi:10.1046/j.1360-0443.2000.95110711.x.

Germain, C. B., & Gitterman, A. (1980). *The life model of social work practice*. New York: Columbia University Press.

Gotterer, R. (2001). The spiritual dimension in clinical social work practice: A client perspective. *Families in Society, 82*(2), 187–193.

Hanson, M., & Gutheil, I. A. (2004). Motivational strategies with alcohol-involved older adults: Implications for social work practice. *Social Work, 49*(3), 364.

Heckman, T. G., Sikkema, K. J., Hansen, N., Kochman, A., Heh, V., & Neufeld, S. (2011). A randomized clinical trial of a coping improvement group intervention for HIV-infected older adults. *Journal of Behavioral Medicine, 34*(2), 102–111. doi:10.1007/s10865-010-9292-6.

Hill, R. (1972). *The strengths of black families*. New York: Emerson Hall.

Hodge, D. R. (2005). Spiritual ecograms: A new assessment instrument for identifying clients' spiritual strengths in space and across time. *Families in Society, 86*(2), 287–296.

Hodge, D. R., & Limb, G. E. (2010). Native Americans and brief spiritual assessment: Examining and operationalizing the Joint Commission's assessment framework. *Social Work, 55*, 297–307.

Illa, L., Echenique, M., Saint Jean, G., Bustamante-Avellaneda, V., Metsch, L., Mendez-Mulet, L., et al. (2010). Project roadmap: Reeducating older adults in maintaining aids prevention: A secondary intervention for older HIV-positive adults. *AIDS Education and Prevention, 22*(2), 138–147.

Jacobson, S. A. (2011). HIV/AIDS interventions in an aging U.S. population. *Health and Social Work, 36*(2), 149–156.

Johnson, J. E. (1997). Insomnia, alcohol, and over-the-counter drug use in old–old urban women. *Journal of Community Health Nursing, 14*(3), 181–188.

Kalichman, S. C., Pellowski, J., Kalichman, M. O., Cherry, C., Detorio, M., Caliendo, A. M., et al. (2011). Food insufficiency and medication adherence among people living with HIV/AIDS in urban and peri-urban settings. *Prevention Science, 12*, 324–332. doi:10.1007/s11121-011-0222-9.

Li, F., Fisher, K. J., Brownson, R. C., & Bosworth, M. (2005). Multilevel modeling of built environment characteristics related to neighbourhood walking activity in older adults. *Journal of Epidemiology and Community Health, 59*, 558–564. doi:10.1136/jech.2004.028399.

Li, Y., Li, D., Ma, C. Y., Liu, C. Y., Wen, Z. M., & Peng, L. P. (2012). Consumption of, and factors influencing consumption of, fruit and vegetables among elderly Chinese people. *Nutrition, 28*(5), 504–508. doi:10.1016/j.nut.2011.07.023.

License for child placement agency, Family Law Article, §5-507(b) Annotated Code of Maryland (2010).

Livingston, G., & King, M. (1993). Alcohol abuse in an inner city elderly population: The Gospel Oak survey. *International Journal of Geriatric Psychiatry, 8*, 511–514.

Marshall, V. W., & Altpeter, M. (2005). Cultivating social work leadership in health promotion and aging: Strategies for active aging interventions. *Health and Social Work, 30*(2), 135–144.

Martin, E., & Martin, J. (2002). *Spirituality and the black helping tradition in social work*. Washington, DC: NASW Press.

Masotti, P. J., Fick, R., Johnson-Masotti, A., & MacLeod, S. (2006). Healthy naturally occurring retirement communities: A low-cost approach to facilitating healthy aging. *American Journal of Public Health, 96*(7), 1164–1170.

Massey, D. S., & Denton, N. A. (1993). *American apartheid: Segregation and the making of the underclass*. Cambridge: Harvard University Press.

McCall, L. (2005). The complexity of intersectionality. *Signs, 30*(3), 1771–1800.

Miller, W. R., & Rollnick, S. (2012). *Motivational interviewing: Helping people change* (3rd ed.). NY: Guilford Press.

Min, J. W. (2005). Cultural competency: A key to effective future social work with racially and ethnically diverse elders. *Families in Society, 86*, 347–358.

Montgomery, P. R., John, P. D., & Tyas, S. (2009). Alcohol misuse, gender, and depressive symptoms in community-dwelling seniors. *International Journal of Geriatric Psychiatry, 24*, 369–375.

National Institute on Alcohol Abuse and Alcoholism. (2012). *Older Adults*. Retrieved March 10, 2013 from http://www.niaaa.nih.gov/alcohol-health/speical-populations-co-occurring-disorders/older-adults

Norris, M. P., & Karlin, B. E. (2006). Public mental health care utilization by older adults. *Administration and Policy in Mental Health and Mental Health Services Research, 33*(6), 730–736.

Oswald, F., Wahl, H. W., Schilling, O., Nygren, C., Fange, A., Sixsmith, A., et al. (2007). Relationships between housing and healthy aging in very old age. *The Gerontologist, 47*(1), 96–107.

Pérez-Escamilla, R., & Putnik, P. (2007). The role of acculturation in nutrition, lifestyle, and incidence of Type 2 diabetes among Latinos. *The Journal of Nutrition, 137*(4), 860–870.

Quinn, A. (2008). Healthy aging in cities. *Journal of Urban Health, 85*(2), 151–153. doi:10.1007/s11524-008-9268-9.

Ramsey, R., Giskes, K., Turrell, G., & Gallegos, D. (2011). Food insecurity among adults residing in disadvantaged urban areas: Potential health and dietary consequences. *Public Health Nutrition, 15*(2), 227–237. doi:10.1017/S1368980011001996.

Richard, L., Gauvin, L., Ducharme, F., Gosselin, C., Sapinski, J., & Trudel, M. (2005). Health promotion and disease prevention for older adults: Intervention themes and strategies used in Québec local community health centres and seniors' day centres. *Canadian Journal of Public Health, 96*(6), 467–470.

Robert, S. A., & Ruel, E. (2006). Racial segregation and health disparities between black and white older adults. *The Journals of Gerontology: Series B: Psychological Sciences and Social Sciences, 61B*(4), S203–S211. doi:10.1093/geronb/61.4.S203.

Ross, E. M. (2000). In the literature: A guide to evidence-based nutrition practice. *Nutrition in Clinical Care, 3*(2), 71–75.

Saleebey, D. (1996). The strengths perspective in social work practice: Extensions and cautions. *Social Work, 41*(3), 296–305.

Sayles, J. N., Ryan, G. W., Silver, J. S., Sarkisian, C. A., & Cunningham, W. E. (2007). Experiences of social stigma and implications for healthcare among a diverse population of HIV positive adults. *Journal of Urban Health, 84*(6), 814–828. doi:10.1007/s11524-007-9220-4.

Shendell, D. G., Johnson, M. L., Sanders, D. L., Nowakowski, A. C. H., Yang, J., Jeffries, C. D., et al. (2011). Community built environment factors and mobility around senior wellness centers: The concept of "safe senior zones". *Journal of Environmental Health, 73*(7), 9–18.

Slack, T., & Jensen, L. (2008). Employment hardship among older workers: Does residential and gender inequality extend into older age? *The Journals of Gerontology, 63B*(1), S15–S24.

Snyder, C., van Wormer, K., Chadha, J., & Jaggers, J. W. (2009). Older adult inmates: The challenge for social work. *Social Work, 54*(2), 117–124.

Uphold, C. R., Rane, D., Reid, K., & Tomar, S. L. (2005). Mental health differences between rural and urban men living with HIV infection in various age groups. *Journal of Community Health, 30*(5), 355–375. doi:10.1007/s10900-005-5517-y.

U.S. Census Bureau. (2010). Retrieved May 1, 2013 from http://www.census.gov/c2ss/www/
Ward, E. G., Disch, W. B., Levy, J. A., & Schensul, J. J. (2004). Perception of HIV/AIDS risk among urban, low-income senior-housing residents. *AIDS Education and Prevention, 16*(6), 571–588.
Wight, R., Seeman, T. E., Miller-Martinez, D., Karlamangla, A. S., Botticello, A. L., & Aneshensel, C. S. (2007). Urban neighborhoods and depressive symptoms among older adults. *The Journal of Gerontology, 62*(1), S52–S59.
Wolfe, W. S., Frongillo, E. A., & Valois, P. (2003). Understanding the experience of food insecurity by elders suggests ways to improve its measurement. *The Journal of Nutrition, 133*(9), 2762–2769.
Yeh, C. J., Inman, A., Kim, A. B., & Okubo, Y. (2006). Asian American collectivistic coping in response to 9/11. *Cultural Diversity and Ethnic Minority Psychology, 12*, 134–148.
Zheng, X., Zhang, J., Song, X., Pang, L., Li, N., & Chen, G. (2011). Risk factors for depression in older adults in Beijing. *Canadian Journal of Psychiatry, 56*(8), 466–473.
Zunzunegui, M. V., Sicotte, M., Tellechea, L., Beland, F., & Alvarado, B. (2007). Social and Gender inequalities in depressive symptoms among urban older adults of Latin America and the Caribbean. *The Journal of Gerontology, 62*(4), S226–S236.

A Call to Action: Dementia Screening of Alzheimer's Disease in Older African Americans

Sheila Black and Gaynell M. Simpson

Abstract In the current chapter, we discuss racial disparities with respect to the detection of dementia. We also discuss reasons for the disparities including cultural bias with respect to dementia screening instruments. Topics include the impact of health and educational disparities on test performance, and how these disparities need to be considered when assessing members of minority groups. Special emphasis is given to the fact that accurate detection of dementia is important. Individuals suffer needlessly when falsely diagnosed with dementia. Thus, it is important that clinicians make a concerted effort to minimize false positive diagnoses. On the other hand, it is important that dementia be detected early. If dementia has progressed before an accurate diagnosis is made, individuals will have missed the opportunity to delay the progression of the disease through medication and other cognitive interventions.

Healthy cognitive functioning of individuals across the lifespan represents a complex process associated with memory, attention, language, and decision-making. Dementia is a clinical syndrome caused by an ongoing loss of structure or functioning of neurons, including death of neurons (Henderson and Jorm 2002). The most common dementia pathologies are Alzheimer's disease (AD), vascular dementia, Lewy body and front temporal dementia. Ethnic and racial group members, such as African American, Latino, and Asian American elders have higher rates of vascular dementia compared to White Non-Hispanic (WNH). However, only African American and Latino elders have increased rates of AD compared to WNH (Froehlich et al. 2001). Given the importance of screening for cognitive impairment as a step towards intervention, it is critical to examine the diagnostic accuracy of screening tools for dementia, specifically as it relates to the disproportionate numbers of elders of color. As Dilworth-Anderson et al. (2012) eloquently point out, it is an injustice that ethnic minority elders, particularly

---

S. Black (✉) · G. M. Simpson
University of Alabama, Tuscaloosa, AL, USA
e-mail: sbpsych31@aol.com

H. F. O. Vakalahi et al. (eds.), *The Collective Spirit of Aging Across Cultures*,
International Perspectives on Aging 9, DOI: 10.1007/978-94-017-8594-5_13,

African American elders are not diagnosed properly with respect to dementia. Today, if a timely and accurate dementia diagnosis is made, interventions are available that can delay the rate of cognitive deterioration and functional decline. Diagnostic inaccuracy continues to disconnect diagnosis and proper intervention for ethnic minority elders.

Utilizing concepts of health disparities, cognitive reserve theory, and acculturation theory as guiding frameworks, the authors examine possible explanations for the differential rates of AD among African American elders; and explore why established diagnostic screening measures for dementia are not necessarily valid for African American older adults. Furthermore, the authors discuss the compounding problems of inaccurately labeling cognitively intact African Americans as demented. The authors conclude with recommendations of potential dementia screening tools that can improve detection of geriatric cognitive impairment for the African American population.

The Role of Educational and Health Disparities

Older, African Americans have been subjected to widespread discrimination with respect to healthcare, educational attainment, and occupational advancement, that have led to greater vulnerability for dementia and AD. The *Healthy People Report of* 2010 defines health disparities as difference in incidence, prevalence mortality, burden of disease and other adverse health conditions/outcomes among specific populations. Special attention is targeted to ethnic minority elders, particularly to the significant health disparities of African American elders. Studies have confirmed that older individuals with vascular diseases such as hypertension, diabetes, and other cardiovascular problems, have a greater likelihood of cognitive decline (Grodstein 2007; Hachinski 2008; Viswanathan 2009), which connects the health disparity among African American elders and their disproportionate rates of dementia. Although older individuals suffering from metabolic disorders such as diabetes are unaware of the neurological consequences of their chronic disease, such diseases are insidiously compromising cognitive functioning through their effect on the brain and other parts of the central nervous system (Viswanathan 2009). Indeed, research has extended our understanding about the complex intersections of health disparities, such as health care access and utilization, region, poverty, and racial discrimination and cognitive decline of community dwelling older African Americans (Crowe et al. 2010).

Similarly, disparities with respect to access to education and intellectually challenging occupations have a deleterious effect on the cognitive functioning of older adults. High education and socioeconomic status serve as protective factors for cognitive functioning for a number of reasons including the process of cognitive reserve, which refers to the enrichment of the brain that occurs because of intellectual stimulation (Stern 2002). These alterations serve to protect the brain from the possible deleterious changes in the brain as a result of illness or aging

(Whalley et al. 2004). Intellectually challenging experiences increase cognitive reserve by changing the strength and the number of synaptic connections between neurons. However, because of historically discriminatory practices, there remain disparities between African Americans and Whites with respect to educational and occupational achievement. As a result of continuous discrimination, older African Americans have had less opportunity to pursue higher education or professional careers that impedes exposure to intellectual stimulation. Many elders educated during the Jim Crow era (1800–1950s)in which schools were segregated were often forced to drop out of school to help with sharecropping and other duties in the family that contributed to their survival. Moreover, there was often only one school designated as the "Colored High School" in a vicinity, and transportation was not readily available for the one "colored" institution in the area. Therefore, African Americans wishing to pursue a high school education often had to go through a number of obstacles to attend school.

Some researchers using dementia screening scales have attempted to circumvent the problems associated with the differences in years of education. For instance, Manly et al. (2003) recruited matched samples of African Americans and European Americans elders in terms of education level. That is, if the mean years of education was 13 for the European American sample, it was ensured that the mean years of education was 13 for the African American sample. Although the number of years of education was equal, it did not mean that African American and European American older adults received the same content or quality of instruction. African American elders were often subjected to dilapidated schools, out of date textbooks, and inadequate training as children. In addition to differences in formal education contributing to differences in performance on dementia screening instruments, the differences in the socioeconomic status and the economic status of one's family of origin must be considered. For example, vocabulary is dependent somewhat on the environment in which one is raised. If one is raised in an environment in which low frequency words such as "egregious" are used on a regular basis, one will learn the meaning of the word just through osmosis. African American elders are less likely than European American elders to have been raised in households in which they would have been exposed to low frequency vocabulary words. Thus, African American older adults are sometimes at a disadvantage because of the vocabulary and the phrasology used in such instruments.

Evidently, although African American and European American older adults may be equated on education level, there may still be differences in reading levels and quality of education because of institutionalized racism in the education system and socioeconomic disparities (Whitfield and Wiggins 2003). Therefore, it is imperative that future studies include quality of education as measured by reading ability, as well as, differences in quality of education when assessing cognitive performance in African Americans with dementia and AD. More importantly, it is critical to avoid confounding measures of reading ability by the effects of dementia (Chin et al. 2012).

Misdiagnosis, Culture, and Acculturation

Thus far, we have discussed factors that might result in African American elderly adults being more vulnerable to dementia. However, as Tappen et al. (2012) point out, to a certain extent, it is difficult to assess the prevalence rate of dementia among African Americans because misdiagnosis is so prevalent. A recent report building on decades of multiple studies examining the prevalence of dementia among the African American elderly have to be cautious in interpreting their findings due to the high rate of false positives (i.e., results indicating that the test-taker has dementia when the test-taker does not) among African American elderly test-takers (Borson et al. 2013). The high rates of false positives among elder African Americans are in large part because of the inherent biases of the test. Manly (2008) has conducted interesting work that indicates that performance on standardized tests of African Americans across age groups is often dependent on acculturation. Acculturation refers to an interpersonal and adaptive process of being exposed to and adopting customs and behaviors of the dominate culture (Manly 2008). Several studies have revealed that test performance is as much dependent on acculturation as education (Arnold et al. 1994; Byrd et al. 2005).

Why would acculturation matter for African Americans? To the extent that African American test takers are not acculturated, with respect to the dominant culture, African American individuals taking a standardized test may have difficulty comprehending test questions due to being unfamiliar with the wording and some of the situations described in tests. Several researchers have proposed that there are so many environmental and cultural differences between European and African Americans, that standardized cognitive tests have low construct validity with respect to African Americans (see Raydin and Katzen 2013). Construct validity refers to the degree to which a test actually measures the psychological concept that it was designed to measure (i.e., dementia). For example, Manly (2008) proposes that racial differences in performance on cognitive screening measures can be attributed to lack of construct validity with respect to the cognitive screening test and confounds differences in culture and environment between European and African Americans.

Finally, another issue that contributes to misdiagosis is the lack of knowledge concerning dementia and its associated symptoms within the African American community (Dilworth-Anderson et al. 2012). Alzheimer's disease is fairly easy to recognize in advanced stages, but not so easy to detect in the very early stages (Nestor et al. 2004). Research has shown that African American families often have only a basic knowledge of dementia and miss some of the subtle signs that are manifested in the early stages of the disease. For example, African American families often attribute pathological memory problems to "old age." Evidence has also shown that physicians of African American patients were also less likely to refer African American clients to neuropsychologists for a neurological workup. In fact, African Americans are often diagnosed much later in the disease stages that their European American counterparts (Dilworth-Anderson et al. 2012).

Early diagnosis of dementia is crucial for a number of reasons. For one thing, there are now treatments available that can slow the progression of dementia symptoms (Thompson et al. 2001), if the treatment is implemented early in the disease process. For example, pharmaceutical treatments such as Aricept have been shown to delay the rate of memory decline among individuals in the mild to moderate stages of dementia. Moreover, research has shown that cognitive interventions (e.g., memory training) are effective in delaying cognitive decline among individuals in the early to moderate stages of dementia (McDougall et al. 2010). In addition, early diagnosis allows for family members to engage in long term planning. Given the importance of screening for cognitive impairment it is critical that we continue to examine the diagnostic accuracy of screening tools for dementia.

Diagnostic Accuracy of Screening and Detection of Cognitive Impairment

In framing the discussion on cultural fairness of specific diagnostic instruments, the concepts of sensitivity and specificity must be considered. The sensitivity of a test refers to the ability of the test to accurately diagnose individuals who in fact suffer from some pathology (i.e., dementia). On the hand, specificity refers to the ability of the test to accurately screen out those individuals who do *not* suffer from the pathology (i.e., dementia). The goal of the individuals who construct a test is to have optimal sensitivity and specificity. These test developers set cutoff scores and create scales in an effort to maximize both sensitivity and specificity (Hamrick et al. 2013). However, there is often an inverse relationship between the two concepts and applicability to certain populations. Generally, cognitive tests have shown lower specificity with respect to African Americans (Fillenbaum et al. 1990).

The mini-mental status exam (MMSE) has been studied more than any other measure in its ability to accurately detect dementia when it exists (Folstein et al. 1975; Hawkins et al. 2011; Mitrushina 2009; Pedraza et al. 2012). More recently, researchers have been interested in its ability to accurately detect dementia among African Americans. There are a plethora of studies that indicate that the MMSE yields an unacceptable degree of false positives, when the diagnostic cutoffs listed in the manual are applied to African American test takers (Albert and Teresi 1999; Hawkins et al. 2011; Folstein et al. 1975). The MMSE examines attention, orientation, calculation, free recall, and to a limited extent, what cognitive psychologists refer to as working memory, or the ability to hold and manipulate information. Albert and Teresi conducted a classic study in which they used the MMSE as a dementia screening instrument to assess the prevalence of dementia among older adults in Harlem, New York in which they found that test performance was not only influenced by education; it was also influenced by the quality of the test takers education. Participants received the Wide Range Achievement Test-Revised (WRAT-R) so that the investigators would have some estimate of the quality of their education. The authors pointed out that nearly a quarter of the participants were reading below the third grade. Reading ability was correlated with MMSE score in that 66 % of participants in the lowest quartile met criteria for impairment but only 13 > 7 % in the highest quartile met the MMSE criteria for impairment. Only 26 % of participants had readings skills commensurate with their level of education. As indicated earlier, MMSE scores were highly correlated with reading levels. This research did not have any independent assessment of cognitive ability to assess the specificity of the MMSE but it did shed light on the notion that quality of education is an important factor to consider when administering the MMSE.

Likewise, because the MMSE is comprised of test items that relate to education (i.e., ability to spell, write sentences), researchers have modified the instrument to increase its sensitivity for individuals with lower levels of literacy. For instance, Hamrick et al. (2013) substituted serial and backward 7s/World subtest with the days of the week. Individuals were instructed to list the days of the week backwards rather than count backwards by 7s or spell the word "world" backwards. Hamrick et al. found that having participants recite the days of the week backwards resulted in fewer false positives than having participants perform the typical working memory tasks associated with MMSE (i.e., serial 7s). Similarly, Shen et al. (2006) point out that healthcare professionals may be able to circumvent the bias associated with psychometric testing by opting for informant interviews. These researchers examined informants' reports of cognitive functioning of two groups—Nigerians and African Americans and found a relationship between informants' reports of cognitive decline and the testers' score on a dementia screening instrument, (i.e., the MMSE). This finding indicate that dementia is essentially an appropriate diagnosis if there has been a change in one's cognitive functioning relative to an earlier portion of one's life.

A screening instrument that has been developed to reduce misdiagnosis among diverse populations is the Community Screening Instrument (CSI) for dementia (Prince et al. 2011). This instrument includes culturally fair test items and evidence shows that rates of false positive results with regard to dementia screening are consistent across ethnic and racial groups. Additionally, the CSI includes an informant interview that queries the informant about the test taker's functional abilities. The instrument has successfully examined dementia among cultures as varied as Nigerians, Cree Native Americans, and African Americans. While touting the CSI's ability to successfully screen for dementia across diverse populations, a number of researchers have complained that the test takes too long to administer—approximately 30 min. Thus, it would not be feasible for primary care physicians to administer the test on a regular basis, given the time constraints associated with office appointments. As a result, researchers have refined the test to include fewer questions, reducing the scale from 32 to 7 items with validity intact, to be administered in 5 min or less at a doctor's office. The brief version is in its infancy stage, and its efficacy in accurately detecting diagnosis across various populations remains to be seen. Developers of the test maintained that nurses and doctors (primary care healthcare professionals) can make accurate diagnosis of dementia given their knowledge of the patient. Tappen et al. (2012) pointed out that it is not always convenient to have an informant, that there needs to be a stand alone assessment tool for dementia that can be administered quickly, and that the test needs to be free of bias or systematic error in the process of measuring.

Tappen et al. (2012) also examined the possibility that the most accurate dementia screening tool is a combination of a functional screening assessment and a cognitive assessment. The Mini-Cog—Functional Activities Questionnaire (MC-FAQ) and the MMSE-FAQ have been widely used as cognitive dementia screening instruments. This study examined the performance of three ethnic groups (i.e., African Americans, Hispanic Americans, and European Americans) on both the Mini-Cog and the MMSE-FAQ. The MC-FAQ is a functional assessment designed to be free of educational or socioeconomic bias (Tappen et al. 2012). This instrument assesses the ability to perform everyday activities with scores ranging from 0 to 3. Data for this project was obtained from individuals who took part in a statewide screening for mild cognitive impairment and dementia. Individuals age 59–95 were included in the data analyses. Three ethnic groups were matched in terms of education, gender, and age. Participant scores were combined as suggested by Steenland et al. (2008) to allow rapid screening classifications. Findings indicate that when the MMSE and Mini-Cog were combined with the Functional Activities Questionnaire there were no differences across ethnic and/or racial groups, although effects of education remained.

The Time and Change Test is another innovative test that was developed to decrease educational and cultural biases associated with dementia screening (Froehlich et al. 1998). The test-takers' tasks revolve around everyday activities such as making change and correctly identifying the time on a clock. The test has high levels of accuracy with respect to correctly classifying individuals of dementia and has relatively low levels of bias.

Overall, the MMSE and other similar tests are still administered quite frequently in an effort to screen for dementia. However, the MMSE and other tests that rely heavily on reading and skills related to education achievements tend to yield a high rates of false positives among African Americans in comparison with Non-Hispanic whites. Researchers have developed several tests designed to be free of cultural bias (i.e., time and change test). However, more work needs to be conducted in this area. Clinicians often grapple with the problem that the norms that have been developed with regard to appropriate test performance were based on performance of white middle class older adults. Researchers need to collect normative data on the cognitive functioning of non-demented African American older adults in order to compare clients to a normative sample similar to them.

Implications of Misdiagnosis for Older African Americans

There are a number of implications associated with misdiagnosis of dementia in the African American community. Older adults, are falsely labeled as demented, must cope with the extra anxiety and discomfort associated with a false label. In fact, there are instances in which family members have started to doubt an older person's competence based on a false label. On the other hand, when older African American adults are inappropriately diagnosed with dementia there are dire consequences. Individuals diagnosed in the later stages of dementia, do not have the opportunity to engage in long term care planning, and do not benefit from interventions that work best in the early stages of the disease.

Early detection is extraordinarily important for a number of reasons. For instance, if treatment is implemented early in the disease process, the progression of dementia symptoms are decreased (Thompson et al. 2001); Pharmaceutical treatments such as Aricept have been shown to delay the rate of memory decline among individuals in the mild to moderate stages of dementia. Cognitive interventions (e.g., memory training) are also effective in delaying cognitive decline among individuals in the early to moderate stages of dementia. Professionals maintain that treatment should start when individuals are diagnosed with mild cognitive impairment (MCI), a syndrome in which cognitive decline is greater than would be expected based on a person's age and educational level. MCI has been identified as the very early stages of dementia (Belleville et al. 2011). Belleville and Bherer (2012) found that the implementation of memory training among individuals with MCI resulted in individuals using new strategies to compensate for decline that had occurred because of cognitive impairment, and this greater compensation was manifested by greater activation in parts of the brain. Because memory training has been effective for individuals suffering from MCI, Belville and colleagues suggested that memory training become routinely administered to individuals suffering from mild cognitive impairment.

Given all of the treatments now available for individuals suffering from dementia, it is an injustice if African Americans suffering from dementia do not receive appropriate treatment or interventions until the disease has progressed to the moderate or severe stage. Likewise, given the psychological trauma associated with a false diagnosis of dementia, it is also an injustice if African American older adults are more likely to receive a false diagnosis. The onus is on aging researchers to develop appropriate screening tools for older adults of all ethnicities and cultures.

References

Albert, S. M., & Teresi, J. A. (1999). Reading ability, education, and cognitive status assessment among older adults in Harlem, New York City. *American Journal of Public Health, 89*(1), 95–97.

Arnold, B. R., Montgomery, G. T., Castaneda, I., & Longoda, R. (1994). Acculturation and performance of Hispanics on selected halstead-reitan neuropsychological tests. *Assessment, 1*, 239–248.

Belleville, S., & Bherer, L. (2012). Biomarkers of cognitive training effects in aging. *Current Translational Geriatrics and Experimental Gerontology Reports, 1*, 104–110.

Belleville, S., Clément, F., Mellah, S., Gilbert, B., Fontaine, F., & Gauthier, S. (2011). Training-related brain plasticity in subjects at risk of developing alzheimer's disease. *Brain: A Journal of Neurology*, 134, 1623–1624.

Borson, S., Frank, L., Bayle, P. J., et al. (2013). Improving dementia care: The role of screening and detection of cognitive impairment. *Alzheimer's and Dementia, 9*, 151–159.

Byrd, D., Sanchez, D., & Manly, J. (2005). Neuropsychological test performance among Caribbean-born and U.S-born African American elderly: The role of age, education and reading level. *Journal of Clinical and Experimental Neuropsychology, 27*(8), 1056–1069.

Chin, A. L., Negash, S., Xie, S., Arnold, S. E., & Hamilton, R. (2012). Quality, and not just quantity, of education accounts for differences in psychometric performance between African Americans and White Non-Hispanics with alzheimer's disease. *Journal of the International Neuropsychological Society, 18*(2), 277–285.

Crowe, M., Sartori, A., Clay, O., Wadley, V., Andel, R., Wang, H., et al. (2010). Diabetes and cognitive decline: Investigating the potential influence of factors related to health disparities. *Journal of Aging and Health, 22*(3), 292–306. doi:10.1177/0898264309357445.

Grodstein, F. (2007). Cardiovascular risk factors and cognitive function. *Alzheimer's Dementia., 3*, 16–22.

Dilworth-Anderson, P., Pierre, G., & Hilliard, T. S. (2012). Social justice, health disparities, and culture in the care of the elderly. *The Journal of Law, Medicine & Ethics, 40*(1), 26–32.

Fillenbaum, G., Heyman, A., Williams, K., Prosnitz, B., & Burchett, B. (1990). Sensitivity and specificity of standardized screens of cognitive impairment and dementia among elderly black and white community residents. *Journal of Clinical Epidemiology, 4*, 651–656.

Folstein, M. F., Folstein, S. E., & McHugh, P. R. (1975). "Mini-mental state". A practical method for grading the cognitive state of patients for the clinician. *Journal of Psychiatric Research, 12*, 189–198.

Froehlich, T. E., Bogardus, S. T., & Inouye, S. K. (2001). Dementia and race: Are there differences between African Americans and Caucasians? *Journal of the American Geriatric Society, 49*, 477–494.

Hachinski, V. (2008). Shifts in thinking about dementia. *Journal of the American Medical Association, 300*(18), 2172–2173. doi:10.1001.

Hamrick, I., Hafiz, R., & Cummings, D. M. (2013). Use of days of the week in a modified mini-mental state exam (M-MMSE) for detecting geriatric cognitive impairment. *Journal of the American Board of Family Medicine, 26*, 429–435.

Hawkins, K. A., Cromer, J. R., Piotrowski, A. S., Pearlson, G. D. (2011). Mini-mental status exam performance of older African Americans: Effect of age, gender, education, hypertension, diabetes, and the inclusion of serial 7s subtraction versus 'world' backward on score. *Archives of Clinical Neuropsychology, 26*(7), 645–652. Retrived from doi:10.1093/arclin/acr054.

Henderson, S. A., & Jorm, A. F. (2002). Definition and epidemiology of dementia: A review. In M. Maj & N. Sartorius (Eds.), *dementia* (2nd ed.). New York: Wiley.

Manly, J. J. (2008). Race, culture, education, and cognitive test performance among older adults. In S. M. Hofer & D. F. Alwin (Eds.), *Handbook of cognitive aging: Interdisciplinary perspectives* (pp. 398–417). Thousand Oaks: Sage Publications Inc.

Manly, J. J., Touradji, P., Tang, X., & Stern, Y. (2003). Literacy and memory decline among ethnically diverse elders. *Journal of Clinical and Experimental Neuropsychology, 25*, 680–690.

Manly, J. J., Jacobs, D. M., Touradji, P., Small, S. A., & Stern, Y. (2002). Reading level attenuates differences in neuropsychological test performance between African American and White elders. *Journal of the International Neuropsychological Society, 8*, 341–348. doi:10.1017/S1355617702813157.

McDougall, G. G., Becker, H., Vaughan, P., Acee, T., & Delville, C. (2010). The SeniorWISE study: Improving everyday memory in older adults. *Archives Psychiatric Nursing, 24*(5), 291–306.

Mindt, M. R., Arentoft, A., Coulehan, K., Byrd, D. (2013). Considerations for the neuropsychological evaluations of older ethnic minority populations. In L. D. Ravdin and Katzen (Eds.), *Handbook on tmhe Neuropsychology of Aging and Dementia Clinical Handbooks in Neuropsychology* (pp. 25–41). Berlin: Springer

Mitrushina, M. (2009). Cognitive screening methods. In I. Grant & K. M. Adams (Eds.), *Neuropsychological assessment of neuropsychiatric and neuromedical disorders* (3rd ed., pp. 101–126). New York: Oxford University Press.

Pedraza, O., Clark, J., O'Bryant, S. E., Smith, G. E., Ivnik, R. J., Graff-Radford, N. R., et al. (2012). Diagnostic validity of age and education corrections for the mini-mental state examination in older african americans. *Journal of The American Geriatrics Society, 60*(2), 328–331. doi:10.1111/j.1532-5415.2011.03766.x.

Prince, M., Ferri, C. P., Jotheeswaran, A., Acosta, D., Guerra, M., Huang, Y., et al. (2011). The association between common physical impairments and dementia in low and middle income countries, and, among people with dementia, their association with cognitive function and disability: A 10/66 dementia research group population-based study. *International Journal of Geriatric Psychiatry, 26*(5), 511–519. doi:10:/gps.2558.

Ravdin, L. D., Katzen, H. L. (2013). Handbook of the neuropsychology of aging and dementia, Clinical Handbooks in Neuropsychology. New York: Springer.

Shen, J., Gao, S., Unverzagt, F. W., Ogunniyi, A., Baiyewu, O., Gureje, O., et al. (2006). Validation analysis of informant's ratings of cognitive function in African Americans and Nigerians. *International Journal of Geriatric Psychiatry, 21*(7), 618–625.

Steenland, N. K., Auman, C. M., Patel, P. M., Bartell, S. M., & Lah, J. J. (2008). Development of a rapid screening instrument for mild cognitive impairment and undiagnosed dementia. *Journal of Alzheimer's Disease, 15*(3), 419–427.

Stern Y. (2002) What is cognitive reserve? Theory and research application of the reserve concept. *Journal of The International Neuropsychological Society*, *8*(3), 448–460. doi:10.1017/S1355617702813248

Tappen, R. M., Rosselli, M. M., & Engstrom, G. G. (2012). Use of the MC-FAQ and MMSE-FAQ in cognitive screening of older African Americans, Hispanic Americans, and European Americans. *American Journal of Geriatric Psychiatry, 20*(11), 955–962. doi:10.1097/JGP.0b013e31825d0935.

Thompson, S., MacDonald, J., & Coates, T. (2001). Improving visual memory with Aricept (donepezil hydrochloride, E2020) in mild-to-moderate alzheimer's disease. *Clinical Gerontologist, 24*(1/2), 55–73.

U.S. Department of Health and Human Services. (2001). *Healthy people 2010.* McLean: International Medical Publisher.

Viswanathan, A., Rocca, W. A., & Tzourio, C. (2009). Vascular risk factors and dementia: How to move forward? *Neurology, 72*(4), 368–374. doi:10.1212/01.wnl.0000341271.90478.8e.

Whalley, L., Deary, I., Appleton, C., & Starr, J. (2004). Cognitive reserve and the neurobiology of cognitive aging. *Aging Research Reviews, 3*, 369–382.

Whitfield, K. E., & Wiggins, S. (2003). The impact of desegregation on cognition among older African Americans. *Journal of Black Psychology, 29*, 275–291.

Reworking the Template: The Financial Crisis and Housing Transitions of Older Americans

Tam Perry

Abstract This chapter highlights the impact of the global financial crisis on older Americans who were planning to voluntarily relocate in its aftermath, illustrating how the instability of the world economy played out in the individual lives of older adults within different social, environmental, and economic contexts. Older Americans are "reworking the template," or reassessing their approaches to analyzing their housing options. Drawing from an ethnographic study (January 2009–May 2012) of older Americans relocating ($n = 81$), this chapter shows the various ways this population addresses housing concerns including (1) reconfiguring relationships between sellers and buyers (2) rethinking home ownership (3) reconsidering moving and (4) redirecting resources. As the effects of the recession may be long-term and continue to influence actions of older Americans in particular ways, the chapter concludes with suggestions for professionals working with older Americans who are facing these concerns and decisions.

For many decades, Americans have been drawn to homeownership, which served personal and financial purposes (Cauley et al. 2007). Homeownership created communities which were stable places to live and raise a family, often planned near neighborhood schools and stores (Rohe et al. 2002). Homeownership also contributed to the reduction of crime and provided opportunities for advancement for those living in the community (Haurin et al. 2002; Rohe et al. 2002). Tax policies have long provided incentives for homeownership (Pynoos and Liebig 2009), and homes often served as investments that could be sold to cover medical and living expenses in old age. Mortgage payments could thus be viewed as "self-imposed savings plans" promoting a "systematic bias" of homeownership (Beracha and Jonson 2012, p. 218). In many ways, homeownership acted as a template, or a source of stability and security in older adulthood, that complemented the potential volatility of the stock market and diminishing corporate-offered pension plans.

T. Perry (✉)
Wayne State University, Detroit, MI, USA
e-mail: teperry@wayne.edu

H. F. O. Vakalahi et al. (eds.), *The Collective Spirit of Aging Across Cultures*, International Perspectives on Aging 9, DOI: 10.1007/978-94-017-8594-5_14,

Arguably, this template worked primarily for those from middle class and upper middle class economic backgrounds (Beracha and Johnson 2012; Xiao and Liu 2007). The homeownership template served as an aspirational plan for people of varying socioeconomic backgrounds.

The global financial crisis of 2008 destabilized expectations of security in older adulthood. Older adults have had to create new plans for the future or "rework" their template for old age. Some have chosen to postpone retirement, or to rejoin the workforce in a part-time capacity (McFall 2011; O'Loughlin et al. 2010). McFall found that older Americans planned to delay retirement by approximately 2.5 months due to wealth loss between July 2008 and May/June 2009. Others have delayed purchases and travel while they reassess their financial well-being. Given the emphasis on the individual rather than corporate planning for retirement in the American context (Herd 2009), older adults and their families must make these decisions.

The full impact of the global financial crisis on older adults' decisions about housing is yet to be fully known. This chapter begins by presenting an overview of what is known about how Americans make choices about their housing in old age. A discussion of various ways that the financial crisis affected the relocation practices older adults follows, including (1) reconfiguring relationships between sellers and buyers (2) rethinking home ownership (3) reconsidering moving and (4) redirecting resources. Older Americans are "reworking the template," or reassessing their approaches to analyzing their housing options, while planning for their housing needs in older adulthood. This chapter highlights the impact of the global financial crisis on older Americans who were planning to voluntarily relocate at the peak of the crisis, illustrating how the instability of the world economy played out in the individual homes of older adults within different social, environmental, and economic contexts. The chapter concludes with suggestions for professionals working with older Americans facing these changing structural contexts.

Background

As they age, older adults examine whether their residences and communities suit their needs. The proximity of their kin to their residence, the cultural norms regarding co-residence with kin or living independently in older adulthood, and affordability all affect these examinations. Cross-cultural research shows differences in navigation of spaces when living intergenerationally (Lamb 1997) as well as different preoccupations in relocation (Marcoux 2001). When considering housing needs, environmental and social contexts are interrelated for older adults. The environmental contexts of the older adult's residence include ways that older adults navigate their physical spaces (e.g., difficulties arising from climbing stairs) and the tasks that older adults execute to maintain their contexts both inside and outside the home (e.g., lawn maintenance). Sometimes lawn care and seasonal tasks like shoveling snow may be outsourced. The location and configuration of residences within neighborhoods and communities as well as community resources and safety are also important in understanding an older adult's environmental context. For example, an older adult living in a community with crime, abandoned homes, and unreliable public transportation may not utilize resources offered in the larger community. In these ways, environmental contexts impact the social contexts of older adults. The social contexts of older adults include accessibility and connectedness to family members, peers and neighbors.

Environmental contexts, often termed the "built environmental" include several elements. The built environment is comprised of "urban design" or the arrangement and appearance of physical elements, "land use" or the allocation of residential and commercial space within a community, and the transportation systems available within this community (Handy et al. 2002). Gerontological scholars have assessed the built environment and its impact on older adults in terms of walkability and physical activity (Cunningham and Michael 2004) and in terms of access to services and amenities like grocery stores (Smith et al. 2013). Research also links the built environment and older adults to predictors of obesity (Michael and Yen 2009), mental health (Evans 2003), and perceptions of social support (Brown et al. 2009). Brown et al. found that architectural features that promoted interactions, such as porches, were positively associated with perceptions of social support.

In assessing their contexts, older persons may reevaluate the benefits and drawbacks of remaining in their current housing (Wiseman 1980; Sergeant et al. 2010). Erickson et al. (2006) found that one reason for staying in one's home includes having an adult child in the area. Those who moved were "pulled" toward family and toward a residence with less maintenance, rather than "pushed" out of their homes because of the difficulty of maintaining them (Erickson et al. 2006; Krout et al. 2002). This pull toward family has been demonstrated repeatedly in the literature and can be thought of as the "joint impact of human and economic resources" (Longino et al. 1991). Human resources might include relatives who will come over with meals, help after medical operations, and mow the lawn. Economic resources can sometimes circumvent the need for family as professionals can be hired for many of these same tasks; however, scholars have shown that older adults experience varying levels of comfort with using non-kin as support for these concerns (Miner 1995).

If older adults plan to stay in their current homes, complementary supports like home modification may help those older adults living alone or living with a spouse/partner to continue to age in place. Examples of modest home modification include installation of grab bars to prevent falls and ramps to promote accessibility. More complex home modification includes creating additions with barrier-free bathrooms and widened doorways. Costs associated with modifications vary greatly, and some communities offer assistance with the modification of grab bar installation and home repairs. These agencies may be supported by local and federal funding and may rely on volunteer labor.

Relocation is an alternative to aging in place, particularly for older adults who are concerned about their emotional well-being, physical health and safety. Relocation is an option that may be supported or even strongly encouraged by kin as they express concern for the well-being of their older family members. Wiseman's model of relocation (1980) addresses triggering events like "age related loss and critical events", push factors such as loss of a spouse, pull factors such as "successful relocation" of a friend that may lead to relocation of the older adult, indigenous and exogenous factors that are considered when making a decision to move such as tangible factors like wealth and cost of living, and intangible considerations like ties to one's community (p. 145). Wiseman also coined the term "involuntary stayers" for older adults who would like to move but who are not able to relocate for various reasons.

Litwak and Longino (1987) identified three types of moves made by older adults. The first is an "amenity" move, where older people relocate to embrace a different lifestyle, usually involving less maintenance of their housing unit, the possibility of participating in leisure activities (such as golf and swimming), or residing in an environment with warmer weather. The second type of move is considered an "assistance" move, where older people move towards family or other relatives who can provide support when needed. Lastly, "nursing home" moves, where older people move to facilities because family assistance cannot address all of their needs. As long-term care options expanded, new models of housing options have combined some of these types of moves into, for example, facilities offering a continuum of care (Geboy et al. 2012; Groger and Kinney 2006; Shippee 2009). Scholars continue to explore "place attachment" or ties to one's home and possessions (Rowles and Watkins 2003; Ekerdt et al. 2012) and designing new models of relocation to incorporate "experience-driven belonging" resulting from place attachment and "behavior-driven agency" based on beliefs about controlling one's environment (Wahl and Oswald 2010). Attending to those older adults who are "stuck in place" (Torres-Gil and Hofland 2012), researchers describe the emotional reframing of their current residence that may take place for older adults when they realize they will not be able to move (Golant 2011). Golant suggests that "residential normalcy" is achieved by reformulating goals, thinking less about present living conditions by focusing on memories in a residence and minimizing problems associated with living in the residence.

Relocation can be either local, allowing older adults to remain in their own community while changing their residence, or non-local, where moving entails a new community in addition to a new residence. Non-local moves often involve development of a variety of new networks and identification of new resources in the community such as selecting new physicians, finding new banks and grocery stores, and learning about trash delivery and farmer's market schedules, among others. Yet, distance may be relative. Moves to another neighborhood within the same town may feel more, less, or equally drastic than moving across states or oceans. Longino et al. (2008) argue that those who had experience with non-local moves, such as having their children move away for education or work, or having themselves moved previously for work, were more likely to undertake non-local relocations in older adulthood. Length of time in their homes also factored into whether they moved or not, as those living in their homes for less time tended to move non-locally.

As discussed above, decisions to relocate are multi-faceted. However, unprecedented changes in the global financial landscape during the crisis that began in 2007 created an uncertain backdrop for older adults as they made decisions about relocating. Eleven trillion dollars of household wealth was lost, "with retirement accounts and life savings swept away" (Financial Crisis Inquiry Commission 2011, p. xv). Some attribute this to overconfidence in the housing industry that permitted loans and financing of both new construction projects and mortgages for persons who should not have been given loans. Others attribute a great deal to the deficiencies in the oversight of the credit markets. Whatever the origins, the ramifications of the turmoil and uncertainty in the international banking sectors transcended the globe.

The Financial Crisis Inquiry Commission (2011) writes, "the profound events of 2007 and 2008 were neither bumps in the road nor an accentuated dip in the financial and business cycles we have come to expect in a free market economic system. This was a fundamental disruption—a financial upheaval, if you will—that wreaked havoc in communities and neighborhoods across this country" (p. xv). The residents of these communities had to react to this global disruption as they made plans for their futures. Many sectors were affected by the global financial crisis. For example, as communities plan for their health care needs, their models of public and private pay service users is uncertain as a result of the recession (Sisko et al. 2009).

Older adults were affected by the global recession in major ways. Their retirement savings became less secure due to diminished assets and a short horizon in which to replace the assets (Butrica et al. 2010). The Great Recession has also been linked to experiences of pain and stress (Deaton 2012). Scholars have noted the psychosocial and personal shock that older adults experienced as a result of such uncertainty (Hicks and Kingson 2009). Also, older adults reported deterioration in their mental health (McInerney and Mellor 2012). In a longitudinal study of older adults in Australia, those older adults who reported being impacted by the financial crisis reported a significantly higher level of depression and anxiety than those who did not (Sargent-Cox et al. 2011). Their study also shows that those surveyed between April and September 2009 reported higher levels of depression than those surveyed afterwards, leading researchers to question the possibility of a protective effect as the financial crisis affected so many that it may have become socially normalized (Sargent-Cox et al. 2011). Framing the issue in moral terms, Moody (2009) links the economic turmoil to issues of trust, blame, and levels of risk tolerance. Furthermore, Moody suggests that the crisis has led to the re-examination of expectations that those who work should pay for those who do not. With these expectations shaken, questions of the reward of delaying gratification (by saving instead of spending) surface (Moody). Communities continue to work on ways to address the mental health concerns resulting from the financial crisis (Wahlbeck and McDaid 2012). Given the extent of the challenges resulting from the financial crisis, we gain a deeper understanding by examining housing issues in an American context.

Relocating Older Adults: A Case Example

To illustrate the impact of the financial crisis on older Americans, a case example is presented. This example is well-placed for understanding the global financial crisis. First, this study was conducted in the state of Michigan, a state known for its auto industry that was particularly hard hit in the financial crisis and experienced the recession earlier and more deeply than the rest of the country (Manning 2009). Second, data collection took place between January 2009 and May 2012, a period capturing the repercussions of the "Great Recession" that peaked in fall 2008.

Sample and Method

This study which comprised of older adults (n = 81), family members (n = 49), and supportive professionals (n = 46) highlights the experiences of older adults (ages 57–91) and their networks as they embark on relocating from their homes to senior living communities, condos, or other types of housing. Over 90 % of the older adult participants were White European Americans living in Michigan. There were fifty-nine women and twenty-two men among the older adults in the study. The researcher conducted interviews, participant observation of the moving process (packing, garage sales, moving day, adjustment) and document reviews at three stages: pre-move planning, move in-process, and post-move adjustment. The ethnographic approach of observing these contexts and participating in the moving process allowed for prolonged engagement through participant-observation, enabling the researcher to gain insight to the phenomenon studied (Geertz 1973). At the data analysis stage, interviews were selectively transcribed, and transcripts, fieldnotes, documents and photographs of spaces and experiences were analyzed for emergent themes. This method allowed the researcher to understand the impact of the financial crisis on the emotional concerns of participants and on their "surroundings," such as bank loan procedures and local housing markets. The original research questions focused on the processes of relocations; themes specifically about financial issues emerged during the project.

Contextual Backdrop

For older adult participants who were planning to relocate post-financial crisis, there were two significant contexts that they had to understand and factor into their need to "rework" the template. They had to understand the meanings and implications of the devaluation of their assets and the insecurity of the housing market on their plans to relocate. Depending on their asset distribution, many older adults had to determine whether their short-term and long-term spending habits needed revision. Others had to reassess whether their mix of assets or the distributions received on a monthly or yearly basis made sense in the new financial climate.

Older adults were also concerned about their housing in specific ways. First, some were concerned about the decreased value of their homes. People planning moves had counted on proceeds from the sale of a house at a certain price based on housing market trends over the previous few years. Coming to terms with housing values that might be much lower than expected was difficult financially. Many older adults' financial equations were built expecting a certain housing value. Reduced proceeds from sales would factor into which down payment plan these individuals would choose when moving to a retirement community or condo. A second concern was the inability to sell one's home or the illiquity of a large asset. Depending on the price range of the home and the number of homes in that price range in a local economy, homes might remain on the market for a long time.

The meanings and implications of these two contexts, devaluation of assets and insecurity of the housing market, lead to older adults experiencing anxiety related to their financial well-being. Many of the study participants had actively and proudly saved for old age, with the intention of not burdening others with either finances or care. The desire not to burden others stems from both Western ideas of individualism and also from the impact of growing up in or hearing stories about the Great Depression and frugality in consumption patterns and consciousness about saving. This recession created great emotional uncertainty about the financial futures and financial decisions related to housing choices for older adults. Feelings of apprehension about the inability to control one's financial futures weighed heavily. Worry about diminished stock portfolios lead to reluctance to commit to the cost of a retirement community if one was planning to relocate, both in terms of down payment and in terms of monthly fees. Given increased longevity and increased costs of long-term care, the timing of the move was re-examined by many movers who participated in this study.

Findings

Given the contextual backdrop of devaluation of assets and housing market insecurity, several themes were identified that described how the global financial crisis impacted the moving practices of older adults in this study. These included: (1) reconfiguring relationships between sellers and buyers, (2) rethinking home ownership, (3) reconsidering moving, and (4) redirecting resources.

Reconfiguring Relationships Between Sellers and Buyers

Uncertainty in the housing market troubled movers as potential buyers' ability to secure financing was questionable. For example, after receiving a bid on their home, one married couple experienced difficulty in completing the sale of their home because of a buyer's delay in securing finances, despite their buyer′s a long standing relationship with his bank and had a prior history of securing mortgages. In other cases, older adults had difficulty with buyers whose income was scrutinized by banks, a consequence of the financial crisis. Banks required potential mortgage holders to go through tedious processes of providing financial documentation. Holding non-traditional jobs or jobs where money earned was not classified as income, such as medical residency fellowships, created circumstances where financing was delayed. This created a tenuousness that led to a reconfiguring of the relationships between buyers and sellers; this reconfiguration possibly depended on the size of the community and the contacts between buyers and sellers. The sellers, older adults planning to relocate, obtained more information about buyers' work and home ownership histories than might have been expected in previous real estate transactions. Furthermore, the buyers' histories, also scrutinized as a result of the housing market insecurity, became a key factor impacting the possibility of relocating for these older adults.

Rethinking Home Ownership

The global financial crisis led some older adults to rethink home ownership in the relocation process. One option was to rent a unit at a retirement community rather than to purchase a residence. Relocating to a retirement community with rental options brought not only freedom from home and lawn maintenance and the possibility of a network of peers for social interaction, it also brought freedom from home ownership. Looking forward to giving up the responsibilities of home ownership, including avoiding maintenance and selling one's home, a participant in this study expressed great anxiety in negotiating with buyers about the condition of his house. The demands of the buyers kept him concerned about the sale. This man looked forward to the completion of the transaction so that he could reduce his anxiety over keeping the house ready for sale on a daily basis. Questioning the need for home ownership also resulted from the very challenges of trying to sell a home in an insecure housing market.

Reconsidering Moving

After the global financial crisis, some older adults who intended to move chose not to relocate due to their changed financial contexts. Despite previous excitement about moving and a belief that moving might increase their ability to function through a reduction in home maintenance and lawn care, older adults reassessed relocation in this new economic context. Attention to the affordability of a residence in a retirement community prompted them to withdraw deposits from retirement communities and emotionally postpone the goal of relocating. One couple, who were excited that a senior residence would open in their town, changed their decision about moving due to the financial crisis. They altered their budget by cutting back on their spending, and in consultation with their financial advisor, chose to receive less monthly income in order to preserve their assets. They hung a sign on their front door, publically displaying the economic impact on their household that said, "No sorry! No No No No Money!". Even after exploring alternative financing options, in the end, they still could not afford to move. When discussing whether they would consider moving in the future, the older man said, "It depends on our health…We could be a burden to our home." Their plan is to remain in their home. He notes that "We try to keep active as much as we can", but then acknowledges, "Of course, our body tells us to slow down. So, we have to listen to it".

Leaving a familiar environment, growing health concerns, and the physical difficulty of moving at an older age were also anxieties that were linked to moving for these older adults. Reconsidering moving meant "reworking" the current template of how to maintain the house, since affordability of hiring assistance for lawn care and snow removal, also weighed on older adults affected by the financial crisis. A suggested way to "rework" the template was to consider different types of residences to relocate to, such as apartments not noted as senior living, or senior housing in less desired locations.

Redirecting Resources

Older adults also "reworked" the template due to the inability to sell a home, which left them without the cash to make a down payment on a unit in a retirement community. Coming to terms with the lack of resources to fund a move led to reexamining of resources. In some cases, older adults had been preparing for the sale of their homes by repainting walls and disposing of possessions. They had expressed a sense of accomplishment at clearing out their spaces. But by reconsidering moving, older adults avoided the need to keep the space ready to show to potential buyers, and they avoided worry about transitions for pets. Some older adults also cancelled plans for moving to help out family members also affected by the financial crisis.

One woman provided an example of how her finances were reconfigured over the process of planning for a move. She said, "So I sort of put it all together, and the market was lousy…And, I just decided, I'm not gonna put it on the market." Instead of incurring the cost of a move and the cost of living in a retirement community, some older adults, like this woman, gave financial gifts such as college tuition to children and grandchildren. Additionally, some older adults aimed to preserve the financial resources for their relatives' inheritance upon their deaths.

As one woman explained, "I am the richest of all the people I love." An older adult may forego a more accessible residence to support relatives also facing tough times after the global financial crisis.

Discussion

Around the globe, the global financial crisis thwarted plans for individuals, families and communities. As addressed in this chapter, older Americans planning to relocate had to reassess their residential options. They had to "rework" their template in many ways: (1) reconfiguring relationships between sellers and buyers, (2) rethinking home ownership, (3) reconsidering moving, and (4) redirecting resources. Sometimes "reworking" a template involved a transformation in the meaning and practices of home ownership where relationships between buyers and sellers were reconfigured, and where long-held principles of homeownership were called into question. Especially in cases where home sale prices were decreased, or homes remained on the market without potential buyers, the "systematic bias" of homeownership (Beracha and Johnson 2012, p. 218) had to be reassessed. Did homeownership really bring security in older adulthood? In what ways did owning a home weigh an older adult with responsibilities rather than stability?

At times, relocation plans were delayed or cancelled, making futures uncertain, especially for those who had decided that their original homes did not suit their needs in terms of accessibility. Deciding to remain in one's home after the anticipation of a more navigable space, or a space requiring less internal and external maintenance, requires an emotional reframing to achieve "residential normalcy" (Golant 2011). Golant argues that when older adults realize that they were unable to relocate, they go through emotional processes to accept their current residence as their place of residence in the future. Lastly, older adults chose to forego moving in order to financially support members of their family as job loss and instability affected all generations. Or they chose to save resources for inheritances to be used for educational or other support for family members.

Examining the impact of the global financial crisis on older adults is important for three reasons. First, we can understand how such a crisis affected relocation plans of older adults. Relocation often involves emotional, health and financial projections (Perry 2012). For the latter, the financial projection is often a result of preparation decades before a move, as employment status and work history all play into this projection. In the uncertain economic times created by the global financial crisis, older adults frequently found that previous financial preparations had evaporated. Their personal financial projections became threads that were a part of unraveling of the fabric made up of financial institutions around the world. The aftermath of this unraveling is yet to be fully understood.

Second, relocation researchers have argued for the need to understand the decision-making processes that lead older adults to actively plan to move (Wiseman 1980; Bekhet et al. 2009). These studies suggest linearity beginning with triggering events and push factors with a goal toward relocation with place attachment and pull factors as key issues. This chapter has shown that the reverse can and did occur. Retreating from an active moving stage in these cases was based on financial reasons. Our models of relocation can become better informed by understanding the lived experiences of those older adults intending to relocate, to anticipate the fluctuations in moving plans resulting from financial, health and emotional variation.

Lastly, as the fluctuations in the economy affected older adults' plans for the future, practitioners serving older adults must redesign their practice interventions appropriately for the new reality. Older adults are delaying retirement, and reengaging in the work force in order to accommodate this new financial reality. With increased longevity, older adults must plan for longer periods of old age, and including health care needs that may accompany their longer life spans. Decisions about housing accompany this new reality. There has been an increase in multigenerational living since the Great Recession. While Bracha and Jamison have established that older adults have greater beliefs in home ownership than younger generations (2013), there has also been a documented shift in beliefs about home ownership (Frederick 2012). The older clients may be concerned about supporting themselves and other family members. Practitioners should explore whether anxiety and depression may be related to the financial concerns of older clients. Communities should also continue to explore various culturally appropriate solutions for older adults relative to the economic crisis, hoping to learn from this historical moment.

Conclusion

There were many ways the Great Recession impacted older adults who were planning to relocate. The effects of the Recession may be long-term and continue to influence decisions of older Americans in particular ways. Practitioners working with older adults and their families need to understand the impact of the Global Financial Crisis on significant decisions such as relocation. Policymakers also need to understand the altered impact on present projections of older adults' financial well-being. On a conceptual level, the country and arguably the world has experienced the need to reexamine the challenges of aging and housing under unprecedented contexts. Older adults share this historical moment on a global scale and will need to plan for its repercussions. There may be comfort or normalcy created to realize that most older adults were impacted by the Great Recession. However, we know that the lived experiences of the recession depends on one's multiple identities, including race, gender, and socioeconomic roles. For older adults, their physical and cognitive abilities intersect with these other identities which affect their navigation of their homes, factoring in whether relocation is considered. Understanding the gradations of impact depends first on acknowledging the issues of intersectionality represented in housing issues. Only then, can we best support older adults as they "rework" the financial template.

References

Bekhet, A. K., Zauszniewski, J. A., & Nakhla, W. E. (2009). Reasons for relocation to retirement communities: A qualitative study. *Western Journal of Nursing Research, 31*, 462–479. doi:10.1177/0193945909332009.

Beracha, E., & Johnson, K. H. (2012). Lessons from over 30 years of buy versus rent decisions: Is the American dream always wise? *Real Estate, 40*(2), 217–247.

Bracha, A., & Jamison, J. (2013). The great recession and confidence in homeownership. *Communities & Banking, 24*(2), 27–30.

Brown, S. C., Mason, C. A., Lombard, J. L., Martinez, F., Plater-Zyberk, E., Spokane, A. R., et al. (2009). The relationship of built environment to perceived social support and psychological distress in hispanic elders: The role of "Eyes on the Street". *The Journals of Gerontology Series B: Psychological Sciences and Social Sciences, 64B*(2), 234–246.

Butrica, B. A., Smith, K. E., & Toder, E. J. (2010). What the 2008 stock market crash means for retirement security. *Journal of Aging and Social Policy, 22*(4), 339–359.

Cauley, S. D., Pavlov, A. D., & Schwartz, E. S. (2007). Homeownership as a constraint on asset allocation. *Journal of Real Estate, Finance, and Economics, 34*, 283–311.

Cunningham, G., & Michael, Y. (2004). Concepts guiding the study of the impact of the built environment on physical activity for older adults: A review of the literature. *American Journal of Health Promotion, 18*(6), 435–443. doi:10.4278/0890-1171-18.6.435.

Deaton, A. (2012). The financial crisis and the well-being of Americans. *Oxford Economic Papers, 64*(1), 1–26.

Ekerdt, D. J., Luborsky, M., & Lysack, C. (2012). Safe passage of goods and self during residential relocation in later life. *Ageing and Society, 32*(5), 833–850.

Erickson, M. A., Krout, J., Ewen, H., & Robison, J. (2006). Should I stay or should I go? Moving plans of older adults. *Journal of Housing for the Elderly, 20*, 5–22.

Evans, G. (2003). The built environment and mental health. *Journal of Urban Health, 80*(4), 536–555.

Financial Crisis Inquiry Commission. (2011). *The Financial Crisis Inquiry Report: Final Report of the National Commission on the Causes of the Financial and Economic Crisis in the United States.* Official Government Edition.

Frederick, R. (2012). Seniors housing. *Long-Term Living: For the Continuing Care Professional, 61*(8), 18–22.

Geboy, L., Moore, K. D., & Smith, E. K. (2012). Environmental gerontology for the future: Community-based living for the third age. *Journal of Housing for the Elderly, 26*(1–3), 44–61.

Geertz, C. (1973). *The Interpretation of Cultures*. New York: Basic Books.

Golant, S. M. (2011). The quest for residential normalcy by older adults: Relocation but one pathway. *Journal of Aging Studies, 25*(3), 193–205.

Groger, L., & Kinney, J. M. (2006). CCRC here we come! Reasons for moving to a continuing care retirement community. *Journal of Housing for the Elderly, 20*(4), 79–101.

Handy, S. L., Boarnet, M. G., Ewing, R., & Killingsworth, R. E. (2002). How the built environment affects physical activity. *American Journal of Preventive Medicine, 23*(2), 64–73.

Haurin, D. R., Parcel, T. L., & Haurin, R. J. (2002). Does Homeownership Affect Child outcomes? *Real Estate Economics, 30*, 635–666.

Herd, P. (2009). The two-legged stool: The reconfiguration of risk in retirement income security. *Generations, 33*(3), 12–18.

Hicks, J., & Kingson, E. R. (2009). The Economic Crisis: How Fare Older Americans? *Generations, 33*(3), 6–11.

Krout, J., Moen, P., Holmes, H., Oggins, J., & Bowen, N. (2002). Reasons for relocation to a continuing care retirement community. *Journal of Applied Gerontology, 21*(2), 236–256.

Lamb, S. (1997). The making and unmaking of persons: Notes on aging and gender in North India. *Ethos, 25*(3), 279–302.
Litwak, E., & Longino, C. (1987). Migration patterns among the elderly: A developmental perspective. *The Gerontologist, 27*, 266–272.
Longino, C. F., Jackson, D. J., Zimmerman, R. S., & Bradsher, J. E. (1991). The second move: Health and geographic mobility. *Journal of Gerontology, 46*(4), S218–S224.
Longino, C. R., Bradley, D. E., Stoller, E. P., & Haas, W. (2008). Predictors of non-local moves among older adults: A prospective study. *Journals of Gerontology. Series B, Psychological Sciences and Social Sciences, 63B*(1), S7–S14.
Manning, R. (2009). *Keeping people in their homes: Policy recommendations for the foreclosure crisis in Michigan*. Madison, WI: Filene Research Institute.
Marcoux, J. S. (2001). The 'Casser Maison' ritual: Constructing the self by emptying the home. *Journal of Material Culture, 6*(2), 213–235.
McFall, B. H. (2011). Crash and wait? The impact of the great recession on the retirement plans of older Americans. *American Economic Review, 101*(3), 40–44. doi:10.1257/aer.101.3.40.
McInerney, M., & Mellor, J. M. (2012). Recessions and seniors' health, health behaviors, and healthcare use: Analysis of the Medicare current beneficiary survey. *Journal of Health Economics, 31*(5), 744–751. doi:10.1016/j.jhealeco.2012.06.002.
Michael, Y. L., & Yen, I. H. (2009). Invited commentary: Built environment and obesity among older adults—Can neighborhood-level policy interventions make a difference? *American Journal of Epidemiology, 169*(4), 409–412.
Miner, S. (1995). Racial differences in family support and formal service utilization among older persons: A nonrecursive model. *Journals of Gerontology. Series B, Psychological Sciences and Social Sciences, 50B*(3), S143–S153.
Moody, H. R. (2009). The moral economy of retirement. *Generations, 33*(3), 27–33.
O'Loughlin, K., Humpel, N., & Kendig, H. (2010). Impact of the global financial crisis on employed Australian baby boomers: A national survey. *Australasian Journal on Ageing, 29*, 88–91. doi:10.1111/j.1741-6612.2010.00425.
Perry, T. (2012). Leaving home later in life: Voluntary housing transitions of older adults as gift giving practices in the Midwestern United States (Doctoral Dissertation). Retrieved from ProQuest Dissertations and Theses.
Pynoos, J., & Liebig, P. (2009). Changing work, retirement, and housing patterns. *Generations, 33*(3), 20–26.
Rohe, W. M., Zandt, S. V., & McCarthy, G. (2002). Home ownership and access to opportunity. *Housing Studies, 17*(1), 51–61.
Rowles, G. D., & Watkins, J. F. (2003). History, habit, heart and hearth: On making spaces into places. *Aging independently: Living arrangements and mobility*, 77-96.
Sargent-Cox, K., Butterworth, P., & Anstey, K. J. (2011). The global financial crisis and psychological health in a sample of Australian older adults: A longitudinal study. *Social Science and Medicine, 73*(7), 1105–1112.
Sergeant, J. F., Ekerdt, D. J., & Chapin, R. K. (2010). Older adults' expectations to move: Do they predict actual community-based or nursing facility moves within 2 years? *Journal of Aging and Health, 22*(7), 1029–1053. doi:10.1177/0898264310368296.
Shippee, T. P. (2009). But I am not moving: Residents' perspectives on transitions within a continuing care retirement community. *The Gerontologist, 49*(3), 418–427. doi:10.1093/geront/gnp030.
Sisko, A., Truffer, C., Smith, S., Keehan, S., Cylus, J., Poisal, J. A., et al. (2009). Health spending projections through 2018: Recession effects add uncertainty to the outlook. *Health Affairs, 28*(2), w346–w357.
Smith, R. J., Lehning, A. J., & Dunkle, R. E. (2013). Conceptualizing Age-Friendly Community Characteristics in a Sample of Urban Elders: An Exploratory Factor Analysis. *Journal of Gerontological Social Work, 56*(2), 90–111. doi:10.1080/01634372.2012.739267.

Torres-Gil, F., & Hofland, B. (2012). Vulnerable populations. In H. Cisneros, M. Dyer-Chamberlain, & J. Hickie (Eds.), *Independent for life: Homes and neighborhoods for an aging America* (pp. 221–232). Austin: University of Texas Press.

Wahl, H. W., & Oswald, F. (2010). Environmental perspectives on ageing. The SAGE handbook of social gerontology, 111–124.

Wahlbeck, K., & McDaid, D. (2012). Actions to alleviate the mental health impact of the economic crisis. *World Psychiatry, 11*, 139–145.

Wiseman, R. F. (1980). Why older people move. *Research on Aging, 2*, 141–154.

Xiao, D., & Liu, Z. X. (2007). The importance of wealth and income in the transition to homeownership. *Journal of Policy Development and Research, 9*, 137–152.

Understanding the Human-Animal Bond: Implications for Practice with Community-Living Older Adults

Ann M. Callahan and Regina Brotherton

Abstract *Understanding the Human-Animal Bond: Implications for Practice with Community-Living Older Adults* begins with an examination of research on why older adults own pets. Research in this area provides insight into what older adults expect to gain from pet ownership. Then the most recent research designed to explore the effects of pet ownership will be systematically reviewed for more robust conclusions. This review is followed by research that clarifies why pets are retained and the conditions under which pets are surrendered for a broader understanding of the potential implications of pet ownership for older adults. The chapter ends with an overview of basic information about pet care which provides gerontologists and other professionals with the knowledge necessary to support the benefits of pet ownership in their work with older adults.

This chapter examines research on reasons for pet ownership by older adults. Research in this area provides insight into what older adults expect to gain from pet ownership. The authors conducted a systematic review of the literature to explore the effects of pet ownership. A thorough discussion on why pets are retained and under what conditions pet owners surrender their animals is provided. From a contextual and gerontological lens, the authors seek to gain a broader understanding of the potential implications of pet ownership for older adults and reasons they surrender their pets. The authors hope that this chapter will provide gerontologists and other professionals with the knowledge necessary to support pet ownership in their work with older adults.

A. M. Callahan (✉)
Middle Tennessee State University, Murfreesboro, TN, USA
e-mail: Ann.Callahan@mtsu.edu

R. Brotherton
Southwest Georgia Technical College,
15689 U.S. Highway 19 North, Thomasville, GA 31792, USA

H. F. O. Vakalahi et al. (eds.), *The Collective Spirit of Aging Across Cultures*,
International Perspectives on Aging 9, DOI: 10.1007/978-94-017-8594-5_15,

The process of aging involves the inevitability of change. Some of these changes involve significant life stressors that can extend beyond one's capacity to cope. Depending on the circumstances, such consequences can be softened by the support of a pet. Studies with older adults have found that pets can reduce anxiety, depression, and loneliness as well as enhance social support and general well-being (Wells and Rodi 2000; Johnson and Meadows 2002; Nagasawa and Ohta 2010; Watt and Pachana 2007; Hara 2007; de Guzman et al. 2009; Chur-Hansen et al. 2009; Connell et al. 2007). However, despite the benefits of pet ownership, the experience of caring for an animal may lead to challenges, including physical, emotional, and financial hardship (Wells and Rodi 2000; Parslow et al. 2005; Chur-Hansen et al. 2009). As such, it is imperative that gerontologists and other professionals understand the human-animal bond (HAB) which could help older adults offset potential negative consequences and maximize the benefits of pet ownership.

Prevalence of Pet Ownership

The current number of pet owners in the United States is 72.9 million. This means three in five Americans own a pet (Harris Interactive 2012). Seventy five percent of households with pets have dogs and/or cats. This amounts to 78.2 million dogs (39 % of pet households) and 86.4 million cats (33 % of pet households) (American Pet Products Association 2012; The Humane Society 2011). Approximately four out of ten pet households own multiple pets (PR Web 2011). This has lead to a significant investment in pet care. As the numbers of pets have increased, so have pet expenditures from $17 billion in 1994 to $53 billion in 2012. The majority of pet expenditures have been for food at $21 billion (American Pet Products Association 2012). In addition to paying for basic necessities, dog owners have spent more on gifts for their dogs increasing from $56 million in 1994 to $73 million in 2012. There are some estimates that suggest pet owners will spend an average of $11,000 per pet during the pet's lifetime (PR Web 2011).

This upward trend in pet ownership and expenditures suggests that pet owners are investing an enormous amount of financial resources to maintain the health and wellness of their animals. Pet owners also show a strong sense of commitment to their pets. For instance, according to a survey conducted by the American Animal Hospital Association (AAHA) in 2004, 93 % of pet owners indicated that they would risk their own lives for their pet. The American Pets Products Association (APPA) survey likewise found a majority of pet owners would take their pets if they had to evacuate their homes (PR Web 2011). Another study confirmed that a number of people refused to evacuate their homes without their pets despite hazardous conditions (Knight and Herzog 2009). This degree of commitment, even at one's own expense, shows that the human-animal bond (HAB) can be a powerful one. For gerontologists and other professionals, it is important to understand the potential implications of pet ownership for older adults.

The social sciences research of HAB has added to our knowledge about older adults and has cited the positive benefits of the HAB. However, there are methodological issues which reduces the generalizability of existing knowledge to older adults who own pets. One problem has been inconsistent controls for demographic characteristics such as the age of pet owners as well as pet-related characteristics (Chur-Hansen et al. 2010; Our pet causes, 2008; Rijken and van Beek 2011; Langfield and James 2009; Power 2008; Risley-Curtiss et al. 2006). HAB research has also largely focused on the efficacy of animal assisted therapy/services provided by professionals with older adults who are ill, disabled, and/or required institutional care. Additional research is needed to capture the experience of pet ownership by community-living older adults.

Toward this end, a systematic literature review of HAB research about the psychosocial impact of pet ownership by older adults was conducted. Studies selected for review were published in peer-reviewed journals between 2000 and 2012 as found in multiple databases through EBSCO, PubMed, reference lists (including Barker et al. 2003), and through Internet sites.[1] The review yielded 14 research articles about the psychosocial effects of pet ownership on older adults living in Australia (n = 6), United States (n = 5), Netherlands (n = 1), Japan (n = 1), and the Philippines (n = 1). The following discussion will help to clarify what older adults should expect from pet ownership, including pet owner satisfaction and pet retention as well as the benefits and/or consequences of pet ownership based on outcomes research.

Literature Review

Reasons for Pet Ownership

Only a couple of studies specifically addressed why older adults chose to own pets. In a qualitative study of 96 older adults living in the Netherlands, Enders-Slegers (2000) found that the reasons for pet ownership fell into the following five categories: (1) love for the pet, (2) need for companionship, (3) no life without pet, (4) antidote to loneliness, and (5) gift from children. Those who previously owned pets indicated that they did not currently own pets because a pet was not allowed in their residence or they did not want the responsibility or extra work involved in pet care. Additional reasons included, dealing with grief and loss issues related to accidental death of pet or euthanasia due to pet illnesses; and lack of funds to pay for pet care. Participants who had never owned pets said they did not like animals or felt indifferent toward them.

[1] A majority of these resources were accessed through HABRI central (http://habricentral.org/resources/journalarticles), Pet Partners (http://www.petpartners.org/humananimalbondresourcecenter), and Center for the Interaction of Animals and Society (http://research.vet.upenn.edu/publications/tabid/1918/default.aspx).

Chur-Hansen et al. (2008) conducted a similar qualitative study, but focused on why eight older adults living in Australia did not currently own pets. The reasons for not owning a pet were similar to those reported by Enders-Slegers (2000). They were broadly categorized as being either emotional or pragmatic. Emotional reasons were related to the desire to avoid the emotional responsibility of pet care, likened to the care of a child. This included the desire to avoid grief associated with pet loss and causing the pet distress should the pet be surrendered. The most common reason for not wanting to own a pet were pragmatic. This generally involved the desire to avoid the inconvenience of pet care; avoid the potential burden on family finding a new home for the pet; desire for freedom to travel; inability to have pets in residence or yard not being conducive; fear of pet causing personal sickness or injury; and pet behavior problems. The most common reasons for not owning a pet were to avoid the loss of independence and time needed for pet care.

Research that included respondents of all ages, reported similar reasons for pet ownership but indicated slight variation based on age (Staats et al. 2008; McNicholas and Collis 2004), gender (Staats et al. 2006, 2008), and marital status (Staats et al. 2006). For example, Staats, Wallace, and Anderson replicated and extended a study by Staats, Sears, and Pierfelice which surveyed pet ownership of students and middle-aged community members in the United States. The top two reasons for pet ownership were to avoid loneliness and to keep active. Consistent with the study by Staats, Sears, and Pierfelice, men were more likely than women to report the desire for pets to help them keep active; however, the desire to avoid loneliness was the main reason reported for pet ownership regardless of gender and marital status. The desire to avoid loneliness was associated with cat ownership whereas to stay active was associated with dog ownership. The only age difference was that older respondents (over 50) were significantly more likely to report a belief in the positive health benefits of pet ownership compared to younger participants.

McNicholas and Collis (2004) conducted a study in the UK to address the relative importance of pet ownership based on life stages from childhood (8–9 years), adolescent (15–19 years), and older adults (65+ years). Respondents were asked to rank order relationships (human or animal) that were most preferred to provide a specific type of support: need for reliable alliance, show nurturance, greater self-esteem, and alleviate emotional loneliness. Results indicated that a larger percentage of older adults (98 %) were more likely than adolescents (36 %) to identify pet ownership as one of their top ten most important relationships. However, both older adult and children identified pet relationships as being important in every scenario. Pets were selected more than humans as being something to show nurturance towards. However, older adults were more likely than other age groups to identify pets as alleviating emotional loneliness and providing a sense of reliable alliance. They also found that respondents chose different types of pets for different types of support needed. Relationships with dogs and cats were most frequently selected as the most important relationships. Dogs were selected more often for showing nurturance and cats were selected more often for supporting self-esteem. In this study, differences based on age emerged more clearly. Older adult pet owners was relative to the type of support needed and type of pet believed to be the most able to meet their need.

There is not enough research to conclusively identify the reasons for pet ownership by community-living older adults in the United States. However, older adults consistently reported anticipated benefits associated with pet ownership. The most consistently reported anticipated benefits of pet ownership included access to emotional support, such as the provision of companionship to prevent loneliness; and instrumental support, such as to inspire an active life style. The reasons for not owning pets were emotional, such as, to avoid the pain of grief and loss; and pragmatic, indicated by the desire to avoid the loss of independence and time needed for pet care. The choice to engage in pet ownership also seemed relative to pet and pet owner characteristics. For example, Enders-Slegers (2000) found that immediately after retirement older adults wanted to get involved in activities that took precedence over owning a pet. Hence, it seems that pet ownership carries with it the potential for benefits and consequences for older adults. Clarification of such implications requires additional research focusing specifically on the human-animal bond (HAB) among older adults.

Impact of Pet Ownership

Of the quantitative research reviewed, studies yielded inconsistent results between pet ownership and psychosocial well-being for older adults. Four studies found that pet ownership had no significant relationship with degree of loneliness (Eshbaugh et al. 2011; Rijken and van Beek 2011), depression (Simmons et al. 2000), social support (Simmons et al. 2000; Winefield et al. 2008; Rijken and van Beek 2011), and life satisfaction (Simmons et al. 2000). Two studies found that pet ownership was significantly related to an increase in boredom, loneliness, and depression scores (Wells and Rodi 2000; Parslow et al. 2005). In contrast, four studies found a significant positive relationship between pet ownership and psychosocial well-being for older adults based on reported sense of life purpose (Wells and Rodi 2000; Johnson and Meadows 2002; Nagasawa and Ohta 2010), social support (Johnson and Meadows 2002; Nagasawa and Ohta 2010), and emotional attachment (Watt and Pachana 2007). These studies also suggested that the type of pet (Wells and Rodi 2000; Rijken and van Beek 2011), reasons for pet ownership (Watt and Pachana 2007), and degree of pet attachment (Krause-Parello 2008) were significantly correlated with and/or moderated the experience of pet ownership for older adults. Qualitative studies were reviewed as well, and a majority of these studies reported a positive relationship between pet ownership and psychosocial well-being for older adults. Four studies found that pet ownership provided older adults a channel for the expression of autonomy through engagement in pet care (Hara 2007; de Guzman et al. 2009) that boosted self-esteem (de Guzman et al. 2009) and sense of life purpose (Chur-Hansen et al. 2009). Pets not only provided a source of companionship, but they also helped older adults initiate social interaction

(Hara 2007; de Guzman et al. 2009) and served as a calming distraction (Connell et al. 2007). Pets were further an outlet for affective and emotional attachment (Chur-Hansen et al. 2009; de Guzman et al. 2009).

Two studies found that pet ownership had negative consequences for older adults as well. Some pet owners refused to go on trips, visit friends, and receive medical treatment that required an overnight stay to avoid the anxiety associated with leaving their pets alone (Wells and Rodi 2000; Chur-Hansen et al. 2009). There were instances where pet owners said life would not be worth living without their pets, leaving them vulnerable to the pain of loss upon a pet's death (Chur-Hansen et al. 2009). In a study by Wells and Rodi (2000), one respondent described distress over being physically unable to dispose of her cat's body after it died at home other than placing it in a trash bag in her garbage can.

Reasons for Pet Retention and Pet Relinquish

Key studies by Neidhart and Boyd (2002) and New et al. (2000) demonstrated the importance of having a "goodness-of-fit" between a pet and pet owner which determines the likelihood of pet retention or relinquishment. Older adults who adopt a pet might live alone and/or adopt a pet for a grandchild. Under these circumstances, pets are more likely to be considered close companions, if not part of the family. Although older adults are less likely to relinquish a pet, they face unexpected life changes and problems with pet behavior that could negatively impact the experience of pet ownership. New et al., more specifically conjectured that psychosocial variables, such as level of attachment including situational factors, such as health issues or change in housing, could further increase the potential for pet relinquishment.

Implications for Practice

Studies about the reasons for pet ownership suggested that older adults adopted pets to gain emotional support and companionship. This carried with it the potential for suffering grief upon pet death and loss of independence/free time given the need for pet care. While the results of outcome studies were mixed, the positive psychosocial outcomes appeared to be consistent. Pet ownership was significantly associated with emotional attachment, social support/interaction, and sense of life purpose. In contrast, there was also evidence that pet ownership either had no significant relationship, or, was associated with lower psychosocial well-being of older adults. Hence, outcomes research began to identify a more complex relationship between factors. It appears as though pets can have a positive or negative effect on the lives of older adults, depending on a variety of circumstances. Given the importance of pet ownership for some older adults, gerontologists and other professionals may need to help them address pet-related challenges. Pet retention research suggested that one way of helping is to support the "goodness-of-fit" between a pet and pet owner through the provision of pet-related information. Although gerontologists are not veterinary professionals, they may still engage in such information gathering for dissemination with older pet owners.

Collaboration with experts in veterinary medicine is required to learn more about the types of concerns older adults are likely to have as pet owners. The remaining section is the product of one such collaboration. The second author, a veterinary professional, will address key pet-related issues such as choosing an appropriate pet, providing basic pet care, avoiding infection from a pet, meeting the needs of elderly pets, preparation for pet death, and coping with pet loss.

Advice From a Veterinary Professional

Appropriate Pets. When individuals choose pets, they generally choose them based on impulse, personal preference, appearance, and various other things rather than the disposition of the breed, life span, animal's temperament, size, and other traits and/or characteristics. Some pets are predisposed to health and/or behavior issues and require a lot of physical activity. This can be challenging for any owner and especially an older adult. Table 1 outlines typical pet life spans, potential therapeutic benefits, environmental factors, and other information that may be of help with an older adult who is trying to choose the best pet. Table 2, originally developed in collaboration with an occupational therapist, further outlines benefits older adults may achieve by interacting with pets relative to the individual's functional capacity. It is important to keep in mind that pets, along with their owners, will change over time. This can lead to new challenges in pet care. For example, assistive devices such as canes, wheel chairs, and walkers can frighten or otherwise threaten a pet and evoke an inappropriate response. As with any pet, some hazards of human-animal interactions may include bites, scratches, and physical injury to the pet owner. It is highly recommended that a behavior screening (see Appendix A) be performed in coordination with a veterinarian or credential veterinary technician prior to the adoption of a pet.

Pet Owner Responsibilities. As part of maintaining a healthy HAB after pet adoption, veterinary professionals can be consulted to provide client education on how to provide routine pet care at home. For example, it is recommended that pet owners provide routine and consistent dental home care. Dental disease prevention may include brushing the pet's teeth, providing appropriate chew toys, and/or administering oral products that can aid in the reduction and production of dental tarter. Pet owners should use toothpaste labeled for pets as fluoride content of human toothpaste can be a concern in pets. Without proper dental care, diseases such as periodontal disease and heart disease may occur. Some diseases that can arise may even be detrimental to older adults as found in Table 3. This will be discussed further in the next section about how to avoid infection from a pet.

Table 1 Considerations in pet selection with older adults

Group of pet/animal	Life span (average)	Factors to consider (environmental, social, and physical)	Average veterinary medical care/animal for 2001	Benefits pets may provide	Zoonotic disease (diseases that can be transmitted to humans from pets)
Canine	10–14+ years	Indoor/Outdoor access Exercised daily Feed and water daily Attention and interaction Socialization daily Rest daily Grooming Vet care Training Insurance riders Temperament	$179	Companionship Physical Social Psychological Emotional	Rabies Leptospirosis Histoplasmosis Blastomycosis Giardia *Cryptosporidium* Round worms Ringworm Hookworms
Feline	15+ years	Indoor access Litter box clean up daily Feed and water daily Grooming Temperament Vet Care	$85	Companionship Social Psychological Emotional	Cat scratch fever Blastomycosis *Cryptosporidium* Roundworm Ringworm Toxoplasmosis
Equine	30+ years	Clean up Grooming Boarding cost Temperament Insurance riders Vet care Training Rest	$112	Physical Mental Social Psychological Emotional	Rabies

(continued)

Table 1 (continued)

Group of pet/animal	Life span (average)	Factors to consider (environmental, social, and physical)	Average veterinary medical care/animal for 2001	Benefits pets may provide	Zoonotic disease (diseases that can be transmitted to humans from pets)
Avian	20–60+ years	Ample cage size Cage supplied with correct items depending on bird type Clean daily Socialization Temperament Feed and water daily Vet care	$10	Mental Psychological Emotional	Histoplasmosis Psittacosis
Exotics (reptiles and pocket pets including ferrets and rabbits)	1–10 years	Daily cleaning Feed and water daily Heat lamps Temperament Vet care Handle properly	Unknown	Companionship Social Psychological Emotional	Ringworm (rodents) *Salmonella* (reptiles) *Giardia* (rodent and ferret) (Rodent and ferret) *Cryptosporidium*
Fish	Variable	Clean tank regularly Feed and water as directed Low maintenance	Unknown	Companionship Mental Psychological Emotional	

Table 2 Pet selection relative to level of functioning and residence

	Alone, own home	Alone, apt. bldg.	With spouse or other family member	Private care facility, assisted living facility
No cognitive, psychosocial, or significant physical functional limitations	Canine Feline Equine Avian Exotics Reptiles Pocket pets Fish	Canine Feline Avian Exotics Reptiles Pocket pets Fish	Canine Feline Avian Exotics Reptiles Pocket pets Fish	N/A
Mild cognitive deficits (dec. sh. term memory)	Canine Feline Equine Avian Exotics Reptiles Pocket pets Fish	Canine Feline Avian Exotics Reptiles Pocket pets Fish	Canine Feline Avian Pocket pets Fish	Canine Feline Avian Pocket pets Fish N/A
Mod/severe cognitive deficits (severe memory loss, dec. judgment, dec. safety awareness)	N/A	N/A	Canine Feline Avian Pocket pets Fish	Canine Feline Avian Pocket pets Fish
Severe cognitive deficits (disoriented, confused, decreased arousal)	N/A	N/A	N/A	Fish Avian
Mild physical limitations (dec. endurance, min. dec. standing balance)	Feline Pocket pets Fish	Feline Pocket pets Fish	Feline Pocket pets Fish	Feline Pocket pets Fish

(continued)

Table 2 (continued)

	Alone, own home	Alone, apt. bldg.	With spouse or other family member	Private care facility, assisted living facility
Mild to mod. physical limits uses cane, walker for ambulation)	Feline Pocket pets Fish	Feline Pocket pets Fish	Feline Pocket pets Fish	Feline Pocket pets Fish
Mod physical limits (uses wheelchair independently, physical limits of UEs not sig.)	Feline Pocket pets Fish	Feline Pocket pets Fish	Feline Pocket pets Fish	Feline Pocket pets Fish
Mod. physical limits (i.e. requires physical assistance to arise from bed, perform basic self-care) May require use of wheelchair	Feline Fish	Feline Fish	Feline Fish	Feline Fish
Severe physical limitations (Seldom moves from bed or other static position)	N/A	N/A	N/A	Fish Avian

Note This table was originally developed by Molly Mika, OT and Regina Brotherton, DVM November, 2004

It is important for pets to receive grooming care. The maintenance of such may be another concern for older pet owners. Pets should be bathed on a regular basis. Pet owners tend to choose small pets like poodles. No matter the size, however, grooming a pet can be challenging. Older adults may need assistance in performing the bath, selecting and collecting items needed, and/or lifting the pet into the tub. Grooming also includes things such as cleaning the ears, trimming nails, and brushing the animal after the fur dries. If the owner elects to take their pet to the groomer for this basic hygiene care, transporting the animal may be of concern for an older adult. When older adults are faced with such challenges, they may feel they have failed in providing for his/her pet needs, which, in turn, may cause stress, depression, or other unwanted conditions.

Older adults that may have memory loss or mental deterioration may also need assistance in meeting day to day care of the pet. Helping the pet owner to keep a chart of daily tasks to be performed in the form of a check off sheet will help to ensure the daily needs of the pet have been met. For instance, pets must receive food and water daily. Depending on the breed, size, and age of the pet, feeding requirements will be variable. It is imperative to feed the pet according to the manufacturer's instructions listed on the pet food bag. It is also important to check the expirations dates of the food so that food offered to the pet is of good quality. Pets need to have routine "potty" breaks to prevent inappropriate behaviors or "accidents." Feeding, watering, and potty breaks can be checked off using the check off sheet and serves as a great reminder to those assisting the elder individual to keep his/her pet in good health. Pets need ample rest as well as a safe environment. To assist, pets should have a place to rest during the day and evening. Places for resting may include a kennel, crate, or another specific area. Regardless of location, pets need to be kept safe and secure at all times. Ensuring the older adult can care properly for his/her pet will help maintain and add value to the HAB.

At times, pet owners may be asked to administer medications to his/her pet. Assistance may be needed to ensure that an older adult is able to administer pet medications at the appropriate time, in the correct amount, for the proper duration, and in the right manner. Pet owners may elect to provide monthly preventatives and maintenance items including heartworm medication such as Heartgard® or Interceptor®, flea medication which can be administered topically or orally, and dewormers. An older adult may need assistance to help restrain a pet until topical or oral products are administered. Again, a calendar may be used to help the pet owner keep up with the administration of pet medication(s). It is important for older adults and those who assist them to be aware of the pet owner's capacity to maintain care. Over time older adults may experience changes in functional capacity that can compromise one's ability to maintain the HAB in a safe and effective manner, which may require additional intervention.

Avoiding Infection from a Pet. A very important issue when working with older adults is their risk for disease. The risk of contracting a disease from one's pet is called zoonoses and a disease that has the potential to spread from animals to humans is referred to as a zoonotic disease. Table 3 contains several zoonoses, relevant species, transmission, and type of infection that may result from each

Table 3 Zoonotic diseases

Disease	Agent/Vector	Transmission
Rabies (Lyssavirus)	Dog, skunk, fox, bat, horse, cattle, mokeys	Bite wounds of infection animal
Leptospirosis (*L. interrogans*)	Dogs, rats, cattle	Exposure to infected water
Plaque (*Yersina pestis*)	Cats. Dogs, squirrels, rodents	Fleas and infected animals
Lyme Disease (*Borellia*)	Dogs, deer, rodents	Tick bites
Rocky Mtn Spotted Fever (*Rickettsia rickettsii*)	Dogs, deer, rodent	Tick bites
Cat Scrath Fever (*Bartonella*)	Cats	Scratches, bites, and licking from cats
Histoplasmosis (*Histoplasma capsulatum*)	Birds	Environmental exposure
Blasmycosis (*Blastomyces dermatitidis*)	Dogs, cats, horses	Environment & animal exposure
Giardia (*G. duodenslis*)	Dogs, cats, rodents, birds, ferrets)	Environment and animal exposure
Cryptosporidiosis (*Cryptpsporidium*)	Dogs, cats, rodent, ferret	Aerosol
Round worms (*Toxacara canis*, cati)	Dogs, cats	Penetration of skin and organs
Hook worms (*Anclostoma caninum*)	Dogs	Penetrate skin
Toxoplasmosis (*Toxoplasma gondii*)	Cats	Litter, dirt, or raw meat
Psittacosis (*Chlamydia psittiaci*)	Birds	Exposure to aerosols
Ring worm (*Microsporum/ Trichophyton*)	Dogs, cats, cattle, rodents	Direct contact with infected animal

zoonotic disease. Pet owners can help prevent their risk for disease by taking pets in for regular veterinary care. Veterinary professionals will usually recommend the following services:

- Annual veterinary visits (highly recommend 6 month visits)

– Complete and comprehensive physical examination

- Vaccinations

– Examples include rabies, distemper, feline leukemia, lyme, and others
– Administered to pets depend on pet health status, environment, and regional laws and regulations

- Diagnostic screenings

– Blood tests that may suggest an infection, organ dysfunction, and electrolyte imbalances

- Fecal evaluation

– Evaluate for internal parasites such as roundworms (*Toxocara*), hookworms (*Ancylostoma*), and tapeworms (*Dipylidium*) infections

- Additional screening may include radiographs (x-rays) depending on the age of the pet.

The cost of veterinary services can range from $75 to $200 annually. Some older adults may have limited income and feel bad if they cannot afford pet care. One option is to contact an area veterinarian to help identify resources (programs offered in the community to assist with low income services for pet owners) or contact the local animal humane society. As previously stated, the goal is to maintain the HAB in a safe and healthy one so that maximum benefits for older adults may be realized.

Meeting the Needs of Older Pets. Pets like humans are plagued with various diseases and disorders such as diabetes, cataracts, hypothyroidism in dogs, hyperthyroidism in cats, and other conditions that may place undue hardships on an older pet owner. In addition, pet owners of a geriatric pet may be faced with kidney failure, neoplastic disease (cancer), heart failure, and other things such as degenerative joint disease all of which require routine veterinary visits, daily medications, and other treatment regimens. The challenge of managing the cost of these conditions can be very daunting, especially if the older pet owner is living on a fixed income. Although this can be a difficult time, care must be taken to ensure that the pet owner does not forego his/her health care to care for one's pet.

Preparation for Pet Death. When some pets reach an inhumane level of suffering, a veterinarian may suggest euthanasia. Euthanasia is referred to as "painless death." The purpose of euthanasia is to relieve the pet from suffering caused by illness or disease. It may be particularly helpful to work more closely with the older adult and veterinarian during this time.

The veterinary health care team will work with those involved to explain the euthanasia process in addition to what happens to the pet after the animal passes. The pet owner will need to decide if he/she would like to have the pet cremated in mass fashion (with other pets) or individually (so that ashes may be collected and returned to the pet owner). If the owner chooses to have the pet's ashes returned, the pet owner will have to choose a box or urn to have the remains returned to them. The cost for these services can vary, but can range from $75 to $150 and upward. Throughout this process, the pet owner may need additional help to understand what they are being told and what is necessary to expedite the process.

Coping with Pet Loss. As noted earlier in this chapter, older adults can become deeply attached to their pets. The loss of a pet can lead to emotional distress that compromises one's well-being. In working with an older adult who loses a pet, it is important to identify what role the pet had in the home. The degree of pet attachment and reasons for attachment can vary among individuals. Pet owners may be attached to a pet out of loyalty, purpose, responsibility, dependency, and love/acceptance. For example, an older adult may have lost a spouse or child in the past and turned to a pet to give them purpose in life through being responsible for pet care. If the pet becomes ill or passes away, the pet owner may feel as if they cannot go on and express overwhelming grief. During this time, an older adult will need additional emotional support and, potentially, a referral for grief counseling.

Pet owners may need to be reassured that they did everything they could to take care of their four-legged friend. They may also need to be reminded that their pet appreciated the love and care the pet owner provided for the pet through the years. Sensitivity and understanding is imperative for a successful transition through this experience.

Future Directions

Given the complexity of this material, collaboration is an important resource for enhancing the confidence and competence level of gerontology and other professionals in addressing pet-related issues with older adults. Research is needed, however, to determine if the provision of pet-related information by such professionals is significantly related to the psychosocial benefits of pet ownership. This line of inquiry not only has implications for the delivery of evidence-based practice, but also curriculum development for training professionals who are interested in blending veterinary and gerontological social work.

One direction for future research is to involve the development of a training module for students and professionals who are likely to work with older adults. Researchers could measure potential psychosocial changes in older pet owners as they work with trained individuals. This type of research may further involve measuring the psychosocial status of older pet owners who only receive veterinary services. The inclusion of methodological controls for individual characteristics of older adults as well as other potentially influential factors will need to be considered as well.

Pet ownership has the potential to enhance psychosocial well-being, but carries with it responsibilities that require adequate resources to support pet care. Research demonstrates how particular conditions, such as over attachment to pets, can have negative consequences for pet owners. However, knowledgeable professionals may help ensure that pet ownership is maintained in a way that is life enhancing for older adults. This chapter suggests that a way to promote the positive benefits of pet ownership is to provide older adults with key pet-related information. In the process, gerontologists and other professionals may collaborate with and draw from the knowledge of veterinary professionals. Future research is needed, however, to ensure such intervention helps older adults access the psychosocial benefits of pet ownership that may likewise bring joy into the life of an animal.

Appendix A

Individual Questionnaire for Pet Selection by Older Adults

1. Do you have any allergies and if so what are they?
2. Are you aware of any compromises in your immune system?

3. Do you ever feel any increased anxiety or agitation in situations? If so can you name those situations?
4. Do you have any problems with visual perception? If so, can you name them? For example depth perception.
5. Do you experience of loss of feeling in any part of your body? Can you name the part of your body you have that loss of feeling?
6. How is your vision?
7. How is your hearing?
8. Do you feel at any time a loss of endurance, strength? If so, can you name that time or times? For example, any cardiac disease, or related cancer fatigue.
9. Where you live now is there anyone who lives with you who would be interested in taking care of your pet if you get ill?
10. Have you ever had a pet before? If so, what kind?
11. Do you have any life threatening illnesses that might cause your pet to out live you?
12. Do you have a preference of that type of pet you want and can you tell us why?
13. Do you have a fear of phobia of pets? Explain.

References

Allen, K. (2003). Are pets a healthy pleasure? The influence of pets on blood pressure. *Current Directions in Psychological Science,12*(6), 236–239.

American Animal Hospital Association (2004). American animal hospital association 2004 pet owner survey. Retrieved from https://www.aahanet.org/PublicDocuments/petownersurvey2004.pdf

American Pet Products Association (2012). *Industry statistics and trends.* Retrieved from www.americanpetproducts.org/press_industrytrends.asp

Arhant-Sudhir, K., Arhant-Sudhir, R., & Sudhir, K. (2011). Pet ownership and cardiovascular risk reduction: Supporting evidence, conflicting data and underlying mechanisms. *Clinical and Experimental Pharmacology and Physiology,38*, 734–738. doi:10.1111/j.1440-1681.2011.05583.x

Barker, S. B., Rogers, C. S., Turner, J. W., Karpf, A. S., & Suthers-McCabe, H. M. (2003). Benefits of interacting with companion animals. *American Behavioral Scientist, 47*(1). doi:10.1177100027642032552l5

Bath, P. A., Deeg, D., & Poppelaars, J. (2010). The harmonization of longitudinal data: A case study using data from cohort studies in the Netherlands and the United Kingdom. *Ageing and Society, 30*(8), 1419–1437 doi:http://dx.doi.org/10.1017/S0144686X1000070X

Callahan, A. M. (2012). Ram veterinary social work: How undergraduate social workers can make a difference. *Journal of Baccalaureate Social Work,17*(17), 149–158.

Carlisle-Fran, P., & Frank, J. M. (2006). Owners, guardians, and owner-guardians: Differing relationships with pets. *Anthrozoos, 19*(3), 225–242. doi:http://dx.doi.org/10.2752/089279306785415574

Chur-Hansen, A., Stern, C., & Winefield, H. (2010). Gaps in the evidence about companion animals and human health: Some suggestions for progress. *International Journal of Evidence-Based Healthcare,8*(3), 140–146. doi:10.1111/j.1744-1609.2010.00176.x

Chur-Hansen, A., Winefield, H. R., & Beckwith, M. (2009). Companion animals for elderly women: The importance of attachment. *Qualitative Research in Psychology,6*(4), 281–293. doi:10.1080/14780880802314288

Chur-Hansen, A., Winefield, H. R., & Beckwith, M. (2008). Reasons given by elderly men and women for not owning a pet, and the implications for clinical practice and research. *Journal of Health Psychology,13*(8), 988–995. doi:10.1177/1359105308097961

Cline, K. M. (2010). Psychological effects of dog ownership: Role strain, role enhancement, and depression. *Journal of Social Psychology,150*(2), 117–131. doi:10.1080/00224540903368533

Connell, C. M., Janevic, M. R., Solway, E., & McLaughlin, S. J. (2007). Are pets a source of support or added burden for married couples facing dementia? *Journal of Applied Gerontology,26*(5), 472–485. doi:10.1177/0733464807305180

de Guzman, A. B., Cucueco, D. S., Cuenco, I. B. V., Cunanan, N. G. C., Dabandan, R. T., & Dacanay, E. J. E. (2009). Petmanship: Understanding elderly Filipinos' self-perceived health and self-esteem captured from their lived experiences with pet companions. *Educational Gerontology,35*(11), 963–989.

Eshbaugh, E., Somervill, J. W., Koten, J. H., Perez, E., Nalan, K. R., Wilson, C. E., et al. (2011). Brief report: Presence of a dog, pet attachment, and loneliness among elders. *North American Journal of Psychology,13*(1), 1–4.

Enders-Slegers, M. (2000). The meaning of companion animals: Qualitative analysis of the life histories of elderly cat and dog owners. In A. L. Podberscek, E. S. Paul, & J. A. Serpell (Eds.), *Companion animals and us: Exploring the relationships between people and pets*. Cambridge: Cambridge University Press.

Friedmann, E., & Son, H. (2009). The human-companion animal bond: How humans benefit. *Veterinary Clinics of North America: Small Animal Practice,39*, 293–326. doi:10.1016/j.cvsm.2008.10.015

Hafen, M, Jr, Rush, B. R., Reisbig, A. M. J., McDaniel, K. Z., & White, M. B. (2007). The role of family therapists in veterinary medicine: Opportunities for clinical services, education, and research. *Marital and Family Therapy,33*(2), 165–176.

Hara, S. (2007). Managing the dyad between independence and dependence: Case studies of the American elderly and their lives with pets. *International Journal of Japanese Sociology,16*(1), 100–114. doi:10.1111/j.1475-6781.

Harris Interactive (2012). Pets aren't just animals; They are members of the family. Retrieved from http://www.harrisinteractive.com/NewsRoom/HarrisPolls/tabid/447/mid/1508/articleId/1076/ctl/ReadCustom%20Default/Default.aspx

Johnson, R. A., & Meadows, R. L. (2002). Older Latinos, pets and health. *Western Journal of Nursing Research,24*(6), 609–620.

Knight, S., & Edwards, V. (2008). In the company of wolves: The physical, social, and psychological benefits of dog ownership. *Journal of Aging and Health,20*(4), 437–455. doi:10.1177/0898264308315875

Knight, S., & Herzog, H. (2009). All creatures great and small: New perspectives on psychology and human–animal interactions. *Journal of Social Issues,65*(3), 451–461. doi:10.1111/j.1540-4560.2009.01608.x

Krause-Parello, C. A. (2008). The mediating effect of pet attachment support between loneliness and general health in older females living in the community. *Journal of Community Health Nursing,25*(1), 1–14. doi:10.1080/07370010701836286

Kruger, K. A., & Serpell, J. A. (2006). Animal-assisted interventions in mental health: Definitions and theoretical foundations. In A. H. Fine (Ed.), *Handbook on animal-assisted therapy: Theoretical foundations and guidelines for practice* (2nd ed. pp. 21–38). New York: Academic Press. Retrieved from http://research.vet.upenn.edu/Portals/36/media/Animal%20Assited%20Interventions%20in%20Mental%20Health.pdf

Langfield, J., & James, C. (2009). Fishy tales: Experiences of the occupation of keeping fish as pets. *British Journal of Occupational Therapy,72*(8), 349–356.

McColgan, G., & Schofield, I. (2007). The importance of companion animal relationships in the lives of older people. *Nursing Older People,19*(1), 21–23.

McNicholas, J., & Collis, G. M. (2004). The role of pets in the social networks of children, adolescents and elderly people. Abstract retrieved from http://www.petpartners.org/document.doc?id=119

McNicholas, J., Gilbey, A., Rennie, A., Ahmedzai, S., Dono, J., Ormerod, E. (2005). Pet ownership and human health: A brief review of evidence and issues. British Medical Journal, 331, 1252-1254. DOI: http://dx.doi.org/10.1136/bmj.331.7527.1252

Nagasawa, M., & Ohta, M. (2010). The influence of dog ownership in childhood on the sociality of elderly Japanese men. *Animal Science Journal,81*(3), 377–383. doi:10.1111/j.1740-0929.2010.00741.x

Neidhart, L., & Boyd, R. (2002). Companion animal adoption study. *Journal of Applied Animal Welfare Science,5*(3), 175–192.

Netting, F. E., Wilson, C. C., & New, J. C. (1987). The human-animal bond: Implications for practice. *Social Work,32*(1), 60–64.

New, J. C., Salman, M. D., King, M., Scarlett, J. M., Kass, P. H., & Hutchison, J. M. (2000). Characteristics of shelter-relinquished animals and their owners compared with animals and their owners in U.S. pet-owning households. *Journal of Applied Animal Welfare Science,3*(3), 179–201.

O'Haire, M. (2009). The benefits of companion animals for human mental and physical health. Retrieved from http://www.rsp

Our pet causes (2008). *Harvard Health Letter, 33*(6), 6–7.

Parslow, R. A., Jorm, A. F., Christensen, H., Rodgers, B., & Jacomb, P. (2005). Pet ownership and health in elderly adults: Findings from a survey of 2,551 community-based Australians aged 60–64. *Gerontology,51*(1), 40–47. doi:10.1159/000081433

PR Web (2011). New survey reveals pet ownership at its highest level in two decades and pet owners are willing to pay when it comes to pet's health. Retrieved from http://www.prweb.com/releases/2011/4/prweb8252684.htm

Power, E. (2008). Furry families: Making a human-dog family through home. *Social and Cultural Geography,9*(5), 535–555. doi:10.1080/14649360802217790

Rijken, M., & van Beek, S. (2011). About cats and dogs…reconsidering the relationship between pet ownership and health related outcomes in community-dwelling elderly. *Social Indicators Research,102*, 373–388. doi:10.1007/s11205-010-9690-8

Risely-Curtiss, C. (2010). Social work practitioners and the human-companion animal bond: A national study. *Social Work,55*(1), 38–46.

Risley-Curtiss, C., Holley, L. C., & Wolf, S. (2006). The animal-human bond and ethnic diversity. *Social Work,51*(3), 257–268.

Sachs-Ericsson, N. J., Hansen, N., & Fitzgerald, S. (2002). Benefits of assistance dogs: A review. *Rehabilitation Psychology,47*(3), 251–277.

Simmons, L. A., Simons, J., McCallum, J., & Friedlander, Y. (2000). Pet ownership is not associated with human health: A nine year prospective study in older Australians. *Australasian Journal on Aging,19*(3), 139–142. doi:10.1111/j.1741-6612.2000.tb00166.x

Staats, S., Sears, K., & Pierfelice, L. (2006). Teachers' pets and why they have them: An investigation of the human animal bond. *Journal of Applied Social Psychology,36*(8), 1881–1891. doi:10.1111/j.0021-9029.2006.00086.x

Staats, S., Wallace, H., & Anderson, T. (2008). Reasons for companion animal guardians (pet ownership) from two populations. *Society and Animals,16*, 279–291. doi:10.1163/156853008X323411

Tedeschi, P., Fitchett, J., & Molidor, C. E. (2005). The incorporation of animal-assisted interventions in social work education. *Journal of Family Social Work,9*(4), 59–77.

The Humane Society of the United States (2011). U.S. pet ownership statistics. Retrieved from http://www.humanesociety.org/issues/pet_overpopulation/facts/pet_ownership_statistics.html

Thorpe, R. J, Jr, Kreisle, R. A., Glickman, L. T., Simonsick, E. M., Newman, A. B., & Kritchevsky, S. (2006). Physical activity and pet ownership in year 3 of the health abc study. *Journal of aging and physical activity,14*, 154–168.

Walsh, F. (2009). Human-animal bonds I: The relational significance of companion animals. *Family Process,48*(4), 462–480. doi:10.1111/j.1545-5300.2009.01296.x
Watt, D. & Pachana, N. A. (2007). The role of pet ownership and attachment in older adults. Australian Journal of Rehabilitation Counseling, *13*(1), 32-43. DOI: http://dx.doi.org/10.1375/jrc.13.1.32
Wells, D. L. (2009). Associations between pet ownership and self-reported health status in people suffering from chronic fatigue syndrome. *The Journal of Alternative and Complementary Medicine,15*(4), 407–413. doi:10.1089/acm.2008.0496
Wells, Y., & Rodi, H. (2000). Effects of pet ownership on the health and well-being of older people. *Australasian Journal on Ageing,19*(3), 143–148. doi:10.1111/j.1741-6612.2000.tb00167.x
Winefield, H. R., Black, A., & Chur-Hansen, A. (2008). Health effects of ownership of and attachment to companion animals in an older population. *International Journal of Behavioral Medicine,15*(4), 303–310. doi:10.1080/10705500802365532

Summary and Future Directions

Nancy Giunta, Halaevalu F. Ofahengaue Vakalahi and Gaynell M. Simpson

Abstract The idea for this book was conceived from a friendship that developed during our tenure as faculty scholars in the Geriatric Social Work Initiative, Hartford Faculty Scholar (HFS) Program. The HFS Program inspired us in so many ways, but particularly, in terms of collectivity and reciprocity, and the notion of giving back and paying forward in a myriad of ways. It particularly inspired us to build knowledge in the rich field of aging. Our passion and respect for our elders from diverse cultures served as the thread that tied together our unique interests and approaches to aging that are presented in this edited collection. The title of this book, "The Collective Spirit of Aging Across Cultures", emerged after much discussion and processing of our ideas, lessons learned, and lived experiences as beings of diverse backgrounds.

The idea for this book was conceived from a friendship that developed during our tenure as faculty scholars in the Geriatric Social Work Initiative, Hartford Faculty Scholar (HFS) Program. The HFS Program inspired us in so many ways, but particularly, in terms of collectivity and reciprocity, and the notion of giving back and paying forward in a myriad of ways. It particularly inspired us to build knowledge in the rich field of aging. Our passion and respect for our elders from diverse cultures served as the thread that tied together our unique interests and approaches to aging that are presented in this edited collection. The title of this book, "The Collective Spirit of Aging Across Cultures", emerged after much discussion and processing of our ideas, lessons learned, and lived experiences as

N. Giunta (✉)
Silberman School of Social Work at Hunter College, City University of New York, New York, USA
e-mail: nancy.giunta@gmail.com

H. F. O. Vakalahi
Morgan State University, Jenkins Building Room 343, Baltimore, MD 21251, USA
e-mail: Halaevalu.Vakalahi@morgan.edu

G. M. Simpson
University of Alabama, Tuscaloosa, USA

H. F. O. Vakalahi et al. (eds.), *The Collective Spirit of Aging Across Cultures*, International Perspectives on Aging 9, DOI: 10.1007/978-94-017-8594-5_16,

beings of diverse backgrounds. The collective, inclusive, and intersectional framework used in this book speaks to the significance of understanding aging across diverse cultures from multiple perspectives, but still as a shared human experience. The underlying message of the book is that although we are unique and different in our aging processes, we are ultimately connected through this physical, mental and spiritual experience of aging. Thus, regardless of whether we are service providers, service recipients, educators or merely fellow human beings, it is important that we approach the aging experience through a collective lens for discovering and sharing resources as we age; honoring the past while simultaneously accepting that the future is here.

Ambitious as it may seem, our desire for this work stemmed from a commitment to address the unique issues of today's aging communities and the need to begin addressing these issues from collective, intersectional, and strengths-based perspectives. Issues including aging and health, mental health, the impact of social, economic, and environmental contexts, spirituality/religiosity, intimacy and sex, and technology. A major component of framing these issues is intersectionality by race, gender, economies, environment, social context, ability, to name a few which speak to the diversity of our aging experiences and our individual and collective outcomes within our human connection. Building on the concepts of diversity, collectivity, and inclusivity, we offer this work as an entry into lengthier and deeper conversations around aging across cultures.

Each chapter in this book presented an important issue for consideration by educators, service providers, policy makers, community activities, and families of the elders. Overall, aging is the platform where all of the aspects of life intersect; time, and thus, aging is a common denominator that does not leave anyone behind. Thus, the individual chapters in this book are provided for the purpose of offering details on this intersection and to highlight the importance of inter-disciplinary collaborations. Furthermore, there is a rich range of conceptual and theoretical frameworks from across cultures which can be utilized in viewing aging and generating culturally relevant solutions for today's elder population. Likewise, there is a rich historical background in aging that began before institutions were established and prior to the emergence of aging and gerontology as a discipline in the U.S. A few select examples of key findings from this collaborative work are provided in the following paragraphs.

First, despite progress in the field, certain issues remain to be addressed. For example, racial and sexual minority elders face more intense challenges of racism and sexism in their daily lives. Elder mistreatment continues to expand given the longevity of the elder population today, and better resources and cross-discipline professionals are needed to address the complexity of this issue. The digital divide renders our elders more vulnerable to financial abuse or other forms of mistreatment than ever before. Disproportionality in mental health problems among ethnic and racial group elders also continues to be a major problem which links to disparities in access to education and healthcare, poverty, acculturative stress as well as inequity and discrimination. Likewise, the impact of the recent economic crisis has significantly impacted housing and life decisions among the elders.

Basically, the experiences of the elders relative to the current social and economic crises are connected to race, gender, and socioeconomic status.

Second, the need for sustaining a sense of independence among the aged and interdependence among supportive systems is warranted. Families and communities interdependently have provided social support and networks that contribute to securing healthy aging among the elders. Families and communities provide strong ties, relationships, attachment, and commitment that are solid resources for supporting needs and addressing the negative effects of discrimination, disparity, and isolation that impact health outcomes of the aged. Respect for the elders as reservoirs of cultural, familial and communal values, beliefs, and practices is acknowledged and honored.

Third, the elders continue to benefit from community-based health and mental health and social service delivery systems that recognize the importance of cultural, structural and individual elder factors and offer culturally competent services that address age-old and emerging challenges for the elders. Such culturally relevant services have become critical in both urban and rural contexts given global migrations and transnational ties. Professionals from across disciplines are integral in this process, particularly in terms of providing ethnic specific, cross-cultural, and bilingual services.

Fourth, spirituality in both indigenous and contemporary perspectives remains important to the development and quality of life of the elders. Spirituality, traditional ways of healing, and religiosity have been resources for addressing historical experiences and traumas. Fifth, technology is an issue that cuts across topics of health, mental health, spirituality/religiosity, intimacy and sex, care, communication, and quality of life, all which are critical in the lives of the elders. Technology has facilitated networking and social supports among the elders, their families and friends, including those in the deaf community, rural areas, and other remote places. At the same time, it also leads to newly emerging risks that may jeopardize autonomy or safety. Sixth, the challenges of old age can also be softened by having an animal companion which can provide support and help reduce anxiety, depression, and loneliness. Finally, sexuality and intimacy is a dimension of aging often overlooked by clinicians despite its important and unique role in the health and well-being of the elders.

Future Directions

In alignment with the issues discussed in the summary section, a few implications for future work are offered. First, further work is needed in developing dynamic and eclectic conceptual frameworks that place indigenous and cultural perspectives including those of LGBTQ individuals in the center and incorporate increasing longevity of the elders. We must ask ourselves, as the elders continue to live longer, how are the systems in place responding effectively to their longevity? The social capital perspective may be utilized beyond its application to minority and immigrant communities as described in this book. This perspective may be useful in examining and addressing disparities and other complex social problems such as elder mistreatment or health disparities.

Second, future work is recommended to address the needs of immigrants and refugees in terms of the migration and transnational experiences of elders who migrate and/or those who are separated from their families and countries of origin. Attention is needed to both physical and mental health of these elder immigrants and refugees. In terms of the U.S. population, attention must be paid to the rapid growth of the Latino community and to the importance of community-based services and involvement of families in serving the elders. Practitioners must continue to draw lessons from innovative models introduced in the Latin@ and other communities which tap the strength of family and social networks. There is a growing need for a closer look at social networks particularly in rural settings.

Third, additional funding is needed for training of professionals, development of curriculum and systems that respond effectively to the needs of the elders. Spirituality is a documented protective factor for older adults, thus, additional funding to support spiritual resources of the elders is needed. Care for the elders needs to be monitored and regulated more closely because families today are forced to "purchase" help for their elders and are not taking care of their own. Quality of the services for the elders must be responsive to the current demand.

Fourth, intimacy and sex in older age is under studied, thus misconceptions are prevalent. Additional research to inform culturally sensitive responses to intimacy and sex among the elders is needed. Ideally, research in this area will help guide practitioners in offering more mindful approaches that incorporate the intersection of physical and mental health with dimensions of age, culture, and sexual desire.

Fifth, the use of technology for social support and networking among the elders, particularly the aging deaf community, must be further explored and supported. As we continue to see innovative uses of technology to support our elders it will be critical to ensure ethical approaches are taken in introducing new technology that balances elder independence and autonomy with minimizing vulnerability to harm. Elder mistreatment is pervasive across cultures and backgrounds. While inroads are being made to recognize elder mistreatment when it is happening, there is a need for training and interventions for breaking down complex barriers to eliminate it.

Sixth, it is imperative to continue to push for visibility of the issues and conditions that are often invisible in the agendas of mainstream aging studies. The discussions on the role of elder African American lesbians in rural society, elders living in urban communities and elders facing housing dilemmas resulting from the national and global financial crises illustrate this point well. Shining the proverbial spotlight on issues less visible in mainstream aging service arenas will help increase access to and utilization of formal support services as well as possibly contribute to better quality of such services. Additionally, shedding light on the valuable resources and strengths that elders from diverse communities demonstrate, including those from urban communities, will lead to opportunities for partnership and collaboration across communities and service sectors. This was illustrated throughout this book, but particularly in the discussion on the role of the Church among African American elders and the elders who are aging in the urban environment. Likewise, the significance of generating innovative ways to rework the housing template for the elders in light of the financial crisis, particularly among those in urban communities, is a crucial topic of interdependency that requires discussion and practice and policy responses.

In essence, there is indeed much work ahead of us as practitioners, scholars, and educators in the field of cross-cultural gerontology and aging. We present this edited collection as a springboard for moving forward the collective, cross-cultural, intersectional body of knowledge and translating that knowledge to solutions for today's diverse aging population. Whether this is through improving research or diagnostic instruments to minimize inaccuracies caused by cultural insensitivity, or through incorporating trans-disciplinary innovations into one's work, we invite you to join us in moving forward the field of gerontology in responding to the collective spirit of aging across cultures.

Index

H. F. O. Vakalahi et al. (eds.), *The Collective Spirit of Aging Across Cultures*, International Perspectives on Aging 9, DOI: 10.1007/978-94-017-8594-5,

MIX
Papier aus verantwortungsvollen Quellen
Paper from responsible sources
FSC® C105338

If you have any concerns about our products,
you can contact us on
ProductSafety@springernature.com

In case Publisher is established outside the EU,
the EU authorized representative is:
Springer Nature Customer Service Center GmbH
Europaplatz 3, 69115 Heidelberg, Germany

Printed by Libri Plureos GmbH
in Hamburg, Germany